Dutch Elm Disease Research

Mariam B. Sticklen James L. Sherald
Editors

Dutch Elm Disease Research

Cellular and Molecular Approaches

With 125 Illustrations

Springer-Verlag
New York Berlin Heidelberg London Paris
Tokyo Hong Kong Barcelona Budapest

Mariam B. Sticklen
Pesticide Research Center
Michigan State University
East Lansing, MI 48824 USA

James L. Sherald
National Park Service
Washington, DC 20242 USA

Library of Congress Cataloging-in-Publication Data
Dutch elm disease research : cellular and molecular approaches / Mariam B. Sticklen, James L. Sherald, editors.
p. cm.
Includes bibliographical references and index.
ISBN 0-387-94061-8 (U.S.). – ISBN 3-540-94061-8 (Germany)
1. Dutch elm disease—Congresses. 2. Ceratocystis ulmi—Control—Congresses. 3. Elm—Disease and pest resistance—Genetic aspects—Congresses. 4. Elm—Breeding—Congresses. I. Sticklen, Mariam B. II. Sherald, James L.
SB606.E5D87 1993
635.9′773962—dc20 93-17484

Printed on acid-free paper.

Production managed by Henry Krell; manufacturing supervised by Jacqui Ashri.
Camera ready copy supplied by the editors.
Cover design by Scheer Graphics and Design, Inc., East Lansing, MI.
Printed and bound by Edwards Brothers, Inc., Ann Arbor, MI.
Printed in the United States of America.

9 8 7 6 5 4 3 2 1

ISBN 0-387-94061-8 Springer-Verlag New York Berlin Heidelberg
ISBN 3-540-94061-8 Springer-Verlag Berlin Heidelberg New York

Preface

The proceedings of the last Dutch elm disease (DED) conference (Kondo et al. 1982) were published over a decade ago. Over the last ten years, a tremendous research effort has been exerted in the United States, Canada, the United Kingdom, France, the Netherlands, Spain and several other countries toward the control of DED. These efforts include investigations of biological control of the pathogen, biological control of the insect which transmits the pathogen, development of kits to identify degree of resistance to DED, bioassays to evaluate the level of resistance of several naturally selected DED-resistant American elms, tissue culture and germplasm preservation of elms, studies of the pathogenicity of DED wilt toxin, molecular genetics of American elms, and molecular genetics of DED pathogen. These subjects have never been reported in a book. This book arose from the 1992 Dutch Elm Disease Workshop, held at Michigan State University August 1–4, 1992.

Scientists have performed research on Dutch elm disease since the 1930s. This book includes a brief examination of the history of Dutch elm disease followed by an up-to-date account of research in molecular and cellular biology, directed toward the control of the disease.

The members of the research community who have directed their efforts toward the control of Dutch elm disease are to be congratulated. While complete disease resistance has yet to be achieved across the range of many *Ulmus* species, we have come far and we are optimistic. Meanwhile, the ranks of DED researchers continue to grow, and like all endeavors, younger minds come to lend their industry and talents to the preservation of a beautiful and majestic sylvan icon.

Acknowledgments

The 1992 Dutch Elm Disease Workshop and this volume, "Dutch Elm Disease Research: Cellular and Molecular Approaches" have been supported by grants from the U.S. National Park Service, National Capital Region, the Michigan State University Foundation, and the MSU Pesticide Research Center.

The editors owe an immense debt of gratitude to the 22 authors who have contributed their time and effort to this volume.

This book reflects the extraordinary skill of Mr. Dean P. Norton who managed the pre-publication editing of the text. Mr. Norton computer scanned, edited, formatted, and assembled the manuscript of each author. Without his assistance and oversight, publication of this book might never have been possible.

Last but not least, the editors would like to recognize the support of their spouses, Jon Sticklen and Lynne Sherald. This book is dedicated to them for their heartfelt support and understanding.

January 4, 1993

Mariam B. Sticklen
James L. Sherald

Editorial Notice

This book is a computer-scanned camera-ready text copy of the authors' manuscripts. The galley proofs of each chapter have been confirmed by the senior author of each report. The scientific content of the book chapters is the responsibility of the authors. However, the editors assume responsibility for informing the authors about any omissions and mistakes which may be found after the book is published. Thus we kindly ask everyone to inform the editors about possible and necessary corrections.

Contents

Contributors

Louis Bernier
Centre de recherche en biologie forestiére
Faculté de foresterie et géomatique
Université Laval Québec
Québec, Canada 61K 7P4

Mark G. Bolyard
Southern Illinois University at Edwardsville
SL 3328 Science Building
Edwardsville, IL 62026-1651 USA

Clive M. Brasier
Forest Research Station
Alice Holt Lodge
Farnahm, Surrey
GU10 4LH United Kingdom

Subhash C. Domir
Nursery Crops Research Laboratory
USDA-ARS
359 Main Road
Delaware, OH 43015 USA

Luc C. Duchesne
Petawawa National Forestry Institute
P.O. Box 2000
Chalk River, Ontario
Canada K0J IJ0

D. M. Elgersma
Department of Molecular and Cell Biology
University of Amsterdam
Kruislaan 318, 1098 SM
Amsterdam
The Netherlands

Ravindra Hajela
Department of Crop & Soil Sciences
B331 Life Sciences Building
Michigan State University
East Lansing, MI 48824 USA

Hans M. Heybroek
Institute for Forestry and Nature Research
Bosrandweg 20
6704 PH Wageningen, Nederland
Postbus 23, 6700 AA
Wageningen, The Netherlands

Francis W. Holmes
Consulting Shade Tree Pathologist
Emeritus Prof., University of Massachusetts
24 Berkshire Terrace
Amherst, MA 01002-1301 USA

Martin Hubbes
Faculty of Forestry
University of Toronto
Ontario, Canada M5S 3B3

Robert S. Jeng
Faculty of Forestry
University of Toronto
Ontario, Canada M5S 3B3

David F. Karnosky
School of Forestry
Michigan Technological University
Houghton, MI 49931-1295
USA

Guillemond B. Ouellette
Laurentian Forestry Centre
Forestry Canada
Quebec Region
1055 rue du P.E.P.S.
Sainte-Foy, Québec
Canada G1V 4C7

Wayne C. Richards
Forestry Canada
Ontario Region
1219 Queen Street East
Sault Ste. Marie
Ontario, Canada P6A 5M7

Frank S. Santamour, Jr.
U.S. National Arboretum, USDA-ARS
3501 New York Avenue, NE
Washington, D.C. 20002 USA

Lawrence R. Schreiber
Research Plant Pathologist
USDA-ARS, Ohio Research Site
USNA, Delaware, OH 43015 USA

James L. Sherald
Plant Pathologist
Center for Urban Ecology
National Park Service
1100 Ohio Drive, SW
Washington, DC 20242 USA

Eugene B. Smalley
Department of Plant Pathology and Forestry
University of Wisconsin-Madison
Madison, WI 53706 USA

Alden M. Townsend
U.S. National Arboretum, USDA-ARS
3501 New York Avenue, NE
Washington, DC 20002 USA

Joan F. Webber
Forest Research Station
Alice Holt Lodge
Farnham, Surrey
GU10 4LH United Kingdom

Susan J. Wiegrefe
Department of Forestry
University of Wisconsin-Madison
Madison, WI 53706 USA

Makoto Yaguchi
National Research Council
Institute for Biological Sciences
Ottawa, Canada K1A 0R6

1
Why Bother About the Elm?

Hans M. Heybroek

Abstract

There are four main reasons why we should not give up on elms:
- their beauty (what is the human value of beauty?)
- their functional properties (resistance to maltreatment and stress, even on difficult sites)
- their cultural history (5000 years of close association with human culture)
- their contribution to biodiversity (in a single area in Europe, 79 species of insects are specialized on elm). It is not a lost battle to fight for the elm. It has been demonstrated that sanitation can work, that induced resistance can be effective, that resistant elms can be developed. This symposium is expected to give more signals of hope and encouragement that the elm can have a future in addition to its fascinating past.

Introduction

Next to *Homo sapiens* himself, who has caused the demise of so many tree species, no agent has inflicted more death and damage to our trees than *Ophiostoma ulmi* and *O. novo-ulmi*, causes of Dutch elm disease. A contender is the chestnut blight, which has come close to exterminating *Castanea dentata*, that major forest tree species of the eastern U.S. Similarly, the blister rust in several five-needled pines, *Phytophthora cinnamomi* in Australian eucalypts, the pine nematode in Japanese pines and some diseases in cocospalm should not be underestimated; but there is little doubt that DED beats them all. The disease does not yet threaten any elm in its existence as a species, but it has killed uncounted millions of trees in three continents, right at the place where they are valued and needed most: close to the people. If one would try to put a value on those trees, using the valuation method of the International Arboricultural Society, the loss incurred might run into hundreds of billion dollars. In view of that loss, we can only say we never realized how much richer we were when the elms were still around!

But it is a gross simplification to speak of the loss merely in terms of dollars. With the disappearance of the elms, human values too are at stake, as we will see later. That means that the protection of the elms cannot be left to the calculator but has to be dealt with by the policymaker. It is evidently useful if one can calculate and predict that it will be cheaper for a community to protect its elms than to let them go; but even in cases where that does not hold or where the financial outcome is uncertain, the wise policymaker in a community may well decide to invest in actions to save the elms as long as possible, and thus to invest in the beauty, the character and the livability of the town. Similarly, a science administrator may well invest in promising research on elm and elm diseases even if direct financial returns for society are unlikely in the short run.

There are many reasons why we should not give up on elms. These can conveniently, though somewhat artificially, be grouped under four headings: beauty, functionality, cultural history, biodiversity.

In defense of beauty

For many people, the elm's beauty is the first reason why they deplore its ill fate and why they are motivated to act in its defense. These stately trees with their delicate twigs in a rhythmic pattern, with their fine haze of flowers in early spring and their mild forms: they have a special grace. This is not the place for a poetic evocation of the characteristic beauties of the tree; but read and see in Wilkinson (1978) how British artists have seen and admired their elms.

What is the value of beauty? It certainly is a powerful factor. In most people, love of nature and "ecological consciousness", now considered vital for the survival of mankind, started when they were touched by nature's beauty, consciously or unconsciously. And even so, beauty is often seen as a "soft" issue and rarely considered a reason for funding. There are two occasions only where people invest eagerly in beauty: first, where it can attract tourists, as in historic centers of cities; second, where it conveys status to persons or institutions. In a way, both cases testify of the appeal that beauty has.

Even that is an underestimate of the importance of beauty. Beauty is one of the ultimate human values. Developing an eye (and ear) for beauty makes a person more sensitive and human. Meeting beauty can dissolve stress. It is surprising that so little study has been made of the effect of beauty on the human personality. No doubt there is a strong positive effect.

Investing in elms is investing in beauty. This may not appeal to many agencies and institutions; but there may be foundations, citizens and testators who look beyond the stock exchange and consumption, who have an eye for the essentials of life and realize that beauty is a serious matter, worth patronizing.

Functionality

The good and useful properties of elms, especially as urban trees, are well-known and require little elaboration. The American elm was found to have the perfect architecture for a shade tree: its rather dense canopy, borne high above the ground, deals with the incoming radiation, while the relative absence of low branching allows free air circulation at ground level (Heisler and Herrington 1976), thus moderating urban climate better than other species. Its fast growth is an important asset, as it makes the waiting period between planted sapling and sizable tree relatively short. It lacks many of the negative properties that can make other trees problematic: it has no brittle branches that come down in any high wind or ice storm, no inconvenient fruits, no honeydew production. But most valuable is the tree's tolerance to a range of unfavorable conditions. This is what made it the number-one urban tree between Atlantic and Rockies as well as in large parts of Europe and Asia. It can stand various maltreatments, including heavy pruning of the crown and even mutilation of the roots: useful traits in an urban tree that has to coexist with many utilities. It is less susceptible to damage by de-icing salts than many other tree species. It can cope with a wide range of unfavorable soils: dry soils, alkaline soils, as well as wet and oxygen-poor clay soils. The latter property is often explained by the fact that the home of many elm species is in the floodplains, where an occasional summer flooding may cause a serious shortage of soil oxygen in the middle of the growing season: Anderson and Herre (1927) found that *U. carpinifolia* and *U. laevis* survived an eight-week flooding in June and July just as well as poplar and willow, and far better than ash, oak, linden, beech and maple (a certain American ash, called F. pubescens, did remarkably well).

It should be added that not all elm species react the same. *U. pumila* seems to be an exception as it does not tolerate "wet feet"; but it is extremely well adapted to near-desert conditions of soil and climate (Loucks and Keen 1973, Webb 1948, Heybroek 1981). In contrast, *U. glabra* and *laciniata* may be less tolerant of dry soil and air than most elms (Heybroek 1982). But several Asiatic elms are highly promising in these respects (Ware 1980, 1984). Species hybrids may surpass their parents, *vide U.* x *hollandica* and the hybrids *U. rubra* x *pumila.* All this leads to the conclusion that elm is perhaps the most versatile and durable urban tree.

Cultural history

Individual trees can serve as stepping stones for the memory for people who want to look back into their past. If in revisiting one's birthplace, one finds an empty spot at the place of a tree that was significant in some way during childhood, than that part of childhood or family life has lost some of its reality. A link with one's past is broken. Similarly, if a tree dies that was witness of an important incident in the history of a community or nation, then that incident recedes into a less real, more cerebral past.

In a comparable way, through the cultural history of a tree species, certain aspects of past human life can be illustrated or elucidated. In cultural history, we study the role the tree had in human culture, but also the history of the cultivation of the tree. The elm's cultural history is a very rich one, comparable to that of oak, apple and olive tree. It goes back a long way.

In the records of history laid down in peat layers by the annual pollen depositions over the last 10.000 years, a sudden drop in the contribution of elm around 5000 years ago can be observed over much of Europe. This so-called elm decline has long puzzled the palynologists. In their chronology, they used it as a marker for the end of the Atlanticum, the period when the forests over Europe were at their top, and for the beginning of the Sub-boreal, supposedly a slightly cooler and dryer period. Consequently many tried to explain the elm decline as a result of a climatic change, but this proved untenable. Some suggested that Dutch elm disease might have had its first outbreak then, nor realizing that the disease was an entirely new phenomenon in Europe in the beginning of this century. Moreover, the elm decline was very evident in Norway too, a country where even the present aggressive strain or *Ophiostoma novo-ulmi* has scarcely managed to get a foothold. So, the explanation had to be sought elsewhere.

At the same time as the elm decline a new wave of farmers appeared in the forests of Europe. They grew crops, but they certainly had cattle. Palynologists suggest these cows were not ranging freely, but were kept at the village and fed with twigs and leaves from trees, especially ash and elm. Probably such twigs, cut in late summer so that the leaves dry on them, constituted the main winter fodder, as they still do in some mountain areas (Heybroek 1963). They were essential: the supply of winter fodder has long been the main limiting factor for the number of cows a European farmer could keep. So, palynologists think that the elm decline can mainly be explained by the activities of prehistoric farmers, feeding their cattle (Troels-Smith 1960; Nordhagen 1954; Brockmann-Jerosch 1918, 1936). The findings of etymologists also suggest that the use of leaf-fodder has been both universal and essential in the period the European languages got their shape (Trier 1962).

There is a sequel to this story. Indications are that for the prehistoric farmers the cow as the great nourisher was the basis of the economy and of human life as they saw it (Lincoln 1981). In the early creation story of the Germanic people even the first living being, called Ymir, during its genesis was fed by the milk of the cow Audumbla hich arose from the same parent material. The cult of the cow in the Mediterranean and India; the mythical fact that the maiden Europa from Phoenicia was abducted to the continent that would bear her name on the back of a cow: it shows the status the animal had. It was a precondition for human life and culture. So the tree that fed the cow had to be special too. This leads us to the creation of mankind. One day the gods Odin, Lodur and Hunir, walking along the shore, found the tree trunks Askr and Embla, that is, ash and elm. Out of these they created the first people, man out of the ash, woman out of the elm. This may mean that since people are dependent on the nourishing cow, and the cow is dependent on ash and elm for its food, so in a poetic wording, people are made out of ash and elm.

Returning to solid history, elm had even more functions than feeding cattle. Pigs too were fattened with elm leaves. Chopped and soaked bark was great for calves. The fibrous bark was widely used for making ropes. The etymologists tell us that elm names like "wych" in English and "vjaz" in Russian are closely linked to words meaning bending, binding, weaving (Friedrich 1970). A local name for elm in Denmark was "Bastved", bastwood (Holten 1956). Elm bark was used for making clothes, and the dried roots for starting fire by the Ainus of northern Japan; elm was irreplaceable for these uses which provided them with conclusive proof for the contention that the elm was the first tree created (Batchelor 1901). In Norway the elm was precious as the bark could serve as emergency food in years of famine such as 1812. Therefore, the right to use the bark of a certain elm tree could serve as a wedding present. As late as 1863, the planting of elms for this purpose was advocated (Landmark 1863). As the tree could save lives in this way, there was hesitance, and even a taboo to cut elms. It was not considered just to make skis of elmwood; a person who had an accident on such skis would not be buried in consecrated earth (Nordhagen 1954; Ropeid 1960).

While the elm had female connotations in the Germanic myths, it was firmly male for the Romans and for generations of poets after them. This was based on the favorite Roman system of wine growing, in which the grapevine was trained upward into the crown of a living tree, preferably an elm. The latter was pruned in summer, and the leaves given to the cattle. This was called "marrying the vine to the elm", and it gave poets from Virgil to Shakespeare a great metaphor for cooperation and marriage. The elm, like man, is strong and grows toward the sun- but it is barren; the vine, like woman, can bear fruit, but by itself it is weak (!), it crawls over the ground, spoiling the fruit; but if the two are twined in marriage, the vine grows to the sun, and the elm bears wonderful fruit. Even now, this story is to be found in the logo of the publisher Elsevier, accompanied by the words "non solus," not alone (Fig. 1).

Figure 1. Elsevier logo.

There is a taxonomic issue involved. The Roman author Columella, who described the vine growing system with zeal, mentioned two types of elm. First is the common local field elm, propagated by seed; second is a special, perhaps superior variety, called *atinia*, which was propagated vegetatively, probably by root suckers (the translators have problems with the technical terms). Now, this vine growing system has died out in Italy in this century, partly because of the DED. It can still be seen, however, in some parts of Portugal. The elm used there is *Ulmus procera*, the English elm. The same elm can be found in other areas where the Romans started wine growing, such as the Rhone valley in Switzerland and the Palatinate in Germany. So I postilate the *atinia* is the same as the English elm *U. procera* (which is also propagated vegetatively), and the early English authors are right who maintained this elm was introduced by the Romans, like the sweet chestnut, and in connection with wine growing.

Elm-wood, surprisingly, was often associated with military uses. Yew gave the best wood for bows; elm the second best. When the extensive yew forests in Europe were plundered for this purpose, decrees were issued stimulating the use of elm. The false keel of the English man-of-war was mostly of elmwood, partly because beams of that size were rare in oak, partly because elm timber was more shockproof. Also some planking under water was of elmwood; above the waterline, elm was not durable enough. Because of its toughness and resistance to splitting, elmwood was indispensable for parts of wheels, (gun)carriages and carts; a reason why the planting of elms was stimulated in France by king Henry IV as well as by the government during the French revolution. Napoleon had elms planted along his new roads primarily to provide shade for his marching soldiers. As late as 1918, elm-wood, preferably from southern Michigan, was considered irreplaceable for making car bodies (Brush 1918).

So, elm has been a close companion of humanity for over 5000 years. We are now in the middle of a new and dramatic chapter in the cultural history of the elm: the battle against Dutch elm disease. It would be a shame if mankind would lose that battle.

Biological diversity

The former chapter grew too long because of the wealth of available information; the present one is too short because of lack of data. There is no doubt the elm species are host to a great number of organisms, many of which are directly or indirectly specialized on elm. Specialization tends to go much further: organisms may prefer some elm species or even individuals over others, they may prefer either juvenile, adult, overaged, sick or dead trees, in the shade or fully exposed, etcetera. In order to sustain these organisms, a variety of genotypes of different ages is needed on a range of sites, in various climatic zones.

Data are scanty on the numbers of species involved. Möller (1992) states that in the "Land" Schleswig-Holstein alone, 79 species of insects are specialized or dependent on elm. In the Netherlands, certain rare epiphyte associations, containing specialized mosses and liverworts, can be found only on the bark of large elms at certain sites (Barkman 1958, Weeda 1988). In the British Isles, 161 lichen species are recorded on elm; because of DED, several are seriously threatened. The development of resistant elms is seen as the only long-term solution (Watson et al. 1988). More such information is needed.

The death of all elms in a certain location will thus decrease local biodiversity; if it occurs over a wide area, species may be threatened in their existence. The survival of many young seedlings or suckers of elm, as may happen in epidemic areas, will not suffice for organisms that are specialized on later stages of the tree's life. It would be interesting to know how the dying out of the chestnut in North America has affected total biodiversity.

Extinction of species is irreversible. Therefore, there is much concern globally about the maintenance of biodiversity. People should realize that the threat to elm is a threat to a host of dependent species too.

Conclusion

The introduction of the DED-organisms in the western hemisphere was a major ecological accident, comparable to the introduction of rabbits into Australia. Much research and effort is needed to cushion the blow and to counteract the effects. Some parts of this research can conveniently be split up in separate and short-term research projects; other parts, however, cannot. Some tree-related research, by its nature, must be long-term. Unfortunately it is increasingly difficult to get stable financing for long-term projects. The late Jonathan Wright, professor of forest genetics at Michigan State University, used to ascribe this to "administrative fatigue" (Wright 1962), though other factors may be involved too. This tendency should be a matter of concern for science administrators, as it may threaten sound, relevant and indispensable research.

There are real accomplishments in the battle against DED. We know that stringent sanitation can contain the disease, especially if complemented by injections; we know that induced resistance can protect elms; we know that resistant clones can be produced. We expect that more reports on real progress in the fight for the elm will turn up at this workshop.

References

Anderson and Herre. 1927. Einwirkung des Hochwassers auf Forstgehölze. Mitteil. Deutschen Dendrologischen Ges. 38, pp. 255-257.

Barkman, J.J. 1958. Phytosociology and ecology of cryptogamic epiphytes. Assen. 628 pp . + 36 annexes.

Batchelor, J. 1901. The Ainu and their folk-lore. London. The Religious Tract Society.

Brockmann-Jerosch, H. 1918. Das Lauben und sein Einfluss auf die Vegetation der Schweiz. Mitt. Geogr. Ethnogr. Ges. Zürich 1917/18, VIII:131-150.

Brockmann-Jerosch, H. 1936. Futterlaubbaüme und Speiselaubbäume. Berichte Schweiz. Bot. Ges. 46:594-613.

Brush, W.D. 1918. Utilization of elm. USDA. Bulletin no. 683. 43 pp.

Friedrich, P. 1970. Proto-Indo-European trees. The arboreal system of a prehistoric people. Chicago. 188 pp.

Heisler, G.M. and L.P. Herrington. 1976. Selection of trees for modifying metropolitan climates. In: F.S. Santamour e.a. (ed.): Better trees for metropolitan landscapes. USDA Forest Service Gen. Techn. Report NE 22, pp. 31-37.

Heybroek, H.M. 1963. Diseases and lopping for fodder as possible causes of a prehistoric decline of *Ulmus*. Acta botanica neerl. 12(1):1-11.

Heybroek, H.M. 1981. Mini-monograph on elms in agroforestry. FAO Technical Consultation on fast-growing plantation broadleaved trees, 1979. FO:FGB-79-8/7, pp. 423-441.

Heybroek, H.M. 1982. The Japanese elm species and their value for the Dutch elm breeding program. Proceedings of the DED symposium and workshop, Oct. 5-9, 1981, Winnipeg, Man. pp. 78-90.

Holten, N. 1956. Erindringer om Danmarks Skove og Skovbrug. Dansk Skovfør. Tidsskrift 41:359-384.

Landmark, J.Th. 1863. Almetraeet. Tidskrift for det praktiske Landbrug 1: 79-84.

Lincoln, Bruce. 1981. Priests, warriors and cattle. A study in the ecology of religions. Berkeley. 242 pp.

Loucks, W.L. and R.A. Keen. 1973. Submersion tolerance of selected seedling trees. Journ. For. 71:496-497.

Möller, G. 1992. Ulmenerhaltung aus der Sicht des Naturschutzes. Proceedings Ulmensyposium 21-22 May 1992 , Hann. Münden. In print.

Nordhagen, R. 1954. Ethnobotanical studies on barkbread and the employment of wych-elm under natural husbandry. Danmarks Geol. Undersøgelse, II. Raekke. No. 80.

Ropeid, A. 1960. Skav. Ein studie i eldre tids fôr-problem. With English summary. Universitetsforlaget, Oslo-Bergen. 387 pp.

Trier, Jost. 1962. Venus. Etymologien um das Futterlaub. Münstersche Forschungen 15. Köln. 207 p.

Troels-Smith, J. 1960. Ivy, mistletoe and elm: climate indicators, fodder plants. Danmarks geologiske Undersøgelse, IV Raekke, Bd 4, Nr. 4.

Ware, G.H. 1980. Little known Asian elms: urban tree possibilities. Journ. Arboricult. 6 (8):197-199.

Ware, G.H. 1984. Coping with clay: trees to suit sites, sites to suit trees. Journ. Arboricult. 10(4):108-112.

Watson, M.F., D.L. Hawksworth and F. Rose. 1988. Lichens on elms in the British Isles and the effect of Dutch elm disease on their status. Lichenologist 20(4):327-352.

Webb, W.E. 1948. A report on *Ulmus pumila* in the Great Plains region of the United States. Journ. For. 46(4):274-278.

Weeda, E.J. 1988. The bryophyte flora on *Ulmus* and other trees in the northern part of the province of Groningen. Gorteria 44:104-115.

Wilkinson, G. 1978. Epitaph for the elm. Hutchinson, London.160 pp.

Wright, J.W. 1962. Genetics of forest tree improvement. Rome, 399 pp.

2
Seven Dutch Women Scientists Whose Early Research Is Basic to Our Knowledge of the "Dutch Elm Disease"

Francis W. Holmes

Reprinted with permission from the American Phytopathological Society Press from *Dutch Elm Disease - The Early Papers.* Selected Works of Seven Dutch Women Scientists. Holmes, Francis W. and Hans M. Heybroek eds.

In this Phytopathological Classic, our Society presents an often-overlooked fact: that the foundation of our knowledge of the Dutch elm disease (DED was laid by women scientists. Six of those represented in this translation performed the earliest research on this disease, and the seventh summarized it. All worked in The Netherlands

At the very outset, let us realize that the name "Dutch elm disease" should be seen as a compliment, a term of honor, recognizing early and thorough Dutch research on this important disease.

Considering the centuries of commercial and naval rivalry between the Dutch and the English, it may have been natural to assume that "Dutch elm disease" was a negative English usage like "Dutch courage," "Dutch treat," "Dutch uncle," or "a Dutchman's chance." This assumption, however, is fallacious. The real origin of the name comes from the fact that all the early reports and studies on this disease came from The Netherlands

When the disease was first noticed in 1919, it had already covered most of Belgium and the Netherlands, plus parts of northern France. Many Dutch, French, and German writers called it just the "elm disease" or "elm death ' *(de iepenziekte; la maladie des ormes; das Ulmensterben).* Only the Dutch, however, gave it much serious scientific study. A long series of useful, fundamental, scientific publications appeared in technical journals, all by Dutch authors.

It was natural, then, that when the disease reached England in 1926, it became known there as the "Dutch" elm disease.

However, the very first use of the name "Dutch" in print, a year earlier, apparently was in Germany (Brussoff, A., 1925, Die hollandische Ulmenkrankheit...eine Bakteriosis, Centralblatt für Bakteriologie 63:256).

The name Dutch elm disease is now accepted even by Dutch scientists especially when writing in English—who called it that only a few years later (Buisman, 1932, Tijdschift der Nederlandse Heidemaatschappij 40(10):338-344). After all, there are a good many other elm diseases, so it's not very helpful to call this one just "elm disease." We felt free, then, in this Classic, to call it the Dutch elm disease and even to use the common acronym DED.

In 1919, Ms. Barendina Gerarda Spierenburg started work at the Dutch Phytopathologischen Dienst (PD) in the town of Wageningen, near the Dutch Agricultural University. At that time, the PD was a combination of a national extension service, plant disease clinic, and plant quarantine service; now it has a very comprehensive task, including matters relating to the Pesticides Act and interaction with other members of the European Economic Community. Dina Spierenburg was the first to report (in 1921) the excessive dying off of elm trees in the 1919 and 1920 growing seasons. The disease apparently arose (or perhaps arrived) while everyone was distracted by the devastation and danger of World War I. In 1920, DED clearly had been around for at least two or three years, because some still-standing, large, dead elm trees had already lost bark and twigs. Most people assumed that the outbreak had begun about 1917 or 1918.

Naturally the first question was: What is killing the elms? Many suggestions were made. In France, drought and/or war-gas poisoning were proposed. In Belgium, the mycologist Biourge considered this a variation of the Nectria canker disease. In Germany, Brussoff isolated a bacterium that he named *Micrococcus ulmi* and that he thought infested the soil. (But no one ever seems to have found *M. ulmi* again!) A proper, thorough, sober, scientific work was undertaken at the Phytopathologisch Laboratorium Willie Commelin Scholten (WCS), which had been formed in Amsterdam in 1895 and moved to Baarn in 1921.

One of the earliest laboratories in the world to be devoted wholly to plant diseases, the WCS came to be associated with three universities: the University of Utrecht, the University of Amsterdam, and the Free University of Amsterdam.

Dr. Ritzema Bos directed the WCS Laboratory from 1895 to 1906 and then left for the Agricultural University at Wageningen The second Director (from 1906 to 1950) was Prof. Dr. Johanna Westerdijk. She had the creative idea of funding DED studies by raising money from the 996 separate municipalities of The Netherlands. She assigned study of the cause of DED to one of her graduate students in 1920, and later she assigned other personnel to DED: to confirm the causal agent, start resistance research, investigate fungal physiology, and the like. She herself made many public speeches describing progress, advocating further research, and passionately defending the discoveries of her staff. She often spoke to lay audiences, to commodity associations of the agricultural industries, and especially to the Dutch Heath Society (i.e., the Dutch Heathland Reclamation Society).

Marie Beatrice Schwarz was, at that time, preparing a doctoral dissertation (on the pathology of peach and willow (tree diseases) under Prof. Westerdijk. We can imagine what she thought when one day her professor told her she must also include the new problem with elm trees. But she loyally added the elm trees to the others. Today, she is famous for her work on elms—to the point that almost no one knows what it was that she learned about peaches or willows!

Schwarz isolated microbes from the diseased elms and inoculated healthy elms with the microorganisms she had obtained. She concluded that one of them—a certain fungus—caused the new disease. She named its imperfect stage (the stage bearing nonsexual spores, which was all that she had found) *Graphium ulmi* Schwarz, new species. By the rules of botanical nomenclature, this name still may be used if one is dealing with only the imperfect stage, as is so often the case with DED. Dr. Schwarz soon went to the Dutch East Indies, where she was employed and where, a few years later, she married T. C. Schol, becoming Mrs. Dr. M. B. Schol-Schwarz.

In the meantime, although Dr. Schol-Schwarz had indeed discovered the true cause of DED, very few people in the plant disease research world believed her. It was, after all, a rare thing for a graduate student to discover the cause of a new disease, especially one that would become world-famous. Many a doctoral degree is given for far less useful discoveries over small points of detail. Apart from her colleagues at the WCS Laboratory, only five others— Prof. Dr. H. W. Wollenweber at the German National Experiment Station in the Dahlem suburb of Berlin; two women scientists in Germany (Prof. Dr. Countess von Linden and Lydia Zenneck, both at the University of Bonn); and two scientists in England, again one a woman ("Malcolm Wilson and Dr. Mary Wilson," as Westerdijk put it) — could perceive that Bea Schwarz was right.

All through the 1920s, this controversy raged, while the disease continued to spread through Europe and warnings began to be sounded in America that it might someday arrive there. So Prof. Westerdijk assigned another of her students to conduct a further study of the cause. Young Dr. Christine Johanna Buisman had just completed a doctoral thesis on root rot of calla lilies, but she spent the rest of her short life studying nothing but DED!

Prof. Westerdijk asked Dr. Buisman to repeat and expand Schwarz's earlier elm study, so as either to confirm or to refute the findings. Buisman embarked at once into an elaborate program of isolations from and inoculations into elm trees at many different times of the year. She promptly found that Schwarz had been right: the fungus *Graphium ulmi* did cause DED!

The reason why Schwarz never could produce real wilt (and why others also had had trouble repeating Schwarz's work), Buisman discovered, was that the period when elms are susceptible to artificial inoculations is limited to about 6 weeks in early summer: June and part of July in the Dutch climate. In addition, freshly potted or transplanted trees will rarely show symptoms upon inoculation. Buisman succeeded, using the same kinds of elm as Schwarz, when she inoculated at the right time of year and in well-established trees.

In 1931, when Dr. Schol-Schwarz and her husband returned to The Netherlands for a leave of absence from work in the East Indies, the fun-loving scientific community at the Willie Commelin Scholten Laboratory planned a little surprise for her. No one had ever told Dr. Schol-Schwarz that her scientific findings had been widely doubted by most European scientists. So her former colleagues said to her, "Come along with us next week, Bea, to the (annual) plant disease conference in Ghent (Belgium)". She went, and she was sitting in the audience when the results of her graduate studies were triumphantly vindicated in a paper presented by Dr. Christine Buisman.

Meanwhile Prof. Westerdijk did not rest content to have only Schwarz and Buisman studying this disease. She assigned another doctoral candidate, Maria Sara Johanna Ledeboer, to study the physiology of the fungus: What acidity did it prefer? What elements did it need for nutrition? What materials were toxic to it? Dr. Ledeboer, in her dissertation in 1934, opened for the first time the question of possible injections into elm of fungicidal materials against the Dutch elm disease fungus. She squarely faced and discussed a problem that still faces us: the phytotoxicity of such chemotherapeutants to the very elm trees they are to protect!

But Dr. Buisman was not limiting her studies to the cause of the disease. In 1929 or 1930 (not all references agree), in conjunction with planning by Prof. Westerdijk, she initiated a far-sighted and far-reaching program to search for disease resistance in all elm species and elm populations and to enhance the resistance by breeding. She gathered elms from every available source and selected among them by inoculation.

However it was left for Dr. .Johanna Catharina Went, who took the torch from Buisman's hands in 1936, to do the first crossings of various (more-or-less resistant) elms with one another, creating hybrids by interbreeding. This program still continues after 55 years. To date, nine named, DED-resistant elm clones have been released to the public through the nursery industry, to replace the centuries-old clone *Ulmus hollandica* 'Belgica', the Belgian elm that formerly grew by the millions along roads and dikes in The Netherlands.

This research has encountered both expected and unexpected difficulties. The Dutch knew that they must combine DED resistance with wind resistance and still retain a beautiful tree form. Salt-laden North Sea winds are a part of everyday life in The Netherlands and the Dutch public has an appreciative eye for a well-formed tree, especially for trees used in intact, symmetrical rows along straight roads.

No one foresaw the fact that clone number 24, found by the Dutch in Spain and issued in 1936 as the Christine Buisman elm, even though highly DED resistant, would prove almost useless to The Netherlands itself. At that latitude (indeed the same latitude as Labrador) it dies back severely from the coral spot canker disease, caused by another fungus, *Nectria cinnabarina* Fr. In addition, its shape and level of wind resistance were unsatisfactory. But this elm clone is a suitable tree in Spain, in Italy, and in the United States at latitudes where the patterns of day-length variation more closely resemble those of its Iberian origin. (This is true despite the fact that in such U.S. areas the winters are far colder than in either Spain or The Netherlands.)

Even this elaborate resistance-breeding program was not enough for Dr. Christine Buisman's energy and capabilities. Meanwhile, she discovered the perfect, or sexual, fruiting stage of *Graphium ulmi,* first in the laboratory and then in nature. She entered the allied field of mycology to describe it and named it *Ceratostomella ulmi* Buisman.

Ceratostomella seemed then to be the logically correct group, but it suffered from an error that had been made a half-century earlier, in 1881, by a man named Winter. Winter had combined two different types of related fungi under the name *Ceratostomella.* Many fungi of both types were put into this genus in the 51 years between 1881 and 1932. Meanwhile there were many failures in the attempts to find a correct genus name for the non-*Ceratostomella* parts of this mixture. Indeed, it was not until 1952 that Buisman's name for the sexual stage of the DED fungus was changed to what some scientists now consider the correct name: *Ceratocystis ulmi* (Buisman) C. Moreau. Other scientists, however—especially those in Europe—now prefer *Ophiostoma ulmi* (Buisman) Nannfeldt. By the botanical rules, Buisman's own name is still attached to either of these Latin binomials, in honor of her initial discovery.

Dr. Buisman's discovery opened a whole field of research into inheritance of variations in this fungus species, a favorite subject of mine. Dr. Buisman showed that two identical-looking but internally different isolates of the fungus had to mate to form the sexual fruiting bodies. She called these two groups "+" and "-" Today they are called "A" and "B." These are now known to be, not two sexes but two mating or compatibility groups. In nature, each of them contains only hermaphroditic cultures (cultures able to act as either male or female), which can mate only with cultures of the opposite mating group. But in a laboratory, male-only cultures have been derived in both A and B - through the loss of the ability to act as females. (As yet, no female-only cultures have been found.)

Dr. Buisman's other articles consist of long, thoughtful annual reports on the DED research being performed. In only eight years (between her doctorate in 1927 and her tragic death after a cancer operation in 1935), this remarkable scientist, whose name was posthumously assigned to the most widely used of the DED-resistant elms she eventually discovered, published 33 technical scientific papers in five language: Dutch, English, French, German, and Italian. She spent a year in Italy as consultant (DED was killing elm trees that were used as supports for grape vines in a major industry there) and another year in the United States, mostly in Massachusetts at Radcliffe College and at the Arnold Arboretum of Harvard University. Somehow she also found time to teach a mycology course in Portugal! She was also unofficially active in helping a young entomologist at the Agricultural University in Wageningen, J. J. Fransen, to complete his doctoral dissertation, in which he showed that elm bark beetles could carry the DED fungus from one elm tree to another. Perhaps Buisman's tremendous efforts tired her too much. Anyway, to the sorrow of many, including myself, she seems not to have been strong enough in 1935 to survive a necessary operation (by some reports a minor operation, by other reports an operation for cancer).

Of Buisman's 33 technical papers, I have translated 32 (plus two memoria about her) into English (the other was already in English). A complete bibliography of her publications can be obtained from me at cost. Funds are still sought to publish the whole collection as a booklet in her honor. She was very far-seeing in her work. Had Buisman lived, DED research might by now be much farther advanced than it is. Despite the some 3,500 papers that have been published since the disease arose (an average of about one a week for the past 65 years!), there is still much to learn.

At once Dr. Johanna C. Went took over management of the valuable program of selection and breeding for DED resistance. She carried this program through the dangerous and difficult occupation years of World War II, still under direction of Prof. Dr. Johanna Westerdijk. Dr. Went ran that program from 1935 to 1954, a total of two decades, and then turned to other research, delving into processes of decomposition of leaf litter of an oak forest floor.

During the World War II occupation, food, shoes, clothes, fuel, and research materials were scarce. Survival itself was doubtful. Work had to be done under continual German surveillance. It was an effort even to force oneself to go to work each day, to say nothing of giving close attention to rigorous scientific disciplines like phytopathology or the genetics of disease resistance.

We all owe a debt for our present-day resistant elms to the courage of the scientists of the Phytopathological Laboratory Willie Commelin Scholten and also to those in the companion Centraalbureau voor Schimmelcultures (CBS), the Dutch Central Agency for Fungus Cultures, then housed in the same building in the town of Baarn, who continued their research despite the German occupation.

CBS had been created, and was directed, by Prof. Westerdijk. It now rates as one of the three principal world collections of fungi in culture, the other two being the Commonwealth Mycological Institute, Kew Gardens, England, and the National Type Culture Collection, Beltsville, Maryland, USA.

Phytopathology owes a special debt to Dr. Westerdijk, who not only founded a dynasty of able and productive plant pathologists (about 50% women and 50% men) and laid a sound basis for understanding and controlling the devastating DED, but also founded a world-renowned fungus culture collection Her persistence during the war and the survival of her research program in the difficult aftermath provide much to admire. In 1952, after 46 years as director of both WSC and CBS, she finally had to surrender the reins of control of both of the famous institutes she had so long led. She died in 1961.

The new director of the Phytopathological Laboratory Willie Commelin Scholten was Prof. Dr. Louise Catharina Petronella Kerling, who had survived World War II in a Japanese concentration camp. (The new director of the Centraalbureau voor Schimmelcultures was Prof. Dr. Agathe L. van Beverwijk. Both were women scientists of repute.)

Prof. Kerling continued to encourage DED research and even increased it. At one point (1962-1963), she had three researchers working on it at once— a whole "DED corridor" in the Lab! In addition, she herself contributed to DED literature, with a publication on its history. She had the insight, too, to appoint a plant physiologist to study DED in her laboratory. Prof. Kerling retired in 1972 and died in 1985, ending the famous and productive body of early work on DED by Dutch women plant pathologists over the half century 1919-1972. Their work was so well done and is so highly regarded that Dutch research on DED continues to hold a leading position in the world.

The fourth director of the Phytopathological Laboratory Willie Commelin Scholten (1972-1985) was Prof. Dr. Koen Verhoeff, and the fifth (1985-) is Dr. Bob Schippers. The WCS Laboratory thus had two women as directors (for a total of 72 years) and three men (for a total of 27 years). In 1987 came the sad news that the WCS Laboratory is scheduled to be closed in 1991, for lack of funding cooperation among universities. For generations of plant pathologists the world around, it has long been a professional Mecca.

After her husband's death under the Japanese occupation, Dr. Marie Schol-Schwarz returned to the CBS at Baarn. I was honored and delighted there in 1963 to do a small study with her on an atypical variant in the coremial stage of the fungus *Graphium ulmi*—the fungus that she herself had named 42 years earlier. Dr. Schol-Schwarz died in 1969, in Baarn.

In this Phytopathological Classic, the spotlight is on these seven women. That is perhaps not entirely fair: even in the 1920s some men also made contributions to DED research, working with Westerdijk. For example, S. Broekhuizen, whose dissertation at Leiden University in 1929 was entitled *Wondreaksies von hout* (Wound Reactions of Wood), included studies of the brown discoloration that characterizes xylem tissue of DED-infected elms. J. J. Fransen (as mentioned above) demonstrated the transmission of this disease fungus by elm bark beetles. Cooperation with men was generally good, as the testified by the reports of the Committee for the Study and Control of the Elm Disease (an independent nationwide group set up to gather funds, authorize research, and publish reports).

But from the research foundation laid by these women phytopathologists grew the huge structure of study that has led to our present knowledge of DED. Others who publish new discoveries on DED probably cite their publications more often than any other references in the scientific literature.

Those in the biological sciences who now seek proper recognition and support for women desiring phytopathological careers should take courage from these famous examples - just as those in the physical and chemical sciences look with pride to the accomplishments and fame of, for example, Marie Curie in discovering radium. But the large role of women throughout the history of plant disease research is little known even to plant pathologists, and it is far less known to today's public. It deserves greater recognition and high respect.

3
The Dutch Elm Breeding Program

Hans M. Heybroek

Abstract

The program is 64 years old. More than 200,000 seedlings and 2000 clones have been tested for resistance to DED and for other properties. Nine clones have been released so far; more are in the final stage of testing. Some observations are given on genetics and expression of resistance, on self-sterility, compatibility between species, polyploidization, resistance to the elm bark beetles, and rootstock matters.

History

The beginnings of a program for finding elms resistant to *Graphium ulmi* Schwarz, as it was called then, can be traced back to 1928. The year before, Christine Buisman had started fresh research on the cause of DED, which led to the development of a dependable inoculation method; this ran parallel to research by Wollenweber in Germany (Buisman 1928, Westerdijk en Buisman 1929). These results elucidated the true cause of the disease, which hopefully would lead to some way of control; at the same time, the inoculation method could serve as a tool to select for resistance. Both tracks were followed by Westerdijk and Buisman: they realized that a control method, though badly needed for the existing elms, would be no more than a partial answer to the disease. Tree planters would want resistant elms for future plantings. There was need of resistant replacements for the widely planted and prized clones *U. x hollandica* 'Belgica' and *U. carpinifolia* 'Sarniensis'.

Therefore, Westerdijk and Buisman, with the help of Mr. S.G.A. Doorenbos, director of parks of The Hague, started in 1928 to collect species, varieties and clones of elm from all over the world. The famous collections of Arnold Arboretum and Kew Gardens provided some material, more came from various nurseries in Europe and from private contacts. Thus the basis was laid for a breeding program that continued until to-day. Over that period of 64 years, it was headed by three persons only: Dr. C.J. Buisman (1927-1936), Dr. Joha C. Went (1936-1953) and the author (1954-1992). A detailed history of the program can be reconstructed from its annual reports, complete from 1929-1988.

For the period 1929-53, these were published in the series Mededeeling, Comité inzake Bestudeering en Bestrijding van de Iepenziekte; as an example, the report over 1936 has been translated (Went 1937). From 1954 on, they appeared as internal reports in The Dorschkamp research institute; at first as separate stenciled reports, from 1968 on in the series Projectverslagen.

In the program, three (overlapping) phases can be distinguished:

1. collecting (and screening) material
2. crossing selected parents and selecting in the offspring
3. field testing clones.

The collecting phase dominated the first ten years of the program: up to 1939, some 16,000 seedlings from many different sources were collected and screened. Most belonged to some form of *U. carpinifolia*, many to *U. glabra* Huds, and their hybrids, and smaller numbers to various species as *U. americana*, *U. pumila*, *U. laevis*, *U. japonica* and *U. rubra*. Most of the clones that were collected and tested were named cultivated clones of European species and hybrid groups. A full list of recognized cultivars can be found in Green (1964); many of those appear in the lists of tested clones in Buisman's annual reports. In addition, as many clones of American and Asiatic species as could be obtained from various collections were grown and tested.

Even after phase 2 (crossing selected parents) had started in 1936, the collection of various materials to broaden the breeding basis continued; partly by correspondence, but mostly in collecting tours, such as:

1957 :E. Germany, 26 clones of surviving field elms along Elbe and Saale;
1958 - 59 :England, Ireland and France, 100 clones, mainly of presumably exposure-resistant trees;
1960 :Himalayan foothills, 64 clones (see Melville and Heybroek, 1971);
1961 :Finland, 10 clones;
1977 :Japan, 9000 seedlings from 160 open pollinated trees from the three main islands.

All together more than 30,000 seedlings and 650 clones were collected and screened as part of phase 1.

Phase 2: crossing selected parents, and selecting in the offspring, started when collected trees began to flower in 1936. In some years, open pollinated seedlings were raised too. Crossings were made every year from 1936 to 1983, including even 1945. As soon as the best and most resistant seedlings of the early crosses started to flower, those were included in the crossing schemes. As a result, more than 178,000 seedlings of the F_1 - F_4 generations were raised for screening up to 1983; more than half of them in the last 17 years.

In screening, seedlings are inoculated annually or biannually from their third year on. Repeated inoculations are needed, not only because of possible escapes, but also because the young seedlings often do not yet react fully to an inoculation (Heybroek 1957): maybe because of a relatively slow early growth, or because of some physiological or anatomical factor. After each inoculation, plants with distinct disease symptoms are discarded. Remaining seedlings of good growth, shape and foliage are propagated clonally and tested again for resistance. In latter years some 50 grafts of each clone, divided over five repetitions, were inoculated in comparison with standard clones. Some 1375 clones from these seedlings were tested.

The last phase in the breeding process is phase 3: testing in field experiments. Clones that exhibited a good resistance as well as a good growth, shape and foliage are planted in field tests together with standard clones, mostly in three repetitions with four trees per clone per plot, at 5 x 5 m. If planted on a good and homogeneous site, the planting will allow a good first judgment on growth, shape, foliage and general health after as few as 6-10 years. In most of these plantings, cases of DED are sporadic and erratic; so, little information on resistance to DED can be expected from them. Instead, the experiments should show whether certain clones are susceptible to other diseases or damages. Some clones are heavily attacked by *Nectria cinnabarina*, others may prove frost-sensitive, or suffer from leaf diseases. It appears that certain species and clones are not healthy as they do not like our oceanic climate with its mild but long and inconstant winters. This applies specially to *U. pumila* and its hybrids; that species is perfectly adapted to a cold desert-like climate and soil (Heybroek 1981), but it is unhealthy in Western Europe, showing repeated twig-dying and a thin crown.

The first repeated field experiments were planted in 1958. Since 1980, many experiments were planted in other member-states of the European Community too (Heybroek 1983). This was made possible by grants of the EC in two research contracts on Dutch elm disease, 1982-1983 and 1986-1990. Plant material was grown in Wageningen and brought to the cooperating agencies by car and trailer. In this way, 42 trials were planted in 9 countries (including The Netherlands), containing 68 clones and some seedling populations of three species. Not all plantings survived, and some did not grow well, but others yielded good and partly unexpected information.

Financial assistance was also obtained from the Forestry Commission in England, 1975-1978, so that a technician could be taken on.

Released clones and new candidates

Nine clones have been released so far (see table 1). The first two clones are no longer on the market, the third one is available in very small numbers only.

The clones 'C. Buisman' and 'Bea Schwarz' were released without being tested in field experiments. The clone 'C. Buisman' soon appeared to incur heavy damage by annual cankers caused by *Nectria cinnabarina*, lacking two resistance mechanisms. The fungus can enter this host not only through wounds, as it can in many tree species, but also through undamaged twigs that were shaded out in the normal process of extension of the crown. And while cankers of *N. cinnabarina* in clones like *U.* x *hollandica* 'Belgica' will not increase in size after budbreak in April, they have a second period of growth far into June in

'C. Buisman', leading to large lesions and to killing of branches by girdling. In addition, the shape of the clone was not very good, its resistance to seawind limited and its foliage not very attractive; so it was dropped without regret. The clone 'Bea Schwarz', though more resistant to *Nectria cinnabarina*, never was a success because of poor growth and shape. 'Commelin', on the other hand, enjoyed an enormous popularity in the late 'sixties, as it had a fast growth on most sites, a good exposure resistance and a quite acceptable shape. Up to 1974, more than 500,000 trees of that clone were sold, 126,000 in the top year 1973 alone. When it became clear the clone, like 'Vegeta', was fully susceptible to the aggressive strain (Gibbs et al. 1972), numbers quickly dwindled to 5,000 in 1980, to 500 in 1990.

Table 1. Released clones

Nr.	name	provenance or parents *)	sown	cloned	in field plots	released
24	'C. Buisman'	Spain	1928	1931	-	1936
62	'Bea Schwarz'	France	1928	1939	-	1948
274	'Commelin'	'Vegeta' x 1	1940	1949	1954	1960
296	'Groeneveld'	49 x 1	1941	1949	1954	1963
454	'Lobel'	202 x 336	1958	1962	1968	1973
494	'Dodoens'	202 selfed	1954	1962	1968	1973
496	'Plantyn'	202 x 302	1954	1962	1968	1973
568	'Clusius'	202 x 336	1958	1965	1968	1983
760	'Columella'	'Plantyn' selfed	1967	1975	1978	1989

*) Descent of the parents:
1 : *carpinifolia* (France)
49 : *glabra* (England)
202 : *glabra* 'Exoniensis' x *wallichiana*
302 : *carpinifolia* 1 x *carp*. 28 (France)
336 : 'Bea Schwarz' selfed

'Groeneveld', released hesitantly in 1960 (Heybroek 1964), proved of better growth than expected and still has a niche in urban plantings, though its resistance is marginal.

'Plantyn', 'Lobel' and 'Dodoens' were candidates for release in 1970. Because of the resurgence of the epidemic in England, release was upheld until cooperative tests in England, Iowa and Holland showed that these clones, though affected, were more resistant to the aggressive strain than the others tested (Gibbs et al. 1975). Ten years later, the comparable clone 'Clusius' was added (Heybroek 1983).

Of the four latter clones, numbers of trees currently being planted range from 2,000-16,000 per clone per year. It has to be awaited how they will satisfy in the long run. They can become diseased and die in case of a massive beetle attack or by infection through root grafts. It is expected however that the percentage of trees killed will be much lower than in 'Commelin', 'Vegeta', 'Belgica' etc. Perhaps the new clones require an attack by more beetles to get diseased. It is hoped that infected trees may recover in subsequent years, in the way Peace described for susceptible trees infected by the old, non-aggressive strain (Peace, 1960).In the nursery, recovery of infected and diseased young trees in the next year is a common phenomenon. The use of the new clones is recommended in situations where a massive infection is not likely, and where the incidental loss of a tree will not be consequential.

With the newest clone, 'Columella', even these reservations do not apply. It is fully resistant and can be planted without fear of DED. In the nursery, in many trials, no twig dying was induced by inoculation with the fungus. Because of its striking columnar shape, its main use will be in urban forestry.

Future candidates for release should have a comparable resistance combined with a wide crown and some exposure resistance, like that ideal elm 'Belgica'. Such clones seem to be within reach now, as some dozens of clones with sufficient resistance and varied shapes of crown are still in field experiments. Out of the 90,000 hybrid seedlings made since 1967, most of which belong to the F_2 or F_3 generation, only 'Columella' has been released up to now; many other clones still need further testing.

Candidates for immediate release are a clone for use as a rootstock, a golden-leaf clone and a clone of *U. wallichiana.*

Some experiences and observations

1. In general, resistance had to be built up in small steps: no satisfactory clones could be found in the P- or F_1-generation.
2. Resistant offspring was primarily found in crosses between parents which had some resistance themselves; resistance in some individuals in the offspring could then be higher than in either parent. Susceptible parents gave susceptible offspring. There were some positive and negative exceptions to this rule, to wit:

   the resistant clone 'C. Buisman' gave mainly susceptible offspring in crosses, as if it were a susceptible clone. One can suspect resistance is based on recessive genes in this case, so that it can be utilized only in an F_2, in backcrosses or perhaps in crosses with trees of the same provenance such as clones 40 and 43. Such crosses are scarcely made, as others were more promising.

   the susceptible clone *U. carpinifolia* 'Hoersholmiensis' often gave seedlings with more resistance than could be expected, and certainly more than the similarly susceptible 'Belgica'.

- the combination *U. wallichiana* (several clones) x 'Commelin' produced very fine seedlings with an unexpectedly high level of resistance repeatedly; unfortunately not high enough for our present standards.

3. Disease expression may differ slightly between clones, suggesting different resistance mechanisms. Some clones react with severe leafdrop but little dieback of twigs, others react the other way around. Several clones, especially those with Asiatic background, develop disease symptoms rapidly in a few weeks upon inoculation, after which these do not increase further in the course of the summer; in most of the European clones, symptoms become worse slowly over the entire summer. The relation between internal and external symptoms may vary too. An extreme case is *Hemiptelea davidii*, which does not show any wilting or dieback upon inoculation, while the twigs grow nearly black internally.
4. Elms are mostly moderately to highly self-sterile (table 2). Generally, random parents give better germinating seeds and more seedlings per pollination bag in crosses than in selfings, on average 23 times as many. Relations vary strongly from year to year however, and certain clones are relatively self-fertile. On average seedlings from selfings are weaker than those from crosses, and more of them die. Based on these data, emasculation of flowers was abandoned in 1955. Instead, next to bags in which cross-pollination was carried out, always some pollinating bags were kept closed to allow self-pollination only, as a check on self-fertility. That method cannot be trusted blindly, however. In some cases self-pollination yielded few or no seedlings, while in cross-pollinated families of that same mother tree many seedlings occurred which for morphological reasons could be suspected to be selfs. Possibly the foreign pollen acted as a mentor pollen, facilitating self-fertilization.
5. All species in the section *Ulmus* of the genus *Ulmus* L. seem to intercross freely. Species hybrids are generally fertile, so that clones occur in which five or more species are represented (Fig 1.). Such inter-species fertility seems to be relatively common in woody plants as contrasted to annuals and perennials (Stebbins 1958). Hybridization between species of different sections runs into trouble more often. Limited attempts to hybridize with *U. americana* (section *Blepharocarpus*) were not successful in the Dutch program, as in other programs. Successful crossings with *U. laevis* of the same section, however, have been reported repeatedly, first by Klotzsch (1855), who maintained that hybrids between *U. campestris* (syn. *carpinifolia*) and *U. effusa* (syn. *laevis*), made by crossings in 1845, showed strong heterosis at age 8. Mittempergher and La Porta (1991) found an interesting case of post-zygotic sterility in the cross *U. laevis* x *villosa*. Santamour obtained hybrids between *U. parvifolia* (section *Microptela*) and various species of section *Ulmus*; in the Dutch program, such crosses were unsuccessful. Considerable differences in compatibility between individuals of a single species are to be expected.

Table 2. Yield of cross-(C) and self-pollinations (S) over 1967-1977.

year		combi-nations	poll. bag	filled seeds	seedlings	% germination	seedlings per bag	C:S per bag
1967	C	86	111	10,803	6,858	63.5	61.7	43
	S	21	40	174	58	33.0	1.45	
1968	C	96	213	14,063		43.1	23.4	42
	S	24	39	74		29.8	0.56	
1969	C	53	88	7,841	4,496	57.3	51.1	93
	S	10	11	23	6	26	0.55	
1970	C	72	171	8,042	4,257	52.9	24.9	12
	S	18	24	572	48	9.2	2.0	
1971	C	72	222	7,927	2,973	37.5	13.4	268
	S	16	19	132	1	0.8	0.05	
1972	C	48	206	5,167	1,364	26.4	6.6	1.7
	S	14	20	394	77	19.5	3.85	
1973	C	65	227	19,680	7,136	36.3	31.4	33
	S	17	21	244	20	8.2	0.95	
1975	C	110	317	9,604	3,966	41.3	12.5	38
	S	33	36	79	12	15.2	0.33	
1976	C	73	216	2,033	589	29.0	2.7	8.7
	S	22	36	44	11	25.0	0.31	
1977	C	95	323		2,270		7.0	39
	S	21	38		7		0.18	
TOT.	C					42.8	23.5	23
	S					18.5	1.02	

6. The production of tetraploid elm clones is simple. When the freshly cut surface of a root or a stem is treated with colchicine, the callus arising from the cambium will produce tetraploid shoots next to diploid ones. Normally, no chimaeras are found, suggesting that the new buds have their origin in a single cell. Tetraploid leaves differ from diploid ones in leaf color, surface structure, shape and venation, and they have larger stomata. As raised polyploidy levels tend to increase the size of cells generally, and as larger vessels increase susceptibility to DED (McNabb et al. 1970), not much effort was spent on polyploidization. Tetraploid 'Groeneveld' tended to be more susceptible than the diploid clone, and triploid descendants of the Swedish asynaptic specimens of *U. glabra* (Ehrenberg 1949) proved highly susceptible.
7. The breeding program aims at producing trees resistant to the fungi *Ophiostoma ulmi* and *O.novo-ulmi*. Resistance to the vectors, i.e., unattractiveness of the tree for twig feeding by the elm bark beetles would serve the purpose also, either by itself or in combination with some degree of

resistance to the fungus. As early as in the 'thirties, the entomologist J.J. Fransen tried to find differences in attractiveness to beetles between different elm species, but had limited success (Fransen 1983). Differences were found, however, by Webber and Kirby (1983). Any breeding for resistance requires a method that allows comparing the resistance level in many individuals in a reliable and relatively easy way. Such a method does exist for screening for resistance to the fungus, not to the beetle.

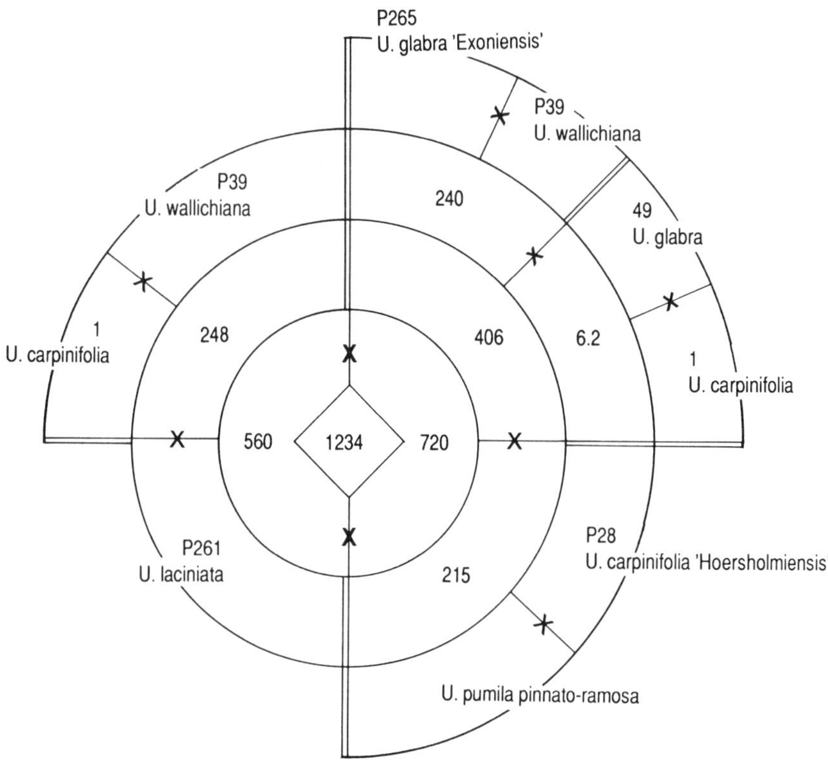

Figure 1. Parents, grandparents, etc. of elm clone no. 1234

8. Rootstock matters have drawn much attention lately. Elm clones are mostly propagated by grafting or budding on clonal *U.* x *hollandica* 'Belgica' or on seedlings of *U. glabra*. Neither produces root suckers, but both are susceptible to DED. For a long time, this was considered unproblematic. Early research had indicated that upon inoculation with *O. ulmi*, grafts on susceptible rootstock were not affected much more than those on a resistant rootstock (Heybroek 1965). This notion has to be reconsidered, perhaps because *O. novo-ulmi* is so much more aggressive, perhaps also because the difference in resistance between rootstock and top part is so much bigger now than in 1965. In the nursery, the fungus often kills the susceptible rootstock upon inoculation of the relatively resistant graft, which is initially little affected. Second-year survival of inoculated clones was much higher when they were grafted on rootstocks of

U. pumila than on 'Belgica', even though the first-year disease symptoms had been only slightly different. Upon inoculation in the roots, tyloses formed much quicker in the vessels of resistant clones than in those of a susceptible clone (Van der Molen et al. 1990). This may reduce the incidence of root transmission of the fungus. Therefore, it seems advisable to grow the resistant clones on their own roots, by rooting cuttings. Another solution is in the release and propagation of the resistant clone 622 for use as a rootstock, as it has the added advantage of a very dense rootsystem, giving more stability to the young tree.

Conclusion

With some perseverance, a range of good elm clones with resistance to DED can be produced.

References

Buisman, C.J. 1928. The cause of the elm disease. Transl. in: Holmes, F.W. and H.M. Heybroek. 1990. Dutch elm disease, The early papers. St. Paul, Minn. pp. 105-110.

Ehrenberg, C. Eklundh. 1949. Studies on asynapsis in the elm *Ulmus glabra* Huds. Hereditas 35 (1): 1-26.

Gibbs, J.N., H.M. Heybroek and F.W. Holmes. 1972. Aggressive strain of *Ceratocystis ulmi* in Britain. Nature 236 (5342):121-122.

Gibbs, J.N., C.M. Brasier, H.S. McNabb Jr. and H.M. Heybroek, 1975. Further studies on pathogenicity in *Ceratocystis ulmi*. Europ. Journ. Forest Path. 5 (3): 161-174.

Fransen, J.J. 1983. The tastiness of different elm species for the elm bark beetles (1939). Translation. Rapport 359, Rijksinstituut De Dorschkamp, Wageningen. 51 p.

Heybroek, H.M. 1957. Elm breeding in the Netherlands. Silvae Genetica 6 (3-4): 112-117.

Heybroek, H.M. 1964. De iep 'Groeneveld'- The 'Groeneveld elm. Plant Disease Reporter 48(3):187-189.

Heybroek, H.M. 1965. Iep en onderstam. (Elm and rootstock). Beplantingen en Boomkwekerij 21(11):172-173.

Heybroek, H.M. 1981. Mini-monograph on elms in agroforestry. FAO Technical Consultation on fast-growing plantation broadleaved trees, FO: FGB-79-8/7, p. 423-441.

Heybroek, H.M. 1983. Resistant elms for Europe. In: D.A. Burdekin (ed.). Research on Dutch elm disease in Europe. For. Comm. Bull. 60, pp. 108-113.

Holmes, F.W. and H.M. Heybroek. 1990. Dutch elm disease - the early papers. Selected works of seven Dutch women phytopathologists. Phytopath. Classics 13. St. Paul, Minn. 154 pp.

Klotzsch. 1855. Die Nutzanwendung der Pflanzenbastarde und Mischlinge. Verhandl. d.k. Preuss. Akademie Wiss. 1854. Berlin. pp. 535.

McNabb, H.S., H.M. Heybroek and W.L. MacDonald. 1970. Anatomical factors in resistance to Dutch elm disease. Neth. Journal of Plant Pathology 76(3): 196-204.

Melville, R. and H.M. Heybroek. 1971. The elms of the Himalaya. Kew Bulletin 26(1): 5-28.

Mittempergher, L. and N. La Porta. 1991. Hybridization studies in the Eurasian species of elm (*Ulmus* spp.). Silvae Genetica, in press.

Molen, S.H. van der, R.J. Scheffer en H.M. Heybroek. 1990. Wortel-en ent-onderstamproblemen bij klonale iepen. Directie Bos-en Landschapsbouw rapport 1990-5. 96 pp.

Peace, T.R. 1960. The status and development of elm disease in Britain. Forestry Comm. Bull. 33. 44 p.

Stebbins, G.L. 1958. The inviability, weakness and sterility of interspecific hybrids. Advances in Genetics 9, pp. 147-215.

Webber, J.F. and S.G. Kirby. 1983. Host feeding preference of *Scolvtus scolvtus*. In: D.A. Burdekin (ed.). Research on Dutch elm disease in Europe. Forestry Comm. Bull. no. 60, p. 47-49.

Went, J.C. 1937. Report on the investigations concerning the elm disease, carried out at the Willie Commelin Scholten Phytopathological Laboratory at Baarn, during 1936. Transl. in Holmes, F.W. and H.M. Heybroek. 1990. Dutch elm disease, The early papers. St. Paul, Minn. p. 130-144.

Westerdijk, J. and C. Buisman. 1929. The elm disease, report on the research conducted at the behest of the Dutch Heath Society. Transl. in Holmes, F.W. and H.M. Heybroek. 1990. Dutch elm disease, The early papers. St. Paul, Minn. p. 87-99.

4
American Liberty Elms and Beyond: Going from the Impossible to the Difficult

E. B. Smalley, R. P. Guries and D. T. Lester

Abstract

'American Liberty' elms are a group of hardy elms from our breeding and selection program which possess an upright vase shape, disease resistance and a vigorous growth habit. These elms constitute a genotypically diverse multiclone variety with similar phenotypes for use in urban areas as a replacement for lost elms killed by Dutch elm disease. Five of the six individuals which comprise the 'American Liberty' elm multiclone were derived from controlled pollinations made in 1969 and 1970 between selected *Ophiostoma ulmi*-resistant parents. One of these (W510) has been patented and named the 'Independence' elm. These individuals were selected from more than 530 survivors of inoculated progeny from controlled pollinations (500 series) which involved more than 43 families and over 3,000 individuals. The sixth individual (M-8) was the single survivor of about 1000 inoculated seedlings obtained from Kansas in 1957. Parents of the 'American Liberty' elms included superior survivors from our American elm screening program, as well as resistant individuals from the New York and the USDA programs. In total, the parents of 'American Liberty' elms are survivors of over 60,000 inoculated American elm seedlings collected from many locations over the northern range of American elm. Selected flowering long-term survivors in the 500 series have been used as parents in further breeding to produce a new (F-2) generation of disease resistant American elms. Certain of these families have yielded nearly 80% resistant individuals following the first field inoculations. Exciting recent research in which the "tetraploid-barrier" of the American elm has been overcome to produce vigorous *U. parvifolia* x *U. americana* hybrids retaining paternal hardiness, leaf size and architecture while carrying maternal genes for disease and insect resistance will be reviewed.

Introduction

American elms, *Ulmus americana* L., resistant to *Ophiostoma ulmi* (Buism.) Nannf. have been the goal of research programs since Dutch elm disease (DED) was first introduced into North America from Europe in the early 1930's. The earliest of these programs began in 1933 cooperatively between Cornell University and the Boyce Thompson Institute for Plant Research and developed extensive collections of American elm (55 lots) from various locations in the northern half of its geographic range (Sinclair et al. 1975). About 21,000 elm seedlings were grown and screened for resistance in the program at the Yonkers, NY nursery. In 1947, 168 surviving trees were moved to Ithaca, NY where screening was completed in 1965. All surviving trees had been inoculated more than seven times and in at least five different years. By 1967, remission of foliar symptoms had occurred in most of the 17 survivors, although six developed bark beetle infestations. Eleven of the remaining non-infested trees included nine very slow growing or stunted trees (probably inbreds), an atypical individual (later identified as *U. laevis*), and a single vigorous "typical" American elm. The latter survivor (R18-2) developed into a well formed, vase-shaped tree and in 1974 was a candidate for release. However, R18-2 succumbed to elm yellows disease (phloem necrosis) during an epiphytotic in 1979; ramets of this tree survive in plantings in Ithaca, NY (Sinclair 1985).

Research by the Division of Forest Pathology, Bureau of Plant Industry, Soils and Engineering (now USDA, Nursery Crops Research) was initiated in 1937 primarily to develop elms resistant to DED and phloem necrosis (elm yellows) (Clapper 1952; Smucker 1944; Whitten and Swingle 1948). American elm seedlings originating over the northern range of the American elm were assembled in nurseries near Morristown, NJ (for DED) and Columbus, OH (for phloem necrosis) and screened for disease resistance. The DED selection program identified two elms resistant to *O. ulmi* from an original population of about 35,000 seedlings. One of these was a dwarfish tree of doubtful usefulness, but the other had the "vase-shaped" characteristics of the traditional American elm. This latter tree was grown to maturity at the National Arboretum in Washington, DC (NA368) and developed into a smallish, spreading tree of moderate vigor. Poor soil conditions at the planting site may have contributed to its small stature although its form is typical of many of our control cross inbreds. Ramets from this tree (now called Delaware #2 or 'Delaware') growing at the USDA Delaware, OH nursery have repeatedly shown high resistance to *O. ulmi* (Townsend and Schreiber 1975). Although the parent tree (NA368) died about 1980 of unknown causes (probably elm yellows), ramets are retained at the USDA nursery at Delaware, OH (Townsend 1985). Another DED resistance screening program was begun in 1950 at the Forest Research Laboratory, l'Assomption Quebec. One of the selections from this program, a slow growing clone called the 'l'Assomption' elm,' possessed high resistance to *O. ulmi* while another selection (56-L-115) with an improved vigor, although developing symptoms following inoculation, recovered successfully from effects of the pathogen (Pomerleau and Bard 1968).

In more recent times certain nurseries such as the Princeton Nurseries, Princeton, NJ have marketed their own "semi-resistant" American elm selection (e.g. 'Princeton Elm') or those from other sources such as the 'Washington Elm' (National Park Service) (Wester 1972) and 'Delaware'(USDA)(Townsend 1985). Another old cultivar, '8630' surviving from selections made more than 35 years ago by Roger Swingle and associates at the Columbus, OH laboratory (USDA), is the most recent candidate for release to nurseries. This seedling was selected in Kentucky from flowering survivors of the initial DED and phloem necrosis epiphytotic. Although not immune, such cultivars in urban plantings have been unexpectedly long-lived and have stimulated interest in further research to discover truly resistant American elms.

Early attempts to introduce resistance genes from diploid (2n=28) elm species into the tetraploid (4n=56) American elm failed (Dermen and May 1966; Collins 1968; Santamour 1970; and others). The only confirmed hybrid recognized is a naturally occurring cross between *U. pumila* and *U. americana* (Santamour 1970) and it's aneuploid trispecific progeny (Santamour 1972). Reproductive isolation in American elm was originally ascribed to an incompatibility due to ploidy differences (Collins 1968; Dermen and May 1966; Lester and Smalley 1969). However, Ager and Guries (Ager and Guries 1982) demonstrated that the crossing barrier is located at the stigmatic surface and even pollen from artificially induced *U. pumila* tetraploids fail to achieve fertilization. To circumvent the "tetraploid barrier" research has sought to induct polyhaploidy by irradiation (Lester 1970), or in screening thousands of germinating seed for the rare occurrence of haploids as indicated by the development of multiple seedlings (Lester and Lee 1974). Others have sought naturally occurring diploid American elms (McGranahan 1982). This latter approach has had some success, and true diploid American elms are recognized (McGranahan 1982; Ware 1992). Their use in our Wisconsin breeding program, however, has not yet resulted in ornamentally useful progeny. These early problems have discouraged breeders from using the tetraploid American elm, and approaches to improving the resistance in America elm have turned to the new biotechnology and molecular techniques such as protoplast fusions between resistant diploid elms and American elm (Townsend and Masters 1984; Sticklen et al. 1991) or through attempts to introduce "resistance factors" into transgenic American elms. These molecular and cellular efforts are discussed in detail by Sticklen and others (Sticklen et al. 1991).

'American Liberty' elms are a group of hardy elms from our Wisconsin breeding and selection programs possessing an upright vase-shape, disease resistance and a vigorous growth habit. These elms constitute a genotypically diverse, multiclone variety (Browning and Frey 1969; Muhs 1982) with similar phenotypes for use in urban areas as a replacement for lost elms killed by Dutch elm disease. Although they are genetically the least resistant of all our Wisconsin elm cultivar releases, they have probably generated the most public and scientific interest. Commercial development of the 'American Liberty' multiclone has been assigned to the Elm Research Institute, Harrisville, NH 03450. Since 1985 more than 186,030 elm trees have been distributed and the Institute anticipates that > 100,00 trees will be distributed annually, with demand exceeding supply (ERI 1992). By contrast, total sales of our more resistant elm cultivars (e.g. Sapporo Autumn Gold, Regal and recently Cathedral) over the past 10 years have barely reach 140,000 (WARF 1992). The activities of the Elm Research Institute over the past few decades are probably responsible for the current resurgence of public interest in replanting the American elm.

American Liberty Multiclone

Origin

Parents of the 'American Liberty' elms (500 series) included superior survivors from our American elm screening program begun in 1958 (Table 1), as well as resistant individuals from the New York and the USDA programs. Five of the six individuals (F-1 or 500 series) which comprise the 'American Liberty' elm multiclone were derived from control-pollinations made in 1969, 1970 and 1971 between selected "putative" *O. ulmi*-resistant parents (Lester and Smalley 1972). More intensive inoculations in 1972 and 1975, however, eliminated many of those original parents, while others which were not used as parents because they had not yet flowered have survived (Table 1 and 3). Seasonal susceptibility trials further defined the resistance of some of the selected parents (Smalley and Kais 1966; Smalley and Lester 1975; Smalley and Lester 1983; Smalley and Lester 1984). In total, the parents of 'American Liberty' elms are the survivors of over 60,000 inoculated American elm seedlings collected from many locations over the northern range of American elm (Table 2). The 'American Liberty' clones were selected from more than 530 resistant survivors derived from controlled pollinations in a partial diallel crossing series begun in 1969 using both putative "resistant" as well as known susceptible parents. In total the series included more than 43 families and over 3,000 individuals (Lester and Smalley 1972) (Table 3). As we now recognize, the 5 selected individuals in the multiclone do not necessarily represent the most resistant individuals in this series since difficulties in propagation have prevented an intensive evaluation of the other remaining survivors.

General description

'American Liberty' elms develop a vigorous upright main trunk in youth with older branches tending to become more horizontal at maturity. Growth of 'American Liberty' elms is vigorous and typical of cross-pollinated seedlings of American elms. General leaf size, shape and color, bark color and texture and other growth characteristics of 'American Liberty' elms are not distinctive and are generally typical of the North American white elms (*U. americana*).

Individually they differ to some extent in architecture, leaf size, growth rate, and in resistance to *O. ulmi.*

Demonstration plantings of the American Liberty clone 'Independence' (W510) on the U. W. campus appear to be developing the traditional American elm architecture. Our knowledge of the ultimate mature form is clearly limited. With the exception of M-8, the oldest ramets, although now 17 years old, are located in a closely spaced (1.2 x 2.4 m) research planting which does not allow normal expression of growth characteristics; additionally, extensive removal of roots for propagation has tended to stunt their growth. M8 grown in a slightly less crowded planting (3.0 x 6.1 m), developed into a typical vase-shaped tree, >3.5 m dbh and was approximately 25 m tall until it's unfortunate demise following accidental removal during plantation thinning. Ramets of M8, however, survive in campus plantings and elsewhere. With the exception of M8, the original 'American Liberty' clones (500 series) were all dug in 1975 for use in propagation of ramets for resistance studies and no original parent trees exist.

In the greenhouse in 12.5 cm clay pots 'American Liberty' rooted cuttings grew at rates varying from 0.85 to 0.95 cm/day (over a 120 day period). Leaves grown under these conditions ranged from 11.5 to 16.5 cm long, 7.0 to 9.5 cm wide with individual petiole lengths varying more widely (0.2 to 0.9 cm). In general, M8, W502 and W505 grew approximately 10% more rapidly than W502 and W510 under the same greenhouse environments.

Table 1. Response of American elm seedlings from North Central U. S. populations to artificial and natural inoculation with *Ophiostoma ulmi.*

Year	Seedling Age	Healthy Trees	Mortality (%)
1959 [a]	3	1150	0.0
1960 [b]	4	1053	3.9
1962	6	966	16.0
1963	7	77	93.3
1969 [c]	13	71	93.8
1972	16	67	94.2
1975[d]	19	39	94.2
1982	26	27	97.7
1985	29	18	98.4
1992	36	15	98.7

[a] Spring planted as 3-year-old dormant seedlings from Lake City and St. Paul, MN, Colton, SD, and Hamburg, IA.

[b] Inoculated 8/11/60, 6/13/62, 6/14/66, 6/8/72 and 6/19/75.

[c] Clonal testing and breeding began.

[d] Continued mortality from natural infection.

Table 2. American Liberty multiclone: geographic origin and pedigrees.

Clone Number	Parent	Origin	DED Resistance	Survival Years	Parent Status
	W185-11 [a]	Iowa	1 of 5 survivors	35	Living [d]
W502	x		out of 202		
	R18-2 [b]	New York	1 of 17 survivors out of 21,000	46	Died 1979 [c]
	W185-22 [a]	Iowa	1 of 22 survivors	19	Died 1973 [d]
W503	x		out of 202		
	R18-2 [b]	New York	1 of 17 survivors out of 21,000	46	Died 1979[c]
	W185-10 [a]	Iowa	1 of 22 survivors	19	Died 1973 [d]
W505	x		out of 202		
	NA368 [b]	New Jersey	single survivor out of 35,000	43	Died 1980 [e]
	CA4 [a]	Wisconsin	(not determined)	50	Removed [f]
W507	x				
	NA368 [b]	New Jersey	single survivor out of 35,000	43	Died 1980 [e]
	W413-1 [a]	Illinois	slight resistance	--	Unknown [g]
W510	x				
	W185-21 [b]	Iowa	1 of 22 survivors out of 202	19	Died 1973 [d]
M8	Unknown	Kansas	1 of 2 survivors out of 1000+	35	Living[h]

[a] Female parent.

[b] Male parent.

[c] Parent from New York program begun in 1933; died of elm yellows disease (phloem necrosis); ramets remain alive.

[d] Seedlings from Interstate Nursery, Hamburg, IA (southwestern Iowa); resistance screening at Arlington, WI began in 1960. Three survivor parents died following severe inoculations in 1972; ramets remain alive.

[e] Parent from USDA program at Morristown, NJ; grown at the National Arboretum, Washington, DC, until its death of uncertain causes, but not Dutch elm disease; ramets survive as Delaware #2.

[f] Parent tree located on University of Wisconsin campus; removed during building construction, but remained free of Dutch elm disease; ramets remain alive.

[g] Parent tree is the widely grown cultivar "Moline" which originated in 1903 in Moline, IL.; Ramets survive at Arlington, WI and elsewhere.

[h] Resistant selection from seedlings donated by the McKay Nursery Co., Waterloo, WI; seedlings originated in Kansas.

Table 3. American Liberty multiclone parents: origin and current status.

Clone No.	Origin	Breeding years	Resistance class	American Liberty progeny	Current status
CA3 [bd]	WI	1979,71	unselected		Removed
CA4 [cd]	WI	1969,70	unselected	W507	Removed
CA1 [d]	WI	1969	unselected		Dead (DED)
W411-1	IA	1969	unselected		Dead (DED)
W411-3	IA	1969,70	unselected		Dead (DED)
W412-1	IA	1969,70	unselected		Dead (DED)
W413-1	IL	1969	unselected	W510	Alive
W427	WI	1970	unselected		Dead (DED)
NA368 [ce]	NJ	1970	selected	W505, W507	Dead (EY)[f]
R18-2 [ce]	NY	1970	selected	W502, W503	Dead (EY)[f]
W181-14	MN	1971	selected		Dead (DED)
W182-3	ND	1971	selected		Alive
W182-5 [c]	ND	1971	selected		Alive
W182-6 [c]	ND	1971	selected		Dead (DED)
W183-4	SD	1970	selected		Dead (DED)
W184-4	MN	1971	selected		Dead (DED)
W184-6	MN	1971	selected		Dead (DED)
W184-7	MN	1971	selected		Dead (DED)
W184-20	MN	1969,70,71	selected		Dead (DED)
W185-5	IA	1969,71	selected		Dead (DED)
W185-9 [abc]	IA	1969,71	selected		Alive
W185-10	IA	1969,70,71	selected	W505	Dead (DED)
W185-11 [ac]	IA	1969,70,71	selected	W502	Alive
W185-12	IA	1969,70	selected		Alive
W185-13	IA	1971	selected		Dead (DED)
W185-14	IA	1971	selected		Dead (DED)
W185-15	IA	1971	selected		Dead (DED)
W185-18	IA	1971	selected		Dead (DED)
W185-19	IA	1969,70	selected		Dead (DED)
W185-20	IA	1971	selected		Dead (DED)
W185-22	IA	1969,70,71	selected	W503	Dead (DED)

[a] Yielded the greatest numbers of vigorous long-term surviving progeny after multiple DED inoculations trials when selfed.

[b] Yielded the greatest numbers of vigorous long-term surviving progeny after multiple DED inoculations trials when used as the male.

[c] Yielded the greatest numbers of vigorous long-term surviving progeny after multiple DED inoculation trials when used as the female.

[d] Random UW-Madison Campus elms originally selected as controls: original trees removed for building construction, grafted trees from CA3 and CA4 survive in the Arlington nursery.

[e] Used only as males.

[f] Died of elm yellows (phloem necrosis).

Table 4. Seasonal resistance in *Ulmus americana*[y]

Clone No.		Rametes without symptoms Dates of inoculation[v]								
		5/25	6/1	6/7	6/14	6/21	6/28	7/6	7/18	8/3
					UNSELECTED[x]					
W411-3	T	3/7	0/8	0/6	0/4	1/12	--	0/6	0/9	3/3
	%	42.9	0	0	0	8.3	--	0	0	100
W412-1	T	1/12	1/9	2/11	0/5	0/12	--	0/10	1/9	2/2
	%	8.3	11.1	18.2	0	0	--	0	11.1	100
					'AMERICAN LIBERTY'--500 SERIES[z]					
W502	T	--	0/2	--	--	--	--	4/6	4/6	--
	%	--	0	--	--	--	--	66.7	66.7	--
W503	T	0/6	0/6	0/4	1/5	2/6	0/2	1/5	3/3	2/2
	%	0	0	0	20	33.3	0	20	100	100
W505	T	--	--	--	--	--	--	5/5	1/1	--
	%	--	--	--	--	--	--	100	100	--
W507	T	5/6	--	--	0/1	1/2	2/2	3/5	2/2	--
	%	83.3	--	--	0	50	100	60	100	--
W510	T	2/11	3/4	0/12	2/8	2/5	6/9	4/9	7/8	1/1
	%	18.2	75	0	25	40	66.7	44.4	87.5	100
					TOPEKA, KANSAS SELECTION[y]					
M-8	T	1/5	1/8	3/5	2/7	6/8	8/8	5/5	7/8	2/2
	%	20	12.5	60	28.6	75	100	100	87.5	100

[v] Mixed inoculum from isolates of *Ophiostoma ulmi* from Wisconsin, Kansas, Massachusetts, Illinois, Nebraska, Minnesota, Connecticut and Maine. (Gkinis, 1977).

[w] Trees planted as rooted cuttings, 3 plants/ clone/ block, 15 clones per block, and 36 to 40 blocks/series. (Many clones are incompletely represented in blocks because of transplant mortality or shortage of plants.) Data is derived from 1971-72, 1974-75 and 1978-79 series and not all clones in the studies are shown.

[x] Data from 1971-72.

[y] Data from 1973-74.

[z] Data from 1978-79.

M-8 is the oldest resistant American elm selection in our collection and was the single survivor in inoculations of about 1000 1.0 to 1.2 m tall seedlings donated by the McKay Nursery Co., Waterloo, WI in 1957, to be used for chemical control investigations. These seedlings were originally purchased by McKay as liners from the Skinner Nursery Company, North Topeka, KS. Parents of M-8 are unknown but the seed was derived from native American elms in the Topeka, KS area. The original 37± year old M-8 tree was a vigorous, vase shaped elm and about 18 m in height. It was also unique in that it only began to flower at 30 years of age. Leaves of M8 have a very short petiole and shortened apex which give the leaves an almost oval shape as compared to the more elongated leaves of the other American liberty clones.

The American elm is difficult to propagate vegetatively as contrasted to Asian elm species such as *U. pumila* or *U. parvifolia*. The 'American Liberty' multiclone (500 series) began in fact as 10 selected clones. Clones designated W500, W501, W504, W506, W506, and W509 have all been discarded because of poor rootability. We have also experienced occasional difficulties in propagating W502 and W505. Clone W510, however, propagates easily and establishes well in the field plantings. Because of its vigor this clone requires careful pruning to allow good basal trunk development before the main scaffold branches become well developed. Some wind damage has developed on a few of these trees in our campus plantings as a result of improper pruning. Sprout leaves develop vestigial petiolar leaflets which often become quite large giving the leaf an almost compound-leaf appearance.

Disease evaluations in the field

'American Liberty' elms exhibit unusual resistance to *O. ulmi* following screening for DED resistance using methods previously described (Gkinis 1977; Lester and Smalley 1972; Smalley and Kais 1966; Smalley and Lester 1975; Smalley and Lester 1983; Smalley and Lester 1984). M-8 was one of our earliest selections to be evaluated and it remained symptomless in screening inoculations in 1958, 59, 60, 62 and 1966. In the 1973-74 seasonal resistance trial vigorous ramets of M-8 were only susceptible through June 1 (Table 4). It also had the earliest decline in high susceptibility (June 7) of any of the American elms in our studies, possibly reflecting its more southerly origin.

Seedlings from controlled crosses in 1969 were grown out-of-doors for one season in a methyl-bromide fumigated seedbed, field planted in 1970 and screened for resistance in 1971. W510 was the single survivor of 10 with symptoms less than 10% crown damage. Seedlings from controlled crosses in 1970 were also grown for one season in a fumigated seedbed, field planted in 1971 and screened for resistance in 1973. W502, W503 and W505 were among 10 seedlings from an original 988 seedlings which developed less than 10% foliar wilt. W507 was one of 475 survivors of the 1971 seedbed inoculations. It was transplanted into a field planting in 1972 and reinoculated in 1973.

Seasonal resistance plots were established in 1976 and inoculated in 1978. The susceptibility periods of clones W502 and 505 were similar, but shorter than the unselected American elm clones (Table 4) (Smalley and Lester 1985). W503 proved to be slightly less resistant than W502 and W505. However, difficulties in propagation of both W502 and 505 along with poor stand establishment make conclusions from this study tentative. W507 along with W510 were the most resistant of the clones in 1978-79 seasonal resistance inoculations (Table 4).

Disease evaluations under controlled environments

More recently, controlled greenhouse and biotron testing have revealed certain new and interesting characteristics of the 'American Liberty' elms. Details concerning the horticultural methods and testing procedures are described by Clark (1984) and Proctor (1990) and elsewhere. Plants for greenhouse experimentation were grown from either year old plants or from freshly rooted elm cuttings in a mixture of compost and coarse sand (3:1) in 12.5 cm clay pots under a 16 hr day (supplemented with high intensity fluorescence illumination). Plants were initially bud-pruned to allow a single dominant bud to develop into a single unbranched shoot. Plants were staked, watered and fertilized regularly, and grown for about 90 days prior to inoculation. In winter trials greenhouse temperatures were maintained at ≤ 24¤ C day and 18¤ C night. In spring and summer trials greenhouse temperatures were maintained at outside ambient levels by shading. Inoculation treatments contrasted susceptibility of the plants to both aggressive (*O. novo-ulmi sensu stricto*)(Brasier 1991) and non-aggressive (*O. ulmi sensu stricto*) strains of the DED pathogen using 10 tree replications/crone/treatment. Plants were harvested 36 to 40 days after inoculation and disease development assessed (Chen et al. 1988; Gkinis 1977; Proctor 1990).

Resistance of the 6 clones which constitute the "American Liberty Elm" multiclone were contrasted with ramets of the first *U. parvifolia* x *U. americana* hybrid (W2213) produced in controlled crosses in 1986, and selected symptomless ramets from DED screened F-2 progeny from resistant parents. American elm controls in the 1990 summer study were equally susceptible to either the aggressive strain (NAN)(70-99) or the non-aggressive strain (NA)(70-116) of the pathogen (Fig. 1). In these trials, however, several of the 'American Liberty' clones (M8, W503, W505, and W507) developed >50% volume of discoloration (VOD) when inoculated with NAN (70-99), although W502 and W510 developed only < 40% VOD. VOD is a calculation based on the areas of xylem discoloration in cross section at 25, 50, and 75% of stem length and the length of discoloration extending from the inoculation site (Proctor 1990). Relatively intense foliar symptoms developed in W503 when inoculated with NAN (70-99) which even exceeded symptom intensity in the controls; W502 and W510 remained virtually symptomless (Fig 1A). In marked contrast to controls, all 'American Liberty' clones were resistant to the NA strain (70-116) and with the exception of W503 and W505, none developed VOD symptoms >30%. Foliar symptoms induced by the NA strain (70-116) were non-existent. We consider elms in this testing procedure to be resistant if they remain symptomless and develop < 50% VOD.

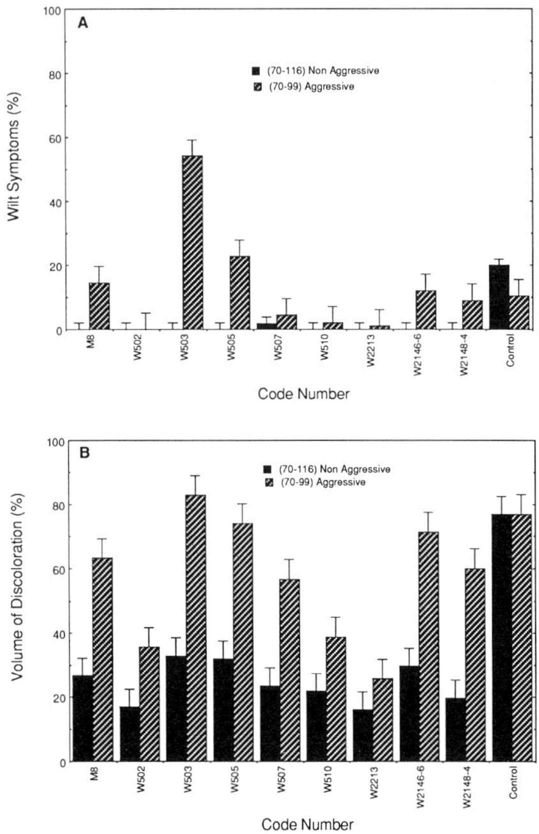

Figure 1. Susceptibility of greenhouse grown 'American Liberty' elms following inoculation with aggressive and non-aggressive strains of *Ophiostoma ulmi.* Wilt symptoms (A) are expressed as mean percent crown wilt of 10 tree replicates / treatment 52 days after inoculation. Volume of discoloration (B)(VOD) is calculated from the length of discoloration from the inoculation point and the areas of discoloration at 25, 50 and 75% of the plant height. The response of clone M8, various 500 series clones, new F-2 progeny of resistant x resistant crosses (W2146-1 and W2148-4) and a *U. parvifolia* x *U. americana* hybrid (W2213) are contrasted with American elm seedlings. Error bars are the SEM of 10 tree replicates/treatment.

General disease and pest resistance

'American Liberty' elms have not been tested for resistance to the elm-yellows mycoplasma (phloem necrosis), but they are probably susceptible since two of their male parents (e.g. NA368 and R18-2) succumbed to this disease naturally. Field observations suggest that 'American Liberty' elms are not unduly susceptible to black leaf spot disease, caused by *Stegophora ulmea* (Schw: Sydow and Sydow) or leaf blister *Taphrina ulmi*. They also appear to be tolerant to *Verticillium dahliae* Kleb (the causal agent of Verticillium wilt) since no symptoms have been observed in any of our plantings, some of which are located in heavily infested soils. Even under "outbreak" conditions 'American Liberty' elms resist the attack of the elm leaf beetle *Xanthogaleruca luteola*, elm and the elm leaf miner *Fenusa ulmi*. The recent elm leaf miner outbreak the past few years has caused major injury to natural populations of elms in the North Central U.S., but only in elms which contain genes from the Fulvae C. Schn and Euglabrae C. Schn. Series (sensu Heybroek) (Heybroek 1976) (e.g. *U. rubra* and *U. glabra*). In our nursery, all of the new DED resistant Dutch clones (e.g. 'Lobel', 'Dodoens', 'Plantyn') as well as 'Regal', 'Pioneer' (Townsend and Masters 1984) and 'Homestead' have been nearly defoliated under conditions in which the 'American Liberty' elms remained untouched.

Beyond 'America Liberty' Elms

The F-2 generation

Beginning in the spring of 1985 we initiated a new American elm diallel crossing series to examine the possibility of developing improved American elm resistance beyond the 'American Liberty' multiclone. Field inoculations of the 1986 series were completed in 1989 and reinoculation of the survivors were completed in 1992 (Table 5 and 6). Results in 1989 were quite dramatic, and indicated distinct differences between certain of the *U. americana* families. Eight families of 13 in the trial contained > 50% resistant individuals, and families with M-8 as female all exceeded 65% resistant individuals. Reinoculation of the survivors, however, reduced the number of resistant individuals. However, seven of the 13 families tested still had >20% surviving resistant individuals. Field inoculations of the 1989 series were completed in 1992. Only two families of 11 in the trial exceeded 50% resistant individuals (Table 7). Completion of the partial diallel plan will still take several more years and final genetic conclusions are not warranted at this time. However, intensive greenhouse testing of selected clones from the best families in the 1986 series has begun (Fig. 1). W2146-1 (M-8 x 185-11) and W2148-1 (185-24 x(CA3x185-4) in the greenhouse were virtually symptomless, but developed VOD levels similar to controls inoculated with the NAN strain (70-99) but less than 25% when inoculated with the NA strain (70-116). However, these clones were selected before the reinoculation and may have been propagated from relatively susceptible trees.

Table 5. Results of 1988 inoculations of progeny of 1986 controlled crosses among selected Dutch elm disease resistant American elms.[x,y]

Family	Genotype	No.	Resistant < 20%CD	Crown Damage (%)
2146	M8x185-11	68	77.9	19.4(2.9)
2147	M8x184-11	46	65.2	32.8(5.1)
2148	185-24x(3x183-4)	58	51.7	48.9(5.6)
2149	185-24xM9	32	35.1	41.6(7.1)
2150	131xMoline	35	25.7	68.9(6.6)
2152	131x(184-15xNA368)	32	28.1	70.3(7.3)
2153	182-3xM8	67	70.1	26.9(4.1)
2154	183-1x182-3	31	41.9	43.5(6.2)
2155	182-5x(4xC18)	47	63.8	30.7(5.3)
2156	(185-llx2)x(184-15xNA368)	30	76.7	22.4(6.5)
2157	(3x183-4)xM9	22	63.6	32.7(8.1)
2158	236Ax(185-11xC18)	42	33.3	65.5(6.3)
2160	(184-15xNA368)x(411-3x185-12)	10	50.0	34.0(9.0)

[x]Trees were inoculated the first week of June, 1988 and final symptoms were recorded in late July 1989.

[y]Details concerning the inoculation procedures are found in the text.

Table 6. Results of 1991 inoculations of progeny of 1986 controlled crosses among selected Dutch elm disease resistant American elms.[x,y]

Family	Genotype	No.	Resistant < 20%	Crown Damage CD(%)
2146	M8x185-11	68	27.9	54.8(3.8)
2147	M8x184-11	46	19.6	62.3(4.7)
2148	185-24x(3x183-4)	58	17.2	70.4(4.4)
2149	185-24xM9	32	21.9	63.8(5.9)
2150	131xMoline	35	8.6	84.0(4.7)
2152	131x(184-15xNA368)	32	6.3	87.5(4.8)
2153	182-3xM8	67	31.3	51.9(4.1)
2154	183-1x182-3	31	9.7	63.7(5.8)
2155	182-5x(4xC18)	47	55.3	40.3(5.5)
2156	(185-11x2)x(184-15xNA368)	30	30.0	46.0(6.5)
2157	(3x183-4)xM9	22	50.0	47.3(8.5)
2158	236Ax(185-llxC18)	42	2.4	82.9(4.1)
2160	(184-l5xNA368)x(411-3x185-12)	10	20.0	68.0(9.8)

[x] Trees were reinoculated the first week of June, 1991 and final symptoms were recorded in late July 1992.

[y] Details concerning the inoculation procedures are found in the text.

Table 7. Results of 1991 inoculation of progeny of 1989 controlled crosses among selected Dutch elm disease resistant American elms.[x,y]

Family	Genotype	No.	Resistant < 20%	Crown Damage CD(%)
1239	(4xC18)x184-1	147	17.0	80.6(2.8)
1242	M8x(185-11x0)	20	65.0	31.0(8.4)
1243	M8x(4x183-4)	11	27.3	67.3(12.7)
1251	182-5x(4x183-4)	7	57.1	42.9(15.1)
1252	(185-11xC18)x184-11	69	13.0	80.7(3.7)
1253	(185-11xC18)xM9	112	12.5	81.6(2.9)
1254	184-1x(3x183-4)	46	15.2	82.6(4.7)
1255	184-lxMoline	45	8.9	86.7(4.1)
1256	184-1x182-3	47	8.5	88.7(4.2)
1259	184-11xM9	30	20.0	77.7(6.3)
1264	(4xC18)xMoline	10	20.0	81.0(11.9)
Amer	U.W. Campus	30	13.3	87.7(5.7)

[x] Trees were inoculated the first week of June, 1991 and final symptoms were recorded in late July 1992.

[y] Details concerning the inoculation procedures are found in the text.

Breaking the tetraploid barrier

Following the investigation of Santamour (Santamour 1970a; 1970b; 1972a; 1972b; 1972c; 1973; 1974; 1989), we began a breeding study in 1984 using the Chinese elm *U. parvifolia* as a source of major genes for disease and insect resistance. Although the Chinese elm is reported to be generally resistant to diseases such as DED, black leaf spot and insects such as the elm leaf beetle (*Xanthogaleruca luteola*), its usefulness in elm breeding has been limited due to its fall flowering habit, small leaf size, and especially its lack of hardiness in the North Central region. We initiated the present breeding study using a hardy flowering selection of Chinese elm (W114-1) in controlled crosses with parents representing some of the best of our selected DED-resistant elms. Crosses made in the fall of 1986 yielded over 1,600 hybrid seedlings from 20 families (including 10 elm species and 10 hybrids or controls). Among the 1986 progeny were 12 seedlings which have now proven to be hybrids with American elm. The best of these hybrids had very large leaves, grew 50% faster than the controls, and morphologically appeared similar to American elm. Others were slightly less vigorous but clearly represented hybrid combinations.

The crossing in the fall of 1987 emphasized the creation of more hybrids between *U. americana* and *U. parvifolia*. A total of 97 new families were created, including five sources of *U. parvifolia* as female, 15 *U. americana* selections and 12 other species or hybrids as males. Crosses yielded a total of 4126 seedlings including 684 seedlings produced using *U. americana* as male and of these, 149 (21.8%) possessed hybrid vigor and had large leaf characteristics typical of *U. americana*. The crosses with diploid males yielded higher percentages of hybrids. Crosses with males of *U. carpinifolia, U. glabra, U. japonica*, and *U. pumila* yielded respectively, 90.3%, 79.5%, 78%, and 67.4% hybrids. However, crosses with the other North American species, *U. rubra*, yielded only 33.3% hybrids.

As a result of these breeding efforts many interspecific progeny have been created and are now being grown at the University Arlington Farms. The older trees in this series have now been screened for DED resistance in the greenhouse. The 1991 greenhouse inoculations contrasted the resistance of various new *U. parvifolia* x *U. americana* hybrids from the 1987 crosses with other elms having known levels of resistance. Selections included 20 *U. parvifolia* x *U. americana* individuals from 9 families; one family (W114-1 x [CA4 x R18-2]) clearly produced the preponderance of superior trees. Most of the *U. parvifolia* x *U. americana* hybrid selections remained symptomless although internal discoloration in certain clones was severe (Fig. 2A). American elm controls (LA4, RR4, American) developed severe foliar symptoms and most individuals died by the end of the study. W510 developed slightly more than 50% VOD but remained virtually symptomless (Fig 2B). 'New Horizon and W114-1, the *U. parvifolia* parent, had < 20% VOD and failed to develop foliar symptoms. There was a great range in variation in the VOD in the 17 Chinese x American hybrids tested. Some such as W2249-1 had VOD almost as intense as the susceptible controls, while others such as W2245-5 had VOD's <40%. Only 2 of the Chinese x American hybrids developed foliar symptoms >40% when inoculated with NAN strains of the pathogen. As with the 'American Liberty' elms, most of the hybrids were more susceptible to the aggressive strain and developed minimal foliar and VOD symptoms when inoculated with the non-aggressive strains of the pathogen. It is clear that resistance genes from *U. parvifolia* were transferred to hybrid progeny with American elm. These progeny both resembled their American elm parent morphologically (e.g. large leaves, highly vigorous, etc.), and physiologically (increased cold hardiness).

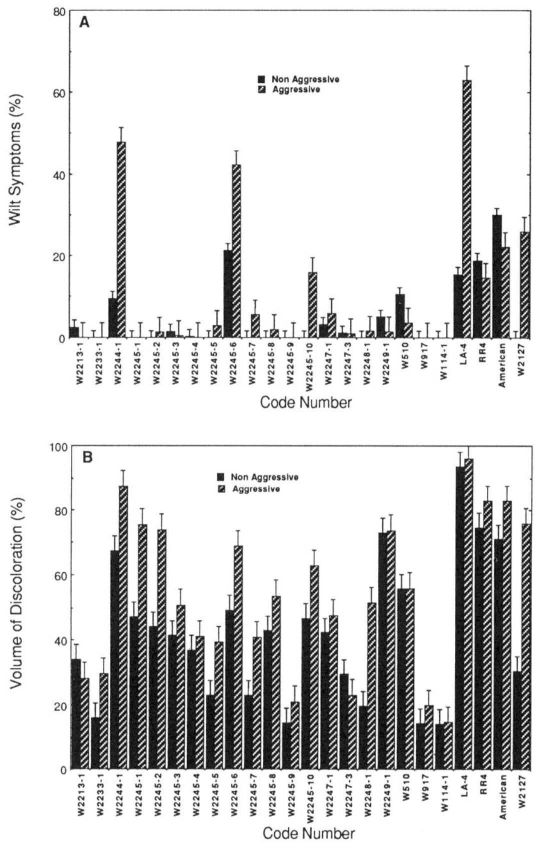

Figure 2. Susceptibility of greenhouse grown selected clones from *Ulmus parvifolia* x *U. americana* controlled crosses following inoculation with aggressive and non-aggressive strains of *Ophiostoma ulmi*. Wilt symptoms (A) are expressed as mean percent crown wilt of 20 tree replicates/ treatment 52 days after inoculation. Volume of discoloration (B)(VOD) is calculated from the length of discoloration from the inoculation point and the areas of discoloration at 25, 50 and 75% of the plant height. The response of clones W2213-1 to W2249-1 are contrasted with susceptible American elm clones (LA4, RR4, American elm seedlings) and *U. procera* (W2127). Error bars are the SEM of 20 tree replicates/treatment.

Asexual reproduction

The 'American Liberty' elms can all be propagated from root cuttings allowed to sprout in the greenhouse in moist sphagnum peat and sand. Greenwood cuttings dipped in a rooting hormone can be transplanted as rooted cuttings after approximately 20 days in a fertile-peat rooting medium under fine intermittent mist. This is the preferred method of propagation; bud-grafting onto highly susceptible American elm seedlings should be avoided.

Availability

Small quantities of rooted cuttings will be available for research only to interested experiment stations upon request. Rights for commercial development of the 'American Liberty' elms in the United States have been assigned to the Elm Research Institute. Inquiries can be directed to The Elm Research Institute, Harrisville, NH 03450.

At the present time we do not recommend the use of the 'American Liberty' multiclone for the UK or Europe where the large elm bark beetle *Scolytus scolytus* (F.) with its more vigorous feeding habits prevails.

Conclusions

Clones comprising 'American Liberty' multiclone in field tests possessed superior resistant to *O. ulmi* when contrasted with non-selected susceptible American elms (Table 4). Greenhouse trials tend to confirm the field results, but indicated an interesting attribute in that all the members of the multiclone possessed resistance to non-aggressive strains of the pathogen, while being more susceptible to the aggressive strain. They were in fact quite similar in this respect to the responses of the English elm *U. procera* to the DED pathogens (Fig. 1 & 2). Their resistance, while clearly greatly improved over non-selected American elms, still remain inferior to resistant cultivars such as 'Sapporo Autumn Gold' or 'Regal' derived from Asian or European sources (Smalley and Lester 1973; 1975; 1983; 1984). However, their resistance to *O. ulmi* represents the highest level that we have thus far achieved by selection and breeding within the tetraploid *U. americana.* The long-time survival and mature plant resistance of parents of the 500 series and the clone M-8 made the 'American Liberty' multiclone worthy candidates for release and public trial. The new Chinese x American hybrids may provide a bridge to introduce major genes for disease and insect resistance which retaining the hardiness, form and other features of the American elm.

W502 and W510 were clearly the outstanding selections among the various members of the 'American Liberty' multiclone, both from the view-point of both the field and the greenhouse tests. The parents of W502 (W185-11 and R18-2) are also unique. W185-11 is still alive after almost 35 years in the Arlington plantation, while R18-2 although not succumbing to DED did finally die from elm yellows disease (Sinclair 1985). In contrast, clone W510 was derived from susceptible parents and emphasizes the fact that about 1% of the natural American elm population is resistant to DED, and that tetraploid inheritance of resistance genes can be very complex.

The lack of response to the non-aggressive strains of the DED pathogen in greenhouse trials is of great interest and may offer clues to the design of improved basic studies on the nature of virulence in *O. ulmi sensu latu* and in the natural development of resistance responses in the American elm.

The multi-clone approach to release of American elms for urban planting may counter the problem inherent in wide-scale planting of identical genotypes with limited genes for resistance to DED (Browning and Frey 1969; Muhs 1982). It is anticipated that as our research progresses, additional and perhaps improved individuals will be added to the 'American Liberty' multiclone.

The problem of reduced resistance in vigorously growing ramets from resistant parents has been a concern to us and has also been observed by other investigators (Sinclair and Brener 1974; Sinclair and Larsen 1980; Sinclair et al. 1974) (Table 4). Under favorable conditions (e.g. wet springs and adequate soil moisture, freedom from weed competition and abundant elm bark beetle populations), rapidly growing young ramets of the 'American Liberty' elms can succumb to systemic Dutch elm disease, although their period of susceptibility under these conditions will be much less than comparable non-selected American elms (Table 4). In the case of the resistant clonal selection M-8, the peak period extended from May 25 to June 1 during the 1974-75 seasonal susceptibility trial (Table 4). In urban plantings, we have not seen major losses from Dutch elm disease in young plantings of 'American Liberty' elm probably because of the young trees limited attractiveness to elm bark beetles. We have rarely, if ever, observed natural bark beetle transmissions of *O. ulmi* in young, vigorously growing elms in our nursery.

Finally, we postulate that under conditions of *Scolytus multistriatus* or *Hylurgopinus rufipes* transmission that vigorously growing ramets which are not infected during the first few years of growth will generally not become infected in later years after reaching maturity. Parents of the 'American Liberty' elms have survived for many years in situations of severe disease and bark beetle pressure. The oldest of the surviving parents is now more than 35 years old and at this time remains healthy and disease free.

Acknowledgments

Research was supported by the College of Agricultural Life Sciences, University of Wisconsin-Madison, and by grants from the Elm Research Institute, Harrisville, NH 03450. Technical assistance of A. Gkinis, C. J. Gale, A. Birkholz-Lambrecht, J. L. Adams, G. Larson, R. G. Showers and the late L. Haakensen and assistance of S. A. Vicen for preparation of the photographs is gratefully acknowledged. We thank Dr. W. A. Sinclair and Dr. F. S. Santamour for pollen collection from R18-2 and NA368, respectively.

Reference

Ager, A. A., and R. P. Guries. 1982. Barriers to interspecific hybridization in *Ulmus americana*. Euphytica 31:909-920.

Brasier, C. M. 1991. *Ophiostoma novo-ulmi* sp. nov., causal agent of the current Dutch elm disease pandemic. Mycopathologia 115:151-161.

Browning, J. A., and R. J. Frey. 1969. Multiline cultivars as a means of disease control. Ann. Rev. Phytopathology 7:355-382.

Chen, M. M., E. B. Smalley, and R. P. Guries. 1988. Comparative resistance of Chinese sources of *Ulmus pumila* to North American and Eurasian strains of *Ceratocystis ulmi*. Proc.Beijing International Symposium of Plant Pathology, Beijing, China, Sept. 1-5, 1988. 3-1 to 3-2.

Clapper, R. B. 1952. Breeding and establishing new trees resistant to disease. Econ. Bot. 6:271-293.

Clark, E. F. 1984. Biochemical aspects of resistance to Dutch elm disease. M.S. Thesis. University of Wisconsin-Madison. 109 pp.

Collins, P. E. 1968. Hybridization studies in the genus *Ulmus*. Ph.D. Thesis, Univ. Minnesota. 84 pp.

Dermen, H., and C. May. 1966. Colchiploidy of *Ulmus pumila* and its possible use in hybridization with American elm. Forest Sci. 12:140-146.

ERI. 1992. Personal communication from the Elm Research Institute, Harrisville, NH.

Gkinis, A. 1977. Factors affecting resistance of elms to Dutch elm disease. Ph.D. Thesis. University of Wisconsin-Madison. 195 pp.

Heybroek, H. M. 1976. Systematiek en nomenclatuur van het geslacht *Ulmus*. Groen 32(8):237-240.

Lester, D. T. 1970. Attempts to induce polyhaploidy in American elm. Forest Sci. 18:137-138.

Lester, D. T. 1971. Self-compatibility and inbreeding depression in American elm. Forest Sci. 17:321:322.

Lester, D. T., and M. J. T. Lee. 1974. Twins and triplets in American elm. Forest Sci. 20:142.

Lester, D. T., and E. B. Smalley. 1969. Prospects for elm breeding in Wisconsin. Proc. 6th Central Forest Tree Improvement Conference. "1968" pages 37-43.

Lester, D. T., and E. B. Smalley. 1972. Improvement of elms through interspecific hybridization with Asian species. IUFRO Genetics-Sabrao Joint Symposia, Tokyo 1972. C-5(V):1-10.

McGranahan, G. H. 1982. *Stegophora ulmea*: Biology and host resistance. Ph.D. thesis. University of Wisconsin, Madison. 151 pp.

Muhs, H. J. 1982. Identifying problems and legal aspects of using multiclonal mixtures. In: Proc. IUFRO Joint Meeting of Working Parties on Genetics about Breeding Strategies Including Multiclonal Varieties. pp. 92-109.

Proctor, R. H. 1990. Phytoalexins and Dutch elm disease resistance. Ph.D. Thesis, University of Wisconsin-Madison. 132 pp.

Pomerleau, R., and J. Bard. 1968. Hardiness and resistance to *Ceratocystis ulmi* (Buism.) C. Moreau of hybrids and clones of European and American elms. Can. Dep. Forest. and Rural Develop., Bimonthly Res. Notes 24:26.

Santamour, F. S. 1970. Natural hybrids between Chinese and cedar elms in Texas. Bull. Torrey Bot. Club 97:72-73.

Santamour, F. S. 1970. A natural hybrid between American and Siberian elm. Forest Sci. 16:149-153.

Santamour, F.S. 1972. A triploid elm *(Ulmus pumila x U. rubra)* and its aneuploid progeny. Bull. Torrey Bot. Club. 98:310-314.

Santamour, F. S. 1972. Aneuploid tri-specific elm hybrids. Forest Sci. 18:290-291.

Santamour, F. S, 1972. Interspecific hybridization with fall- and spring-flowering elms. Forest Sci. 18:283-289.

Santamour, F. S. 1973. Resistance to Dutch Elm disease in Chinese elm hybrids. Plant Dis. Reptr. 57:997-999.

Santamour, F. S. 1974. Resistance of new elm hybrids to Dutch elm disease. Plant Dis. Reptr. 58:727-730.

Santamour, F. S. 1989. Flowering and fertility of hybrids between spring- and fall-flowering elms. HortScience. 24:139-140.

Sinclair, W. A. 1985. Personal communication.

Sinclair, W. A., and W. D. Brener. 1974. Dutch elm disease in clones from white elms resistant and susceptible to *Ceratocystis ulmi.* Phytopathology 64:675-679.

Sinclair, W. A., and A. O. Larsen. 1980. Localization of Dutch elm disease in 10-year-old white elm clones from resistant parents. Plant Disease 64:203-205.

Sinclair, W. A., J. P. Zahand and J. B. Melching. 1975. Localization of infections in American elms resistant to *Ceratocystis ulmi.* Phytopathology 65:129-133.

Sinclair, W. A., D. S. Welch, K. G. Packer and L. J. Tyler. 1974. Selection of American elms for resistance to *Ceratocystis ulmi.* Plant Disease Reporter 58:784-788.

Smalley, E. B., and A. G. Kais. 1966. Seasonal variations in the resistance of various elm species to Dutch elm disease. In: H. D. Gerhold, E. J. Schreiber, R. E. McDermott and J. A. Wenieske (eds.). Breeding Pest Resistant Trees. Pergamon Press, NY. pp. 279-292.

Smalley, E. B., and D. T. Lester. 1973. 'Sapporo Autumn Gold' elm. HortScience 8:51-515.

Smalley, E. B., and D. T. Lester. 1975. Elm tree. U.S. Plant Patent #3780. Patented Sept. 9, 1975. Assignor to Wisconsin Alumni Research Foundation, Madison, WI.

Smalley, E. B., and D. T. Lester. 1983. 'Regal' elms. HortScience 18(6):

Smalley, E. B., and D. T. Lester. 1984. Elm tree called Regal. U.S. Plant Patent #5335. Patented Nov. 13, 1984. Assignor to Wisconsin Alumni Research Foundation, Madison, WI.

Smalley, E. B., and D. T. Lester. 1985. Unpublished research.

Smucker, S. J. 1944. Rebuilding the American elm. Am. For. 50:104-107, 137-138.

Sticklen, M. B., M. G. Bolyard, R. K. Hajela, and L. C. Duchesne. 1991. Molecular and cellular aspects of Dutch elm disease. Phytoprotection 72:1-13.

Townsend, A. M. 1985. Personal communication.

Townsend, A. M., and W. O. Masters. 1984. 'Pioneer' elm. HortScience 19:900.

Townsend, A. M., and L. R. Schreiber. 1975. Recent progress in the breeding and selections of elms. Proc. 1975. Central States Forest Tree Improvement Conference 9:1-6.

Ware, G. 1992. Personal communication.

WARF. 1992. Personal communication from the Wisconsin Alumni Research Foundation, Madison, WI.

Wester, H. 1972. Investigations of American elm clones based on physiological vigor for crown resistance to Dutch elm disease. Plant Dis. Reptr. 56:725-727.

Whitten, R. R., and R. U. Swingle. 1948. The status of research on two epidemic elm diseases. Proc. Nat. Shade Tree Conf 24:113-120.

5
Progress in the Development of Disease-Resistant Elms

Alden M. Townsend and Frank S. Santamour, Jr.

Abstract

An extensive program designed to genetically enhance elms is being carried out by the U. S. National Arboretum, USDA-ARS, Washington, DC. The emphasis is on breeding for greater resistance to the fungus, *Ophiostoma ulmi* (Buism.) C. Nannf., that causes Dutch elm disease. The cultivars released to date are diploid in chromosome number and are 'Urban', 'Dynasty', 'Homestead', 'Pioneer', 'Frontier', 'Prospector', 'Pathfinder', and 'Ohio'. Advanced generation crosses have been made in order to combine tolerance to Dutch elm disease with resistance to the elm leaf beetle and have resulted in over 700 seedlings, three of which have been sent out for nursery evaluation because of their superior pest resistance. With American elm, more than 20 promising clones are being compared for superiority in disease resistance in replicated field trials. Future emphasis will be on the evaluation and release of exceptional American elms and advanced generation hybrid clones.

Introduction

The modern urban environment represents a formidable challenge for landscape and street trees. Diseases, insects, drought, compaction, deicing salts, and air pollution frequently are present to bombard the newly planted tree, which tries to survive and grow as best as it can.

We know that wide variation exists among tree genera in their ability to withstand the rigors of metropolitan sites. Elms generally have proven superior in this regard; their inherent vigor, hardiness, and tolerance to stresses are unsurpassed by most tree genera. And if elms are bred to incorporate insect- and disease-resistance, they will require less maintenance and spraying than do other tree groups.

Because of their superior adaptability and aesthetic attributes, the elms are the focus of a long-term breeding program of the U.S. National Arboretum. Objectives are to produce new research information and to release new disease-

and insect-resistant cultivars with variability in growth rate, form, bark, leaf color, and leaf size. This paper briefly summarizes the work to the present.

Breeding Work and Variation in the Pathogen

Breeding work in the 1970's was concentrated on determining hybridization potential among diploid (non-American) elm species and hybrids and screening hundreds of seedlings with diverse parentages for resistance to *Ophiostoma ulmi*, the fungus which causes Dutch elm disease. Barriers to gene exchange generally were not present between most diploid species, although the success of many crosses depended on which species was used as the female and which was used as the male (Townsend 1975). Unfortunately, attempts to cross an artificially induced tetraploid Siberian elm (*Ulmus pumila* L.) with American elm failed to yield any hybrids.

In order to develop higher levels of disease-resistance, we have tried to use as parents those clones and species which had proven themselves as resistant or had unusual horticultural qualities. Selections of many Asiatic and American species, along with many important clones from the Netherlands, were used to create more than 60 combinations of hybrid elms (Townsend 1979). Many of the controlled crosses have yielded progenies which are superior to the parents in disease resistance, growth rate, and form. When crossed with *U. pumila*, five resistant hybrid clones varied in the amount of resistance to Dutch elm disease transmitted to progeny (Townsend and Schreiber 1976). The transmission of varying degrees of resistance from equally resistant elm clones suggests that resistance to Dutch elm disease is polygenic, therefore lending support to the hypothesis made by other investigators (Arisumi and Higgins 1961; Lester and Smalley 1972; Santamour 1973). We have also found that specific combining ability is more important than general combining ability in transmitting resistance to *O. ulmi*; and that some clones function more effectively as female parents than as male parents in transmitting resistance genes to progeny (Townsend 1979).

In 1980, we initiated advanced generation breeding following a diallel scheme to make a multitude of crosses among the very best of our selections. Also included as a parent was a superior selection of *U. wilsoniana* Schneid., which was used as a source of resistance to the elm leaf beetle (*Xanthogaleruca luteola* (Muller)). Over 700 new hybrids were created as a result of this major crossing effort, and many have shown both insect- and disease-resistance (for example, see Hall and Townsend 1987). Also included in the 1980's were controlled pollinations made between superior clones of American elm.

Variation in the pathogen is an important consideration in any plant breeding program. We have found significant differences in foliar and vascular symptoms after inoculation with different isolates of *O. ulmi* from different geographic areas (Schreiber and Townsend 1976). Isolates generally were either aggressive or nonaggressive, and intermediates were rare. Even when different plant progenies from a wide variety of genetic backgrounds were inoculated, the fungal strains showed consistency in aggressiveness or nonaggressiveness. This lack of specific host-pathogen interaction is good evidence that resistance to *O. ulmi* is horizontal, not vertical. We currently use a mixture of aggressive and nonaggressive isolates for testing resistance of our selections.

Selection Phase

We will briefly describe the process of selection for disease resistance. After crosses are made, fruit are collected and planted in the greenhouse; and the resulting seedlings are grown on and later planted in replicated progeny tests outdoors. After 3 years, seedlings are inoculated in the bottom of the trunk with a mixture of aggressive and nonaggressive strains of the fungus. Those trees which show few or no disease symptoms and also exhibit superior horticultural attributes are propagated and the propagules are planted in replicated plots. After several years of growth, different propagules are inoculated at different times during the growing season to determine the relative degree of disease resistance. Both foliar symptoms and branch dieback are used as measures of resistance. Clones which pass this rigorous process are further propagated and distributed to cooperators nationwide to evaluate environmental adaptability and both horticultural desirability and marketability. Clones are released if they ultimately show high biological and economic superiority.

Cultivars Released

The following cultivars have been released by the U.S. Department of Agriculture. Brief descriptions are given.

'Urban'. Released by L. R. Schreiber and H. V. Main (Schreiber and Main 1976), this cultivar is a hybrid between *U. pumila* and a Dutch selection, N148 (*U.* x *hollandica* 'Vegeta' x *U. carpinifolia* Gleditsch). It has proven itself to be highly resistant to *O. ulmi* and to environmental stresses, but is somewhat susceptible to the elm leaf beetle. It develops a pyramidal crown form and shows a beautiful golden yellow autumn leaf color.

'Dynasty'. Released by F. S. Santamour, Jr. in 1984 (Santamour 1984), this cultivar resulted from a controlled cross between two *U. parvifolia* Jacq. parents. It has demonstrated high resistance to Dutch elm disease and to the elm leaf beetle, and an attractive vase-shaped crown.

'Homestead'. This cultivar has a parentage of a superior selection of *U. pumila* used as the female parent and a complex hybrid between two Dutch hybrids, (*U.* x *hollandica* 'Vegeta' x *carpinifolia*) x (*U. pumila* x *U. carpinifolia*) used as the male parent (Townsend and Masters 1984). A rapid grower, it has a pyramidal crown, a high level of resistance to Dutch elm disease, and excellent tolerance to environmental stresses.

'Pioneer'. This globe-shaped cultivar resulted from a cross between selections of *U. glabra* Huds. (female) and *U. carpinifolia* (male) (Townsend and Masters 1984). It is quite resistant to Dutch elm disease but like 'Homestead', has some susceptibility to the elm leaf beetle. Its crown is dense and provides deep shade.

'Prospector'. This is a seedling selection of *U. wilsoniana*, with a vase-shaped crown like American elm but becoming pendulous at a much lower height (Townsend et al. 1991a). It is resistant to Dutch elm disease and to the elm leaf beetle.

'Frontier'. 'Frontier is derived from a selection of *U. carpinifolia* crossed as a female with a superior selection of *U. parvifolia* (Townsend et al. 1991b). It has a pyramidal crown and red-purple autumn leaf color, moderate resistance to the elm leaf beetle and good resistance to Dutch elm disease.

'Pathfinder'. Released in 1991, this is a superior selection of *U. parvifolia* with a broadly upright crown, high disease and insect resistance, and an attractive greyed red autumn leaf color combined with "lacebark" bark qualities.

'Ohio'. 'Ohio' was released in 1992 and is also an exceptional new cultivar of *U. parvifolia*, with a moderately vase-shaped crown, red autumn leaf color and attractive bark characteristics.

Future Prospects

Future emphasis in our program will focus on evaluation and release of some of the advanced generation clones discussed earlier along with intensive screening and release of the very best American elms. In May 1992, we inoculated about 700 trees representing more than 20 clones in a seasonal susceptibility test growing at Glenn Dale, Maryland. Included in the test are many American elms as well as some diploid hybrid clones with *U. wilsoniana* in their parentage. Several of these advanced generation diploid clones are showing a desirable blend of characteristics, including a vase-shaped crown and good resistance to both *O. ulmi* and to the elm leaf beetle. Analysis of data collected in this study in 1992 and 1993 will help us determine which of these clones we should propagate and send out for further evaluation nationwide.

We have identified more than 20 American elm clones that have shown good resistance to *O. ulmi*. Some of these have resulted from controlled pollinations we have made; whereas others have shown recovery from natural or artificial inoculations. Among the most promising of these are American elm clones 2, 180, 680, 8630, and 11. Many of the 20 clones were inoculated in May, 1992, but in order to develop a comprehensive test of all of our promising American elm clones, we plan to establish in 1993 a large replicated plot of these for later intensive inoculation and evaluation. We hope this test will show us definitively which American elm clones express the highest level of resistance to *O. ulmi*.

References

Arisumi, T., and Higgins, D. J. 1961. Effects of Dutch elm disease on seedling elms. Phytopathology 51:847-850.

Hall, R. W., and Townsend, A. M. 1987. Suitability of *Ulmus wilsoniana*, the 'Urban' elm, and their hybrids for the elm leaf beetle, *Xanthogaleruca luteola* (Muller) (Coleoptera: Chrysomelidae). Environ. Entomol. 16:1042-1044.

Lester, D. T., and Smalley, E. B. 1972. Response of backcross hybrids and three-species combinations of *Ulmus pumila*, *U. japonica*, and *U. rubra* to inoculation with *Ceratocystis ulmi*. Phytopathology 62:845-848.

Santamour, F. S., Jr. 1973. Resistance to Dutch elm disease in Chinese elm hybrids. Plant Dis. Rptr. 57:997-999.

Santamour, F. S., Jr. 1984. 'Dynasty' Chinese elm. 1984. HortScience 19:898-899.

Schreiber, L. R., and Main, H. V. 1976. 'Urban' elm. HortScience 11:517-518.

Schreiber, L. R., and Townsend, A. M. 1976. Variability in aggressiveness, recovery, and cultural characteristics of isolates of *Ceratocystis ulmi*. Phytopathology 66:239-244.

Townsend, A. M. 1975. Species crossability patterns and morphological variation among elm species and hybrids. Silvae Genetica 24:18-23.

Townsend, A. M., and Schreiber, L. R. 1976. Resistance of hybrid elm progenies to *Ceratocystis ulmi*. Phytopathology 66:1107-1110.

Townsend, A. M. 1979. Influence of specific combining ability and sex of gametes on transmission of *Ceratocystis ulmi* resistance in *Ulmus*. Phytopathology 69:643-645.

Townsend, A. M., and Masters, W. 0. 1984. 'Homestead' elm. HortScience 19:897-898.

Townsend, A. M., and Masters, W. 0. 1984. 'Pioneer' elm. HortScience 19:900.

Townsend, A. M., Schreiber, L. R., Masters, W. O. and Bentz, S. E. 1991a. 'Prospector' elm. HortScience 26:81-82.

Townsend, A. M., Schreiber, L. R., Masters, W. 0. and Bentz, S. E. 1991b. 'Frontier' elm. HortScience 26:80-81.

6
An Old Problem; A New Approach

Lawrence R. Schreiber

Abstract

Dutch elm disease caused by *Ophiostoma ulmi* (Buisman) Nannf. has been one of the most destructive tree diseases in North America. The earliest attempts at control involved selection resistant germplasm from within *Ulmus Americana* L., American elm, followed by use of resistant selections from The Netherlands. Finally elm breeding programs were initiated in North America. Tissue culture technology is being studied to reduce the time and cost to identify and incorporate resistant germplasm into new selections of elms. The North American control program emphasized sanitation and chemicals to protect the thousands of highly valued American elms still in the landscape. Environmental considerations limited use of chemicals and encouraged exploration into non-chemical or biological control. Traditional use of biological control, that is pitting an organism or its toxic products against the pathogen, met with varying but generally limited success. With the advent and continuing progress in biotechnology, many of the problems associated with biological control may be overcome and lead to applied disease control.

Background

Dutch elm disease (DED) was first discovered in North America, in Ohio, in 1930 and later near the port of New York in 1933. The pathogen *Ophiostoma ulmi* (Buis.) Nannf., was assumed to have been introduced from Europe on elm logs used for veneer. The disease was soon noted along railroad right-of-ways between eastern ports of entry and midwestern states. In the United States the disease has since spread throughout the host range, natural and introduced, of susceptible elms which encompasses about 42 of the 48 contiguous states.

The disease has caused extremely heavy losses of trees and has significant economic impact because of the widespread planting of the highly susceptible American elm (*Ulmus americana* L.). This elm species was frequently planted in the urban landscape as a monoculture and is also found as wild populations throughout the countryside.

It was widely recognized by scientists in the U.S. Department of Agriculture, that the most effective means of disease control was the selection and planting of disease resistant trees. As a result, American elm seed was collected from entire host range of the species, planted in test nurseries and inoculated with the hope of selecting DED-resistant individuals. As a result of this program, a few trees were selected that possessed both disease resistance and acceptable horticultural form. However, it became obvious from this program that there was little hope of selecting many resistant American elms with naturally high levels of resistance. Elm breeding and selection programs, for control of DED, had been underway and were productive in Europe, particularly in The Netherlands. It was initially believed that the products of these ongoing programs could be utilized in North America and that research for control strategies should be directed to the protection of the numerous, valuable and susceptible trees still in the landscape.

Unfortunately, the supposition that resistant elm germplasm from Europe could satisfy the needs for trees in North America proved erroneous. Elm selections introduced from The Netherlands proved not to be hardy throughout the natural range of the American elm. In addition, aggressive strains of the pathogen found in North America and later introduced into Europe, overcame the resistance in the earlier Dutch selections that survived inoculations with nonaggressive European fungus strains.

Chemical Control

Emphasis for control shifted, in the United States, from use of disease resistance to chemical control of the insect vector and of the pathogen to protect susceptible elms in the landscape. These chemical control programs, along with sanitation of diseased trees to limit elm bark beetle breeding sites and inoculum reservoirs, constituted the main efforts in limiting the spread of the disease.

Literally hundreds of chemicals have been tested for control of the DED pathogen. Application methods involved trunk injections. This technique required solubilized material to assure uptake and distribution in the tree and thus the greatest opportunity for successful control. With most fungitoxicants tested there was a small difference in concentration levels that were efficacious and those that were phytotoxic. The greatest promise for chemical disease control occurred with the advent of solubilized benzimidazoles.

Benzimidazole fungicides showed high levels of toxicity toward many plant pathogenic fungi. They also process low levels of phyto- and mammalian toxicity. Because of their systemic qualities they move through a diseased plant to the sites of infection. Because of these characteristics, they have been used extensively for plant disease control.

With increased use, reports appeared of failure to control diseases that had been previously effectively controlled (Bertrand and Saulie-Carter 1978; Bollen 1971, Bollenand Scholten 1971; Goldberg and Cole 1973; Warren et al. 1974; Littrell 1974; Smiley and Craven 1977). In some cases, as with apple scab, control failures appeared only three years after successful use (Tate and Samula 1976). These failures were associated with increased tolerance of the pathogen to the chemical. The possible cause of this sudden appearance of tolerance may result from mutation, physiological adaptation or selection within a mixed population of sensitive and tolerant strains. Studies at our laboratory indicated that the latter possibility was the most likely (unpublished data).

Mutations may also play a role since benzimidazoles may act as mutagenic agents (Dassenoy and Meyer 1973; Seiler 1972). Physiological adaptation does not appear to play a role in the development of tolerance since the characteristic remains stable in the absence of the chemical.

Fungal tolerance to nonsystemic chemicals occurs at a far lower rate than with systemics due to differences in modes of action. Nonsystemics are not as specific in their activity and control depends upon activity of multiple genes. Systemic fungicides on the other hand act at specific metabolic sites (Davidse 1974). Thus, high levels of fungus resistance may result from single gene mutations (Ben-Yepht et al. 1974; Brasier and Gibbs 1975; Kondo 1978). However, there is no evidence of mutagenic activity at concentrations used for disease control. Thus, these chemicals act in a selective rather than in a mutagenic capacity in mixed populations of both sensitive and tolerant individuals.

The development of, and the role that tolerance might play in control of DED was of great interest since only the injection of solubilized benzimidazoles into elms has offered any practical control of DED. We serially exposed sensitive strains of *Ophiostoma ulmi* (Buis.) Nannf. to low concentrations of MBC-HCl. After 10 exposures isolates were found with tolerances between 10 ppm and 500 ppm (unpublished data). This too indicated a selective rather than an adaptive mechanism since the latter would have produced levels of tolerance closer to the exposure concentration.

While there is general agreement about the mechanisms by which tolerance develops, its presence in nature and the role this would play in chemical control is not clear. Schreiber and Townsend (1976) reported benzimidazole tolerance naturally occurring up to 1,000 ppm from areas of no known benzimidazole use. Others reported similar results (Kondo 1978).

Benzimidazole tolerance develops when large populations of a pathogen have been exposed to the chemical under field or storage conditions. Such conditions do not occur in control situations with DED since relatively few diseased trees would have the aesthetic or economic value to warrant chemical treatment. Also, contact between the chemical and pathogen is limited to the confines of the tree.

Treatment of healthy trees, for prophylaxis, further limits contact between chemical and fungus needed for selection for tolerant strains. However, within these limitations the opportunity may exist for the selection of resistant mutants.

Fungitoxicity of benzimidazoles, as with other fungitoxicants, depends upon the time and concentrations of exposure. At lower concentrations and shorter times, the effect is fungistatic; at higher concentrations and longer exposure times the effect is fungicidal. At concentrations used for DED treatment, toxicity occurs at the fungistatic level. Schreiber et al. (1986) reported that based upon the factors of aggressiveness, competitiveness and stability of tolerance of tolerant isolates, compared with sensitive wild types, tolerant mutants could impact negatively on disease control.

Biological Control

Biological control of plant diseases has long been sought as a alternative to chemical control. The emphasis for such an approach has been accelerated by a greater awareness of possible detrimental effects of chemicals to the environment, greater restrictions on and decreased availability of chemical control agents and the need for continuous reapplication to achieve control.

Biological control may be defined in terms of a microorganism or of its metabolites that act in an antagonistic manner to a plant pathogen to prevent disease development. In a broader sense, biological control may be any naturally occurring mechanism or interaction resulting in the production of toxicants that reduce or inhibit disease development. Such a definition would include products of a resistant reaction such as phytoalexins or those naturally occurring in resistant plants. Biological control involving plant resistance or use of cultural practices has been used successfully for many years while successes involving antagonists and/or their toxic metabolites have been slower to develop (Baker and Cook 1974). Biological control has been studied extensively for the control of soil borne pathogens. The biological and environmental aspects of using antagonists are complex and their interactions are critical to chances of successful control. Some of the principles learned from other biological control systems may be applicable to use in control of vascular wilt organisms.

One of the chief goals of biological control is the reduction or restriction of formation of pathogen inoculum. This is accomplished through the use of hyperparasites or predators which may reduce inoculum density below a critical threshold necessary for disease (Papavizas and Lumsden 1980) or reduction in virulence, aggressiveness or survival ability by agents such as microviruses (Ghabrial 1980; Warren et al. 1974). Microviruses or virus-like agents are common and generally do not effect the host. Fungus infection by such agents can cause hypovirulence, loss of vigor or death of the fungus (Ghabrial 1980; Rawlinson and Buck 1981).

Such an example is hypovirulence of the chestnut blight fungus (*Cryonectria parasitica*) caused by a transmissible dsRNA factor in the cytoplasm (Tate and Samuel 1976; Day et al. 1977). Hyphae of the virulent and hypovirulent strains anastomose with a transfer of the hypovirulent factor to the virulent strains the canker heals (Grente and Sauret 1969a, 1969b).

In the soil situation, various techniques are directed at control by changing the equilibrium between the antagonist and pathogen to suppress the latter. This approach may involve the mass introduction of a large population of the antagonist with or without a change in the soil environment such as the food base that favors the antagonist and/or is detrimental to the pathogen. Natural soils are not conducive to artificial changes in microbial equilibrium and without environmental changes such as food sources (Garrett 1956) such biological imbalances are short lived. Even with manageable changes in the food base, the soil system will soon return to its natural, biological equilibrium (Baker and Cook 1974; Papavizas and Lumsden 1980). Under such conditions, biological control may be effective with annual, but not perennial crops, or perennials that need protection during short periods of susceptibility i.e. during and shortly after seed germination.

In rare cases, biological control can be enhanced through combination with chemicals that adversely effect the pathogen but not the antagonist or reduce competitors of the antagonist (McKeen 1975).

Techniques used to select potential antagonists effect the probability of their success in disease control. Garrett (Garrett 1956) noted that antagonists selected in *in vitro* tests as being antagonistic generally failed in this role in *in vivo*.

Wild type strains of naturally occurring antagonists may be limited in their effect as biocontrol agents because of difficulties in manipulating the equilibrium of the soil system in favor of the antagonists against the pathogen. These problems would be reduced by the selection or creation of mutant forms that would be superior to wild types in survival, reproduction and toxin production. Some studies related to increased cellulase production in *Trichoderma* spp. have been reported (Mandels et al. 1971). Papavizas (Papavizas and Lewis 1981) reported the selection of fungicide-tolerant biotypes of the biocontrol fungus *T. harzianum* through exposure to increasing concentrations of various fungicides. Some of these selections exhibited increased survival ability and antagonism to soilborne pathogens. Changes in soil survival, sporulation, and antifungal activity occurred at random. In some cases the new biotypes were more effective than wild type in suppressing soilborne pathogens.

What are the prospects for effective biocontrol of vascular tree pathogens, particularly *O. ulmi*, and what lessons can we benefit from in the area of biological control of soil pathogens? The area for host-parasite interaction in the tree xylem and surrounding parenchyma is spatially more restricted compared to a similar area in the soil. Chemical, physical and biological factors in the tree environment are more amenable to analysis and thus to testing as to their influence on host parasite interaction and upon interactions with potential antagonists. They may however, be less amenable to controlled change.

Silver maples (*Acer saccharium* L.) are recognized as the least susceptible of the important *Acer* species to Verticillium wilt (Smiley and Craven 1977). Schreiber and Hock (Schreiber and Hock 1975) reported the presence, in silver maple seedlings, of naturally occurring toxicants to *O. ulmi* in bioassays. The source of these fungi-toxicants was not identified but the presence of antibiotic-producing, xylophytic microorganisms was considered a possible source. Studies by Hall et al. (1986) indicated that some naturally occurring isolates of *Bacillus subtilis* (Ehrenberg) Cohn from maples, but not other *Bacillus* spp., reduced *Verticillium* infection in silver and Norway maples. Antagonism of isolates *in vitro*, was not correlated with their activity in reducing infection *in vivo*.

Biological control of DED has been attempted by introducing *Pseudomonas syringae* van Hall (Myers and Strobel 1983), *T. viride* Pers. ex S. F. Gray (Ricard 1983) and *Stretomyces* spp. (O'Brien et al. 1984) into American elms with limited to no control. In most cases, prophylactic treatments as with chemical control procedures resulted in the highest levels of control. In field studies using *P. syringae*, Scheffer (1989) reported significant reduction in DED incidence. *In vitro* bioassay with a number of microorganisms have demonstrated antifungal activity against *O. ulmi* (Desrochers et al. 1987; Gregory et al. 1986; Krause et al. 1987). However, *in vivo* tests (Seiler 1972) were not successful in reducing tree mortality. *B. subtilis* has been studied as a potential biocontrol agent and was found to produce, *in vitro*, antimycotic agents to a number of vascular pathogens including *O. ulmi* (Gregory et al. 1984; Krause et al. 1987; Schreiber et al. 1986).

The step from identification of microorganisms and their toxicants, inhibitory to vascular pathogens including *O. ulmi*, to the use of these organisms in biocontrol has been only moderately successful at best. The goal in such a system is the identification and introduction of a microorganism that can colonize the elm xylem, produce antifungus toxicants at levels necessary to inhibit the pathogen and reproduce continually to maintain effective control and avoid the need for retreatment. In essence the potential control agents adaptability to the internal environment of the elm must equal that of the pathogen. One major problem in establishing an antifungal microorganism, even one that may occur naturally at concentration sufficient to maintain control, is overcoming the equilibrium of population concentrations existing in the host's areas of infection and disease development. To date, high concentrations of antifungal, microbial cells injected into elms had been maintained for relatively short periods. Hall et al. (1986) reported recovery of *B. subtilis*, injected into maple, for 17 wks. The necessity for continued microbial retreatment negates one of the advantages of biocontrol over use of chemicals. The complex interaction of pathogen, host, and microbial control agent, where the pathogen and host have evolved together over centuries, has limited success of this strategy.

Genetic engineering may provide the technical vehicle to circumvent the obstacles for successful control. Thus, the *in vitro* work on the isolation and identification of antimicrobial agents may now be followed by the identification of the loci for their production. Mechanisms and techniques of biotechnology may allow transfer of these genetic factors to the host. This approach, while undoubtedly complex, is theoretically possible. It would overcome the obstacles to use of microbial fungitoxicants in biocontrol that occur when making direct use of the microbe.

While these new techniques offer the possibility for solutions to old problems, they also call to the fore new challenges. These involve the unique problems associated with techniques of vegetative propagation necessary to derive intact plants from transformed protoplasts or callus tissue of American elms, the most likely target of such research. Progress has been made in regeneration of American elm by *in vitro* explant techniques (Bolyard et al. 1991)). Further work will be necessary to meet the needs dictated by the specific techniques of transformation.

Biotechnology has opened new possibilities for the successful resolution of the problems that have hindered improved control of DED. As with any new technology its potential may outstrip its accomplishments. Yet, the opportunity for additional advances exist in the areas of genetic engineering of resistant plants, vegetative regeneration, and biological control. Enough is known about the existing problems for control of DED and the potentials for biotechnology to be optimistic regarding the chances for significant future accomplishments.

References

Baker, K. T. and Cook, R. J. 1974. Biological control of plant pathogens. W. H. Freeman, San Francisco.

Ben-Yepht, Y., Henes, Y. and Dinoor, A. 1974. Genetic studies on tolerance of carboxin and benomyl at the sexual phase of *Ustilago hordei*. Phytopathology 64:51-56.

Bertrand, P. F. and Saulie-Carter, J. L. 1978. The occurrence of benomyl-tolerant strains of *Penicillium expansum* and *Botrytis cinera* in the mid-Columbia Region of Oregon and Washington. Plant Dis. Reptr. 62(4):302-305.

Bollen, G. J. 1971. Resistance to benomyl and some chemically related compounds in strains of *Penicillium* species. Neth. J. Pl. Path. 77:187-193.

Bollen, G. J. and Scholten, G. 1971. Acquired resistance to benomyl and some other systemic fungicides in a strain of *Botrytis cinerea* in cylamen. Neth. J. Pl. Path. 77:83-90.

Bolyard, M. G., Srinivasan, C., Cheng, Jianping, and Sticklen, M. 1991. Shoot regeneration from leaf explants of American and Chinese elm. Hortscience 26(12):1554-1555.

Brasier, C. M. and Gibbs, J. N. 1975. MBC tolerance in aggressive and nonaggressive isolates of *Ceratocystis ulmi*. Ann. Appl. Biol. 80:231-235.

Dassenoy, B. and Meyer, J. A. 1973. Mutagenic effect of benomyl on *Fusarium oxysporum*. Mutation Res. 21:119-120.

Davidse, L. C. 1974. Mode of action of methyl benzimidazole 2-yl carbamate (MBC) and some biochemical aspects of acquired resistance against this fungicide in *Aspergillus nidulans*. Pestic. Biochem. Physiol. 3:317-325.

Day, P. R., Doods, J. A., Elliston, J. E., Jaynes, R. A. and Anagnostakis, S. L. 1977. Double-stranded RNA in *Endothia parasitica*. Phytopathology 67:1391-1396.

Desrochers, P., Benhamou, N. Gardiner, R., and Ouellette, G. B. 1987. Inhibition of *Ophiostoma ulmi* by an ascomycete. Can. J. Plant Pathol. 9:276 (Abstr.).

Garrett, S. D. 1956. Biology of root infecting fungi. Cambridge Univ. Press. Cambridge, MA.

Goldberg, C. W. and Cole, H. 1973. *In vitro* study of benomyl tolerance exhibited by *Sclerotinia homeocarpa*. Phytopathology 63:201 (Abstr.).

Ghabrial, S. A. 1980. Effects of fungal viruses in their hosts. Ann. Rev. Phytopathol. 18:441-461.

Gregory, G. F., Schreiber, L. R. and Ichida, J. 1984. Microorganisms antagonistic to or producing antibiotics inhibitory to *Ceratocystis ulmi*. Phytopathology 74:804-805.

Gregory, G. F., Lewis, R., Schreiber, L., Roberto, N., Ichida, J. and Thomas, J. 1986. A *Pseudomonas* species isolated from live oaks antagonistic to several tree pathogens. Phytopathology 76:652-653.

Grente, J. and Sauret, S. 1969a. L'hypovirulence exclusive, phenome original en pathologie vegetale. C. R. Acad. Sci. Paris, Ser. C. 268:2347-2350.

Grente, J. and Sauret, S. 1969b. L'hypovirulence exclusive estelle controlee par des determinants cytoplasmiques. C. R. Acad. Sci. Paris, Ser. D. 268:3173-3176.

Hall, T. J., Schreiber, L. R., and Leben, C. 1986. Effects of xylem-colonizing *Bacillus* spp. on Verticillium wilt in maples. Plant Disease 70:521-524.

Hastie, A. C. and Georgopoulos, S. D. 1971. Mutational resistance to fungitoxic benzimidazole derivatives in *Asperqillus nidulans*. J. Gen. Microbiol. 67:371-373.

Kondo, E. D. 1978. Scope and limitations of carbendizin H2PO4 injections in Dutch elm disease control. J. of Arboriculture 4(4):80-86.

Krause, C. R., Ichida, J. M., Schreiber, L. R. and Gregory, G. F. 1987. Morphological and cytological effects on *Ceratocystis ulmi* of an antifungal antibiotic produced by *Bacillus subtilis*. Phytopathol. 77:1771 (Abstr.).

Littrell, R. H. 1974. Tolerance of *Cercospora arachidicola* to benomyl and related fungicides. Phytopathology 14:1377-1378.

Mandels, M., Webber, J., and Parizek, R. 1971. Enhanced cellulase production by a mutant of *Trichoderma viride*. Appl. Microbiol. 21:152-154.

McKeen, C. D. 1975. Colloquium on integration of pesticide-induced and biological destruction of soilborne pathogens: Summary and synthesis on pages 203-206 in G. W. Bruehl, ed. Biology and Control of Soil-Borne Plant Pathogens. Amer. Phytopathol. Soc., St. Paul, MN..

Myers, D. F, and Strobel, G. A. 1983. *Pseudomonas syringae* as a microbial antagonist of *Ceratocystis ulmi* in the apoplast of American elms. Trans. Br. Mycol. Soc. 80:389-394.

O'Brien, J. G., Blanchette, R. A. and Sutherland, J. B. 1984. Assessment of *Streptomyces* spp. from elms for biological control of Dutch elm disease. Plant Dis. 68:104-106.

Papavizas, G. C. and Lumsden, R. D. 1980. Biological control of soilborne fungal propagules. Ann. Rev. Phytopathology 18:389-413.

Papavizas, G. C. and Lewis J. A. 1981. Induction of new biotypes of *Trichoderma harzianum* resistant to benomyl and other fungicides. Phytopathology 71:247-248 (Abstr.).

Rawlinson, C. J. and Buck, K. W. 1981. Viruses in *Gaeumannomyces qraminis* and *Phialophora* spp. in P. J. Shipton and M. J. C. Asher, eds. Biology and Control of Take-all. Academic Press. London.

Ricard, J. L. 1983. Field observations on the biocontrol of Dutch elm disease with *Trichoderma viride* pellets. Eur. J. For. Pathol. 13:60-62.

Scheffer, R. J. 1989. *Pseudomonas* for biological control of Dutch elm disease. III. Field trials at various locations in The Netherlands. Neth. Plant Pathol. 95:305-318.

Schreiber, L. R. and Hock. W. K. 1975. Effects of benomyl and thiabendazole in growth of several plant species. J. Am. Soc. Hortic. Sci. 100:309-313.

Schreiber, L. R. and Townsend, A. M. 1976. Naturally occurring tolerance in isolates of *Ceratocystis ulmi* to methyl 2-benzimidazole hydrochloride. Phytopathology 66:225-227.

Schreiber, L. R., Conaway, E. E. and Peacock, J. W. 1986. Aggressiveness, competitiveness and stability of tolerance of benzimidazole-tolerant strains of *Ceratocystis ulmi*. Plant Disease 70:154-158.

Schreiber, L. R., Domir. S., Ichida, J. and Krause, C. 1988. Factors affecting host-pathogen interactions between elm callus cultures and *Ceratocystis ulmi* Phytopathology 78:1618 (Abstr.).

Seiler, J. P. 1972. Mutagenicity of benzimidazole and benzimidazole derivatives. I. Forward and reverse mutations in *Salmonella typhimurium* caused by benzimidazole and some of its derivatives. Mutations Res. 15:273-276.

Shi, J. L. and Brasier, C. M. 1986. Experiments on the control of Dutch elm disease by injection of *Pseudomonas* species. Eur. J. For. Pathol. 16:280-292.

Smiley, R. W. and Craven, M. M. 1977. Control of benzimidazole-tolerant *Fusarium roseum* in Kentucky bluegrass. Plant Dis. Reptr. 61(6):484-488.

Smith, L. D. and Neely, D. 1979. Relative susceptibility of tree species to *Verticillium dahliae*. Plant Dis. Rep. 63:328-332.

Tate, K. D. and Samuel, G. J. 1976. Benzimidazole tolerance in *Venturia inaegualis* in New Zealand. Plant Dis. Rept. 60(8):706-710.

Van Alfen, N. K., Jaynes, R. A., Anagnostakis, S. L. and Day, P. R. 1975. Chestnut blight: Biological control of transmissible hypovirulence in *Endothia parasitica.* Science 189:890-891.

Warren, C. G., Sanders, P., Cole, H. 1974. *Sclerotinia homeocarpa* tolerance to benzimidazole configuration fungicide. Phytopathology 64:1139-11422

Webber, J. F. 1987. Influence of the d^2 factor on survival and infection by the Dutch elm disease pathogen *Ophiostoma ulmi*. Plant Pathol. 36:531-538.

7
Demands and Opportunities for Selecting Disease-Resistant Elms

James L. Sherald

Abstract

The American elm is the unifying element of the Monumental Core of our Nation's Capital. The National Mall, the focal point of the Core, is composed of 600 elms framing the vista between the Washington Monument and the Capitol. The National Park Service (NPS) has a mandate to sustain the American elms of the Core. Although Dutch elm disease (DED) has been a threat to this resource since 1947, the NPS has successfully managed the disease and continues to plant elms. The most promising long-term approach for sustaining elms is to plant DED resistant American elms or other elms having similar forms. Several new selections are now being incorporated into the Core including two NPS selections believed to be resistant to DED, NPS 3-178 and NPS 3-487. In order to hasten the screening process and increase the number of selections available, the NPS has entered into a Cooperative Agreement with Michigan State University to develop cellular and molecular rapid screening techniques for detecting DED resistance. Approaches include measurement of the *in vitro* response of elm calli to the DED toxin, cerato-ulmin, and identification of nucleotide sequence variations between resistant and susceptible elms.

Introduction

The American elm was once the premier species planted in urban landscapes throughout the Northeast and Midwest. In no city, however, has the elm played as significant a role as in our Nation's Capital. The American elm was planted extensively throughout the early development and expansion of the city. Records of the Columbia Historic Society (1957-1959) reported that by 1958 the city's elm population had risen to over 38,000 trees. The graceful effect and long-term success of this tree on streets and in city parks prompted planners to select the American elm as the backbone of the federal park lands of the Monumental Core. In 1902 the Burnham Park Commission presented a development plan for the City of Washington (Olszewski 1970). This plan was inspired by the original concept developed by Major Pierre L'Enfant in 1791, whose center piece was the

"Grand Avenue" or National Mall as it is referred to today. The Mall provides a symbolic link between the executive branch of government represented by the Washington Monument and the legislative branch represented by the Capitol. In planning the implementation of L'Enfant's Grand Avenue, the Commission visited outstanding landscapes throughout the United States and Europe. Upon completion of their tour and analysis of the site they concluded that the most fitting design for the National Mall was an open greensward a mile long and three hundred feet wide bordered on each side by four rows of American elms spaced fifty feet apart (Figs. 1 and 2). In the Commission's report it was stated that:

> *The American elm was chosen not only because of the architectural character of its columnar trunk and the delicate traceries formed by its wide spreading branches, but also because the tree was found at its best in the District of Columbia, notable examples being found in the city parks and in the grounds of the Capitol* (Olszewski 1970).

Figure 1. Elms bordering the National Mall between the Washington Monument and the Capitol as viewed from the Washington Monument.

Figure 2. Vista of the Washington Monument flanked by American elms on the National Mall.

Since the American elm also has such a grand American tradition as a landscape tree, its prominent presence on the National Mall and throughout the Monumental Core provides a uniquely American cultural statement.

In addition to its aesthetic and cultural charm, the American elm has tremendous vigor. Few species are as tolerant of the poor soils and physical abuses so commonly encountered in the urban environment. The ability to tolerate abuse, particularly root disruption, and its rapid growth, so important in sustaining the visual integrity of uniform stands, are preeminent qualities encouraging continued use of the species in Washington and other municipalities.

Although the development plan for the Mall was first presented in the early 1900s, the first major planting did not occur until 1936, when the Mall development became a public works program. Dutch elm disease was first detected in the United States in the early 1930s and began to spread throughout the East just as the Mall was being planted. The elms of Washington remained free of Dutch elm disease until 1947, when the first case of the disease was detected on the grounds of the Lincoln Memorial by Horace V. Wester, Plant Pathologist of the National Park Service. For about 30 years Dutch elm disease claimed an annual loss of only 1% or less of the city's elm population. This early success was a consequence of a city-wide disease management program involving intensive sanitation and insecticide applications for the elm bark beetle. During the 1970s, however, city elm losses began to increase reaching an alarming rate of loss of 5 - 7% in the early 1980s. Over the last 45 years Dutch elm disease has claimed almost 2,000 elms throughout the Monumental Core. On the National Mall, over 200 trees have succumbed to the disease since its first occurrence there in 1952 (Fig. 3).

The sudden increase in disease incidence in the late 1970s prompted the organization of the Save-the-Elms Task Force in 1983 and the development of the "Management Program for the Perpetuation of the American Elm Tree in the Nation's Capital" in 1985. The Task Force consists of representatives of the six major land management agencies within the District of Columbia. The Management Plan specifies the full range of tactics to be used in a city-wide integrated disease management program. The most critical tactic consists of efficient and thorough sanitation.

The central tenet of the Management Program is that the American elm character must be sustained within the Monumental Core. The significance of the elm character within the Monumental Core has recently been reinforced and expanded by the "Interagency Initiative for National Mall Road Improvement." The Initiative plans to enhance the roads, walks, and street plantings of the Monumental Core over the next six years. As a consequence of this initiative the current elm population in the Monumental Core will increase from 2,500 to approximately 4,000 elms. Another major addition to the Monumental Core will be the construction of the Franklin Delano Roosevelt Memorial in West Potomac Park where existing and new elms are planned as a prominent major feature of the Memorial, just as they have been at the Jefferson and Lincoln Memorials.

Since the initiation of the Management Program in 1985, the National Park Service has been able to contain the Dutch elm disease incidence within the Monumental Core at about 1% in accordance with the stated objective of the Plan. Although Dutch elm disease losses are low, other factors also contribute to elm demise. Meeting the demands created by tree loss and expanded plantings have created a new challenge. Until recent years the wide commercial availability of American elms has enabled the National Park Service to replant elms and thus sustain the visual integrity of the resource. Municipal fear of Dutch elm disease and the subsequent lack of demand has all but eliminated commercial sources of American elm. This presents a major challenge to maintaining the overall integrity and continuity of the elm population. Much of the demand for elms will have to be met by production of seedlings and selected clones at National Park Service nursery facilities.

Although successful disease management has been achieved by the National Park Service through rigorous sanitation and occasional therapeutic fungicide injections, the most promising long-term approach for achieving a vigorous, aesthetically acceptable planting is through the use of multiple disease resistant selections with forms characteristic of the American elm. The principal tactic of the Save-the-Elms Management Program is to plant Dutch elm disease resistant selections of American elm, or other elm species with forms similar to the American elm, within the Monumental Core. Throughout the rest of the city managers are encouraged to dilute the elm population by converting elm streets to species other than elm where practical. The objective is to maintain the overall effect created by the American elm form, while reducing the risk of disease. The aesthetic objective is paramount and the challenge is to develop a biologically sound approach for achieving the objective.

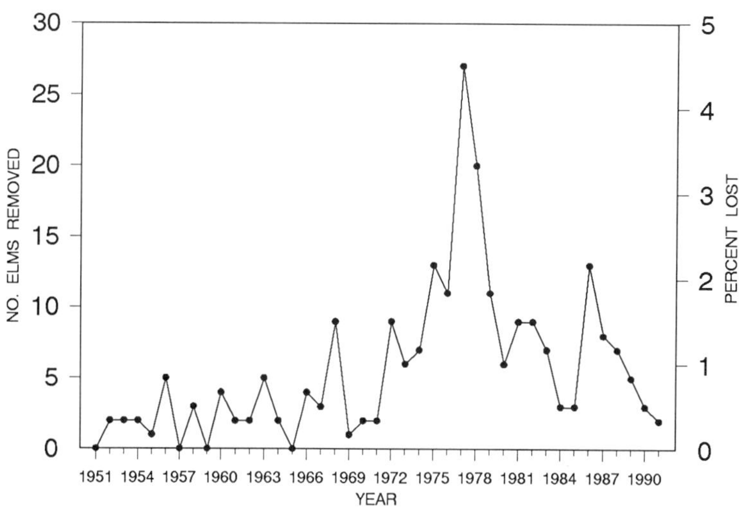

Figure 3. Incidence of Dutch elm disease on the National Mall from 1951 to 1991.

Although the Monumental Core is planted in predominantly American elm, over the years many cultivars and other species have been incorporated throughout the planting. Currently there are 11 species and 19 cultivars (Table 1). Many of these selections are resistant to Dutch elm disease and thus dilute the susceptible elm population. Although some do not have the traditional American elm form, their specific forms are generally lost in the overwhelming dominance of the American elm.

Table 1. Elm species and cultivars planted in the Monumental Core of Washington, D.C.

American elm	*Ulmus americana* L. Princeton Dedfree Augustine Ascending Delaware II Liberty Series NPS 3-487 NPS 3-178 'Washington' elm
September elm	*U. serotina* Sarg.
Smooth-leaved elm	*U. carpinifolia* Gleditsch Bea Schwarz Superosa
Dutch elm	*U.* x *hollandica* Mill. *U. carpinifolia x U. glabra)* Commelin Groeneveld Vegeta Dauvessei Blanford Belgica NPS 1409 NPS 443
Siberian elm	*U. pumila* L. arborea
Chinese elm	*U. parvifolia* Jacq.
English elm	*U. procera* Salisb.
Scotch or Wych elm Camperdown Elm	*U. glabra* Huds. Camperdownii
	U. wilsoniana x *U. japonica*
Sapporo Autumn Gold elm	*U. pumila* x *U. japonica*
Urban elm	(*U.* x *hollandica* 'Vegeta' x *U.* carpinifolia) x *U. pumila*

It is currently proposed that the projected elm population of 4,000 trees in the Monumental Core will be composed of both resistant selections of desirable form as well as seedling American elms. By continuing to use a limited number of seed propagated American elms the overall American elm appearance will be sustained even if some of the resistant clones do not fully achieve the ideal form.

Two of the cultivars currently being planted were selected from the original planting on the National Mall. Both selections were first noted by Horace Wester. These trees caught his attention because of their early spring foliation and late leaf retention in the fall. Early unpublished studies by Wester showed these trees to be resistant to crown inoculations of *Ophiostoma ulmi* (Buisman) Nannf. Both selections have spreading crown habits at maturity and are acceptable for the Monumental Core. Research is underway to confirm the resistance of these trees to Dutch elm disease and to determine the nature of their resistance.

One selection, NPS 3-178, is currently offered by Princeton Nurseries, Princeton, NJ, as the 'Washington' elm (Fig. 4). Unfortunately, the cultivar name 'Washington' has been used previously in naming clonally propagated trees from the American elm under which George Washington assumed command of the Colonial Troops at Cambridge, MA (Green 1964) and NPS 3-178 is not related to this tree.

The unnamed selection NPS 3-487 has been propagated and recently tested for resistance to Dutch elm disease (Fig. 5). Of 104 twig-crotch inoculations, only 15 (14%) became infected compared to 57 of 90 (55%) in wild-type American elm. None of the 22 inoculated 3-487 trees became systemically infected beyond the inoculated branch, while 8 of the 18 wild-type trees succumbed to the disease.

In addition to coming into leaf earlier than wild-type elms and retaining foliage later, 3-487 is also distinguished by its smooth bark when young. Recent chromosome counts performed on root tip cells at Michigan State University indicate that this tree is a triploid (M. B. Sticklen, unpublished). The parental derivation of NPS 3-487 is currently unknown. The tall spreading habit of the parent NPS 3-487 on the Mall and its resistance to Dutch elm disease holds promise for this selection to be one of the principal future clones used throughout the Monumental Core. Both NPS 3-487 and NPS 3-178 can be propagated by softwood cuttings.

Along with these two selections, other cultivars currently considered acceptable for use within the Monumental Core include the 'American Liberty' elm series from the University of Wisconsin and available through the Elm Research Institute, Harrisville, NH, and NA 8630 currently under evaluation by the National Arboretum. Hybrids derived from Asiatic and European species, while fine trees, do not have forms acceptable for widespread planting within the Monumental Core.

Figure 4. Dutch elm disease resistant American elm selection NPS 3-178

Figure 5. Dutch elm disease resistant American elm selection NPS 3-487

The chance occurrence of resistant elms throughout natural and cultivated American elm populations has been a source of encouragement since the threat of Dutch elm disease was first appreciated. Although resistance occurs in American elm, the opportunity for finding resistance is limited by the conventional screening approaches now in use. For example, the selection and testing of putative resistant trees from otherwise decimated elm populations relies on the traditional approach of selection, propagation, production, inoculation and evaluation. This long-term process requires test blocks of nursery grown trees. The cumbersome nature of the process prohibits massive screening of large numbers of trees. Consequently, the few resistant trees found through conventional screening may never be sufficient to provide lasting resistance to an ever changing pathogen and may also increase the risk of other diseases and insect pests taking a major toll on a few extensively propagated clones.

There is a need, therefore, to develop quick and reliable procedures for detecting and quantifying resistance so that many resistant selections can be found and made available. The National Park Service, recognizing the need for rapid screening technology, is cooperating with Michigan State University in the investigation of both cellular and molecular diagnostic systems.

One approach being examined at Michigan State University is the rapid screening of elm callus lines in culture media amended with the Dutch elm disease toxin cerato-ulmin (CU). Hopefully, the growth response of calli subjected to CU will rapidly reflect the susceptibility or resistance of the source tree. Preliminary studies have shown a direct correlation between calli response to CU and resistance (Pijut et al. 1990; Sticklen et al. 1990).

In the past CU has been produced from fungal cultures after extensive purification. Recently, Michigan State University produced DNA which codes for the toxin and can be expressed in *E. coli* (Bolyard and Sticklen 1991; Sticklen et al. 1990). The CU produced in *E. coli* must now be compared with wild-type CU. The availability of large quantities of pure and consistently uniform CU from *E. coli* would greatly facilitate the development of a "kit" system for assessing resistance *in vitro*. Problems, however, still remain such as determining the most reliable tissue from which to begin callus cultures and the ability to standardize the callus lines from a wide range of elm selections. The CU system must also be verified with a wide range of known resistant and susceptible clones.

Since the exclusive role of CU in pathogenesis has not been confirmed, the possibility exists that a CU response may not be universally indicative of resistance. Other toxin(s) or mechanisms may also play significant role(s) in pathogenesis. A consistent correlation, however, in callus response to CU of known resistant and susceptible selections would help substantiate the role of CU in pathogenesis and thus confirm the utility of this screening procedure. The unlimited availability of CU provided by the *E. coli* recombinant will now facilitate evaluations of the *in vitro* screening procedure and promote studies of CU's specific role in Dutch elm disease.

Another approach currently being explored through the Cooperative Agreement between the National Park Service and Michigan State University is the direct identification of nucleotide sequence variations between Dutch elm disease resistant and susceptible elms. The availability of DNA probes derived from resistant elms would provide the most convenient and possibly reliable avenue for rapid detection of resistant elms. Naturally occurring elms, elm seedlings subjected to mutagens, and progeny of known crosses could be screened rapidly for the sequence variations denoting resistance.

Although these approaches are high risk, and may ultimately prove futile, the timing for this pursuit is appropriate given the rapid advances currently being made in tissue culture and genetic engineering. Progress with Dutch elm disease may also lead to advances in understanding and resolving other major problems affecting elm such as elm yellows, bacterial leaf scorch and elm leaf beetle. Regardless of how effective these or other techniques may be in detecting resistance we must assure that the elms ultimately selected are not just resistant to Dutch elm diseasex, but also retain the aesthetic qualities and vigorous nature we have come to expect and appreciate in the American elm and which makes this species uniquely suited for the Nation's Capital and other municipal environments.

References

Bolyard, M. G. and Sticklen, M. B. 1991. Expression in *E. coli* of a functional Dutch elm disease toxin. In press. Nucleic Acid Research.

Green, P. S. 1964. Registration of cultivar names in *Ulmus*. Arnoldia 24:41-80.

Olszewski, G. J. 1970. History of the Mall, Washington, D.C. Office of History and Historic Architecture, Eastern Service Center, Washington, D.C. 115 pp.

Pijut, P. M., Domir, S. C., Lineberger, R. D., and Schreiber, L. R. 1990. Use of culture filtrates of *Ceratocystis ulmi* as a bioassay to screen for disease tolerant *Ulmus americana*. Plant Sci. 70:191-196.

Sticklen, M. B., Bolyard, M. G., and Chang, J. 1990. Methods for *in vitro* selection of Dutch elm disease resistant American elms and expression of synthetic DNA encoding a Dutch elm disease toxin in *E. coli*. In: V. Steenackers and P. Smets (eds.). Proc. International Energy Agency (IEA) Symposium: International Exchange of Genetic Material. Geraadsbergen, Belgium. pp.44-56.

8
Cytological and Biochemical Aspects of Elm Improvement

Frank S. Santamour, Jr.

Abstract

As far as we know, the American elm (*Ulmus americana* L.) is the only natural tetraploid species in the genus, with 2n=4x=56 chromosomes. It was thought that doubling the chromosome number of Siberian elm (*U. pumila* L.) would allow interspecific hybridization between these two species. However, no hybrids with American elm were developed by using the colchiploid Siberian elms as either the male or female parent. Two naturally occurring triploid elms, one a putative hybrid between *U. americana* and *U. pumila* and the other a derivative of *U. pumila* and *U. rubra* Muhl. parentage, have been reported in the literature. Both of these triploids produced viable pollen grains and were used as male parents in crossing on *U. parvifolia* Jacq. Progenies of these tri-specific hybrids, as well as open-pollinated progeny of the *U. pumila* x *U. rubra* triploid, included a wide range of aneuploids, most of which did not have acceptable tree form or were susceptible to Dutch elm disease. Recently, another triploid elm (of unknown ancestry) and two *diploid* American (?) elms have been discovered. Detailed cytological analyses of these trees and their progenies would be desirable. Furthermore, there is a need for biochemical data that will help to make more accurate assessments of the botanical origins of these atypical plants.

Introduction

Our current knowledge of elm chromosome numbers indicates that American elm (*Ulmus americana* L.) is the only tetraploid (2n=4x=56) species in the genus. This count is recorded in the 1955 "Chromosome Atlas" of Darlington and Wylie (1955), along with diploid counts for 13 other taxa. Since 1955, diploid numbers (2n=2x=28) have been reported for *U. alata* Michx., *U. crassifolia* Nutt., *U. macrocarpa* Hance, *U. parvifolia* Jacq., and *U. serotina* Sarg. (Santamour 1969); *U. elliptica* K. Koch (Santamour 1971a); *U. villosa* Brandis ex Gamble (Santamour 1979); *U. wallichiana* Planch. (Mehra and Gill 1974); and *U. davidiana* Planch., *U. glaucescens* var. *lasiocarpa* Rehd., and *U. sukaczevii* Andr. (Mukherjee and Ware 1979). The tetraploid number of *U.*

americana has been confirmed also in *U. americana* var. *floridana* (Chapm.) Little (Santamour 1969) and in the straight species by several authors (Lester 1970; Lester and Lee 1974; Santamour 1969), including the elegantly illustrated meiotic analysis of Karrfalt and Karnosky (1975).

Early attempts to cross American elm with diploid Asiatic species resistant to Dutch elm disease failed and the idea that this sexual incompatibility was the result of disparate chromosome numbers became prevalent. Dermen and May (1966) used colchicine treatment of *U. pumila* L. (Siberian elm) seedlings to induce tetraploidy in that species. They produced and isolated a number of chimeral plants in which the three histological layers of the apical meristem showed different ploidy levels. In some plants, the cells in the L-II layer (which gives rise to reproductive tissue) were tetraploid. Rooted cuttings of several trees with tetraploid L-II layers, as well as other seedlings of the colchiploidy experiment, were established at the USDA/ARS facility in Delaware, Ohio. Alden M. Townsend (personal communication) reported that he was unsuccessful in crossing these colchiploid Siberian elms, as either the male or female parent, with American elm. Furthermore, Ager and Guries (1982), using pollen of a tetraploid Siberian elm supplied by Townsend, were unable to achieve crosses with American elm.

Another approach to accomplishing crosses between American elm and diploid species would be to reduce the chromosome number of American elm to the diploid level. Lester (1970) utilized irradiated pollen of American elm in combination with viable pollen of several diploid elms in an attempt to induce polyhaploidy (diploidy) in American elm. Although a few abnormal seedlings with narrow leaves were produced, all of the progeny had the typical tetraploid chromosome number. In some plants, multiple seedlings that develop from a single seed may occasionally have different ploidy levels. Lester and Lee (1974) published the only cytological analyses of "twin" seedlings of American elm and found none that deviated from the tetraploid number. Similar results were obtained by the author (unpublished) and David F. Karnosky (personal communication). Attempts to develop diploid American elms through the isolation of meiotic protoplasts have also failed (Redenbaugh et al. 1980).

Diploid American elms

Are there any *natural* diploid American elms? And, if there are, would they cross with the diploid species? Actually, a diploid chromosome count for American elm was included in the "Chromosome Atlas" (Darlington and Wylie 1955), although the citation was incorrect. Santamour (1969) doubted that the plant of "Ulmus americana" reported in 1931 by Krause (1931) as a diploid was truly an American elm because of the commercial origin of the plant material and the fact that Willdenow, rather than Linnaeus, was given as the botanical authority for the scientific name. Another potential problem is that the European white elm (*U. laevis* Pall.), a diploid species, is very similar to *U. americana* in many morphological characteristics and Santamour (1972a) had identified an "American" elm being studied at Cornell as *U. laevis* on the basis of leaf flavonoids and chromosome number. In recent years, however, researchers at two American institutions have discovered diploid trees of supposedly true *U. americana*, and the following discussion is based on personal communication and information generously provided by the scientists involved. George H. Ware of the Morton Arboretum in Lisle, Illinois found a tree on the campus of Washington University in St. Louis, Missouri that produced diploid seedlings after open-pollination. Using the Missouri tree as the male parent, Ware was unable to make successful crosses with tetraploid American elms, but did obtain putative hybrid seedlings from crosses on *U. pumila* and some other Asiatic diploid elms. Another tree, which also gave rise to diploid open-pollinated progeny, is located on the campus of the University of Wisconsin in Madison and is being studied by Eugene B. Smalley and Raymond P. Guries. This tree apparently crosses easily with *U. laevis*, but not with tetraploid *U. americana*. It is somewhat noteworthy that the diploid nature of both the Morton Arboretum tree and the University of Wisconsin tree was determined from open-pollinated seedlings rather than from meiotic studies of the parent tree or from rooted cuttings of the parent. Although it is extremely likely that the parent trees are indeed diploid, a critical examination of meiosis might prove interesting. Furthermore, the apparent inability of these diploid American elms to cross with tetraploid American elms deserves additional morphological, physiological, biochemical, and genetics studies.

Triploid elms

Only two triploid elms (2n=3x=42) that involve species with potential resistance to Dutch elm disease have been extensively investigated. One of these trees was determined to be a natural hybrid between American and Siberian elms (Santamour 1970). Chromosome pairing at meiotic metaphase was invariably 14 bivalents and 14 univalents, which would be expected if the American elm chromosomes paired autosyndetically and none of the Siberian elm chromosomes were homologous with any of the American elm chromosomes. This triploid was, however, susceptible to inoculation with the Dutch elm disease fungus (Santamour 1974). Although this tree did not produce any open-pollinated progeny during the years it was under study, it did have viable pollen and was used as a male parent in certain crosses.

The other triploid was determined to be a probable F_2 segregate of *U. pumila* x *U. rubra* Muhl. parentage (Santamour 1971b). Chromosome pairing at meiotic metaphase was extremely irregular with 12-22 univalents, 8-12 bivalents, and 0-4 trivalents. Although some of the "lost" univalents were not incorporated into telophase II nuclei and formed "dwarf" microspores, potential pollen viability was 96 percent. Open-pollinated progeny, produced in only one of 3 years of observation, had aneuploid chromosome numbers ranging from 2n=29 to 2n=52.

Both of these triploids were used as male parents in crosses on Chinese elm (*U. parvifolia*). Early assessments of the susceptibility of these progenies to Dutch elm disease showed considerable variability (Santamour 1974). However, the plants of only the *U. parvifolia* x (*U. americana* x *U. pumila*) progeny were thoroughly analyzed. Chromosome counts on 23 seedlings had aneuploid numbers from 2n=34 to 2n=54 and varied widely in leaf size and growth characteristics (Santamour 1972b). Those aneuploids with the largest leaves and the highest degree of resistance to Dutch elm disease were established in permanent test plots, but none have grown satisfactorily. Furthermore, it appeared that none of these aneuploids would be useful in advanced breeding work. Santamour (1989) found that although hybrids between *U. parvifolia* and several diploid species flowered from October to November, and many produced a relatively high percentage of good pollen, the tri-specific aneuploid hybrids involving American elm never flowered or shed pollen in a normal fashion.

Recently, at least one, and perhaps two, new triploid elms have come to light. Both of these trees were originally selected by the late Horace V. Wester of the National Park Service in Washington, D.C. for early foliation and later leaf retention. Both trees subsequently showed a reasonably high degree of resistance to twig inoculations with Dutch elm disease. Both trees are presently considered to be basically American elms, and are currently under study by James L. Sherald of the Park Service. One of these trees (selection no. 3-487) is definitely a triploid, and the chromosome count was made by Mariam B. Sticklen on a rooted cutting of the original tree in 1991. J. L. Sherald is currently growing open-pollinated seedlings from seed produced on this tree in 1992.

The reason I *suspect* there is another triploid in the Park Service domain is the extreme variability in leaf size and growth of seedlings grown from seed delivered to me by Mr. Wester a number of years ago. Unfortunately, the data concerning that seed collection were somewhat confused, but the range of seedling variation was strongly reminiscent of those aneuploid progenies alluded to earlier. Perhaps further study is warranted.

Biochemical investigations

The thorough and painstaking analyses of elm leaf flavonoids by Sherman and Giannasi (1988), albeit on only six species native to the United States, has certainly cast doubts on earlier reports (Bate-Smith and Richens 1973; Santamour 1972a). American elm contained 16 different flavonoid compounds, many of which were unique to that species among those studied. Thus, flavonoid data might be used to confirm the identification of the "diploid" American elms noted earlier. Unfortunately, similar data are not available for *U. laevis* or any of the other European or Asiatic elms. When, and if, detailed analyses are performed on these exotic species, flavonoids could be useful in determining the parentage of both diploids and triploids of suspected hybrid origin. There is a need for other biochemical criteria that can be used for species identification and hybrid verification.

Summary

Triploid elms may be sexually functional and be useful in producing genetic combinations (in aneuploids) that would not be possible at euploid levels. Meiotic analyses of triploids can be useful in determining whether the tetraploid American elm was involved in their parentage. Meiotic studies should also be undertaken in those presumed American elms that have been determined to produce diploid progeny. Biochemical, or other nonmorphological, criteria are needed to aid in species identification and hybrid verification. Both cytological and biochemical analyses can contribute to a balanced genetic program to develop superior elms for landscape planting.

References

Ager, A. A., and Guries, R. M. 1982. Barriers to interspecific hybridization in *Ulmus americana*. Euphytica 31:909-920.

Bate-Smith, E. C. and Richens, R. H. 1973. Flavonoid chemistry and taxonomy in *Ulmus*. Biochem. System. Ecol. 1:141-146.

Darlington, C. D., and Wylie, A. P. 1955. Chromosome atlas of flowering plants. George Allen & Unwin Ltd., London, 519 p.

Dermen, H., and May, C. 1966. Colchiploidy of *Ulmus pumila* and its possible use in hybridization with *U. americana*. Forest Sci. 12:140-146.

Karrfalt, R. P., and Karnosky, D. F. 1975. Meiotic pairing and chromosome morphology in American elm. Forest Sci. 21:123-127.

Krause, O. 1931. Zytologische Studien den Urticales unter besonder Beruckichtigung der Gattung *Dorstenia*. Planta 13:29-84.

Lester, D. T. 1970. An attempt to induce polyhaploidy in American elm. Forest Sci. 16: 137-138.

Lester, D. T., and Lee, M. J. T. 1974. Twins and triplets of American Elm. Forest Sci. 20:142.

Mehra, P. N. and Gill, B. S. 1974. Cytological studies in Ulmaceae, Moraceae, & Urticaceae. J. Arnold Arb. 55:663-677.

Mukherjee, A., and Ware, G. H. 1979. In: IOPB Chromosome number reports LXIII, A. Love, Ed. Taxon 28:265-279.

Redenbaugh, M. K., Westfall, R. D., and Karnosky, D. F. 1980. Protoplast isolation from *Ulmus americana* L. pollen mother cells, tetrads, and microspores. Can. J. For.Res. 10:284-289.

Santamour, F. S., Jr. 1969. New chromosome counts in *Ulmus* and *Platanus*. Rhodora 71:544-547.

Santamour, F. S., Jr. 1970. A natural hybrid between American and Siberian elms. Forest Sci. 16:149-153.

Santamour, F. S., Jr. 1971a. In: IOPB Chromosome number reports XXXII, A. Love, Ed. Taxon 20: 349-356.

Santamour, F. S., Jr. 1971b. A triploid elm (*Ulmus pumila* x *U. rubra*) and its aneuploid progeny. Bull. Torrey Bot. Club 98:310-314.

Santamour, F. S., Jr. 1972a. Flavonoid distribution in *Ulmus*. Bull. Torrey Bot. Club 99:127-131.

Santamour, F. S., Jr. 1972b. Aneuploid tri-specific elm hybrids. Forest Sci. 18:290-291.

Santamour, F. S., Jr. 1974. Resistance of new elm hybrids to Dutch elm disease. Plant Dis. Reptr. 58:727-730.

Santamour, F. S., Jr. 1979. Resistance of Himalayan small-leaved elm to Dutch elm disease. J. Arboric. 5:110-112.

Santamour, F. S., Jr. 1989. Flowering and fertility of hybrids between spring- and fall-flowering elms. HortScience 24:139-140.

Sherman, S. L., and Giannasi, D. E. 1988. Foliar flavonoids of *Ulmus* in eastern N. America. Biochem. System. Ecol. 16:51-56.

9
Elm Improvement Via Biotechnological Methods

David F. Karnosky and Gopi K. Podila

Abstract

Rapid progress has been made over the past two decades in *in vitro* culture and manipulation of forest tree tissues, cells and protoplasts. It is now possible to micropropagate elms from many different types of cells and tissues; cryogenically preserve rare or unusual elm genotypes; isolate, culture and obtain whole trees from elm protoplasts and produce somatic elm hybrids via protoplast fusion. This paper will provide an overview of the significant advances in elm *in vitro* culture and will speculate as to how the rapidly evolving genetic engineering technologies will be used in genetic improvement for elms. Over the next decade, we envision significant biotechnological progress in: 1) determining the targets and receptors for the fungus-host responses in Dutch elm disease via *in vitro* disease challenge studies, 2) isolating molecular mechanisms for resistance to Dutch elm disease, and 3) introducing Dutch elm disease-resistant genes into ornamentally desirable but disease-susceptible elms.

Introduction

Elms *(Ulmus* species) have been important components of forests, hedgerows, windbreaks and urban areas around the world. Unfortunately, diseases such as Dutch elm disease and phloem necrosis have devastated elms in Europe and North America (Karnosky 1979). Extensive efforts over the past 70 years to develop improved, disease-resistant elms have been hampered by several factors, including: 1) the long time periods needed to turn over breeding generations, 2) the sexual incompatibility of some elm species such as the American elm; and 3) the rapid mutability of *Ophiostoma ulmi* (Buism.) Nannfeldt.

Several recent biotechnological breakthroughs have occurred that offer new opportunities for elm improvement. This paper discusses some of these new methods and speculates as to how they may be useful in elm improvement.

Tissue Culture

Elm species are generally quite amenable to *in vitro* culture (Karnosky and Mickler 1984, 1986). Elms have been regenerated *in vitro* from juvenile tissues such as hypocotyls and cotyledons (Durzan and Lopushanski 1975) and from tissues taken from mature trees, including excised roots, stems and leaf sections (Jacquiot 1949, Bolyard et al. 1991a). Whole plants have been regenerated from callus (Chalupa 1975; Karnosky et al. 1982), cell suspension cultures (Durzan and Lopushanski 1975) and protoplasts (Sticklen et al. 1986; Dorion et al. 1991). To our knowledge, whole plants have not yet been regenerated from male or female haploid reproductive cells. It should be noted that there is considerable genotypic variability in *in vitro* culturability of elms. Thus, considerable effort may be needed to develop successful systems for recalcitrant genotypes.

Micropropagation

As new cultivars of elms are being developed, micropropagation could be used to decrease the time necessary to scale up the numbers of plants necessary to test, distribute and commercially release new elm cultivars. Micropropagation is being widely used for this purpose in the horticultural industry.

In Vitro *Challenge*

In vitro challenge of host cells, tissues or organs with pathogenic fungi or toxins offer many opportunities for disease screening with trees (Diner and Karnosky 1987). Elm callus cultures have been challenged with *O. ulmi* (Szczegola et al. 1987) and with the toxin cerato-ulmin produced by *O. ulmi* (Pijut et al. 1988; Pijut et al. 1990; Domir et al. 1992). These studies have found a strong positive correlation between *in vitro* and *in vivo* response.

Somaclonal Variation

Somaclonal variation, resulting from *in vitro* culture, has yet to be tapped as a source of variation in elms. However, evidence for somaclonal variation in elms derived from protoplasts has been reported by Domir et al. (1992). Protoplasts or cell suspension cultures grown for several generations in increasing concentrations of the toxin cerato-ulmin should provide an opportunity to obtain toxin-resistant somaclones. Because of the complexity of Dutch elm disease resistance, toxin-resistant somaclones should be challenged *in vivo* to test for improved resistance.

Ploidy Manipulation and Protoplast Fusion

American elm is the only tetraploid elm. It has proven to be extremely difficult to sexually hybridize with diploid elms (Bob et al. 1986). The culture of microspores or ovules offer opportunities to reduce the chromosome number of

American elm. Reduced chromosome number plants or microspores and/or ovules from tetraploid elms could be used as sources of protoplasts for somatic cell hybridization with diploid, disease-resistant elms to recreate tetraploid elms with improved disease resistance. Protoplast fusion has been reported by Redenbaugh et al. (1981) but whole-plant regeneration has not yet been obtained for elm somatic cell hybrids.

Molecular Aspects

Detection and Isolation of Resistance Genes

Sticklen et al. (1991) have discussed the need for identifying and isolating genes responsible for Dutch elm disease resistance. There are several types of resistance genes that are prime targets for research. Genes coding the phytoalexin mansonone compounds (Duchesne et al. 1985; Dumas et al. 1986) and genes involved in lignification and/or suberization processes involved in compartmentalizing *O. ulmi* should be studied.

The *in vitro* challenge utilizing callus from resistant and susceptible elms grown in the presence of *O. ulmi* or its toxin cerato-ulmin provide excellent opportunities for gene isolation experiments. Messenger RNA could be extracted from resistant and susceptible elms both before and after challenge. cDNA libraries could be then prepared and differential and subtractive library screening could be used to detect those genes involved in resistance responses. Genes isolated in this way could be sequenced to see if they are unique genes or if they match any of the known defense genes.

Similar studies could be done using *O. ulmi* challenge of whole elm plants, although the process of mRNA extraction is more complex and the number of genes involved in compartmentalization responses are probably greater than for the phytoalexin responses. However, Bugos et al. (1991) recently isolated the gene for a lignin-bispecific O-methyltransferase (OMT) from cambial tissue of *Populus tremuloides* Michx. Thus, the technology is available to search for genes even in complex tree tissues.

Genetic Engineering

The processes of movement of foreign genes into *in vitro* cultured trees followed by regeneration of whole, transgenic trees have now been accomplished with several tree species (Fillatii et al. 1987; James et al. 1989; Huang et al. 1991; Mante et al. 1991; McGranahan et al. 1988; Smigocki and Hammerschlag 1991) or particle-gun bombardment (Christou 1992; Ellis et al. 1991). To our knowledge, no one has yet regenerated a genetically engineered elm tree. However, work in this area is underway (Bolyard et al. 1991b; Bolyard and Sticklen 1992). Given the culturability of elms *in vitro,* we suspect that these methods will be highly successful for genetic manipulation of elms.

While no known Dutch elm disease resistance genes are yet available for gene transfer, the processes of gene transfer and plant regeneration can be researched using readily available reporter genes such as the GUS (ß-glucuronidase) gene (Jefferson 1989) and antibiotic resistance genes (Peach and Velten 1991; Umbeck et al. 1989).

Once the genes are isolated that control the various mansonones, then increased levels of the individual mansonones will be possible through manipulation of the promoter. The cauliflower mosaic virus 35S promoter has been shown to enhance transcriptional activity about tenfold when constructed in tandem duplication as compared to a single 35S promoter (Kay et al. 1987). The 35S promoter is also amenable to tissue-specific gene expression (Benfey and Chua 1990). Thus, it is possible to create a gene construct that would be expressed only in cambial tissues where the effect on *O. ulmi* would be maximized. Fungal defense genes such as chitinase and ß-1,3-glucanase (Broglie et al. 1991; Neale et al. 1990) are also worth examining either separately or in combination with improved mansonone production.

Summary

Because of their general ease of *in vitro* culturability, elms are good candidates for improvement via the newly emerging biotechnological methods. We predict that these methods will play a significant role in elm improvement over the next decade. These methods will be most beneficial when they are fully integrated with traditional elm improvement methods.

References

Benfey, P.N. and N.-H. Chua. 1990. The cauliflower mosaic virus 35S promoter: Combinatorial regulation of transcription in plants. Science 250:959-966.

Bob, C.F., B.L. Redmond and D.F. Karnosky. 1986. On the nature of intra- and interspecific incompatibility in *Ulmus*. Amer. J. Bot. 73:465-474.

Bolyard, M.G. and M.B. Sticklen. 1992. Strategies for the production of DED-resistant American elms. 1992 International Dutch Elm Disease Workshop (In Press).

Bolyard, M.G., C. Srinivasan and M.B. Sticklen. 1991a. Shoot regeneration system for DED-susceptible and putatively resistant American elms. J. Cell Biochem. Suppl. 15A:A202 (Abstract).

Bolyard, M.G., R.K. Hajela and M.B. Sticklen. 1991b. Microprojectile and *Agrobacterium mediated* transformation of Pioneer elm. J. Arboric. 17:34-37.

Broglie, K., I. Chet, M. Holliday, R. Cressman, P. Biddle, S. Knowlton, C.J. Mauvais and R. Broglie. 1991. Transgenic plants with enhanced resistance to the fungal pathogen *Rhizoctonia Solani*. Science 254:1194-1197.

Bugos, R.C., V.L.C. Chiang and W.H. Campbell. 1991. cDNA cloning, sequence analysis and seasonal expression of lignin-bispecific caffeic acid/5-hydroxyferulic acid O-methyltransferase. Plant Mol. Biol. 17:1203-1215.

Chalupa, V. 1975. Induction of organogenesis in forest tree cultures. Commun. Inst. For. Czech. 9:39-50.

Christou, P. 1992. Genetic transformation of crop plants using microprojectile bombardment. Plant J. 2:275-281.

Diner, A.M. and D.F. Karnosky. 1987. Tissue culture application to forest pathology and pest control. In: Bonga, J. and D. Durzan (Eds.) Tissue Culture in Forestry. Martinus Nijhoff Publishers. pp. 351-373.

Domir, S.C., M.B. Sticklen, M. Bolyard and R.D. Lineberger. 1992. Micropropagation and regeneration of DED-resistant elms using protoplast and tissue culture techniques. 1992 International Dutch Elm Disease Workshop (In Press).

Dorion, N., P. Danthu, S. Ohki, C. Preneux, B. Godin and C. Bigot. 1991. Plant regeneration from leaf protoplasts of common elm *(Ulmus compestris* Mill.). C.R. Acad. Sci., Paris, 313:467-473.

Duchesne, L.C., R.S. Jeng and M. Hubbes. 1985. Accumulation of phytoalexins in *Ulmus americana* in response to infection by a non-aggressive and an aggressive strain of *Ophiostoma ulmi.* Can. J. Bot. 63:678-680.

Dumas, M.T., G.M. Strunz, M. Hubbes and R.S. Jeng. 1986. Inhibition of *Ceratocystis ulmi* by mansonones A,C,D,E,F and G isolated from *Ulmus americana.* Eur. J. For. Path. 16:217-222.

Durzan, D.J. and S.M. Lopushanski. 1975. Propagation of American elm via cell suspension culture. Can. J. For. Res. 5:273-277.

Ellis, D.D., D. McCabe, D. Russell, B. Martinell and B.H. McCown. 1991. Expression of inducible angiosperm promoters in a gymnosperm, *Picea glauca (white* spruce). Plant Mol. Biol. 17:19-27.

Fillatti, J.J., J. Sellmer, B. McCown, B. Haissig and L. Comai. 1987. *Agrobacterium mediated* transformation and regeneration of *Populus.* Mol. Gen. Genet. 206:192-199.

Huang, Y., A.M. Diner and D.F. Karnosky. 1991. *Agrobacterium rhizogenes-mediated* genetic transformation and regeneration of a conifer: *Larix decidua.* In Vitro Cell. Dev. Biol. (Plant) 3:201-207.

Jacquiot, C. 1949. Observations sur la neoformation de bourgeons chez le tissu cambial d' *Ulmus compestris* cultivé *in vitro.* C.R. Acad. Sci. Paris 229:529-530.

James, D.J., A.J. Passey, D.J. Barbara and M. Bevan. 1989. Genetic transformation of apple *(Malus pumila* Mill) using a disarmed Ti-binary vector. Plant Cell Reprt. 7:658661.

Jefferson, R.A. 1989. The GUS reporter gene system. Nature 342:837-838.

Karnosky, D.F. 1979. Dutch elm disease: A review of the history, environmental implications, control and research needs. Environ. Cons. 6:311-322.

Karnosky, D.F. and R.A. Mickler. 1984. Propagation and preservation of elms via tissue culture systems. In: M.L. Duryea and G.N. Brown (Eds.) Seedling physiology and reforestation success. Martinus Nijhoff/Dr. W. Junk Publishers. Dordrecht, The Netherlands. pp. 29-36.

Karnosky, D.F. and R.A. Mickler, 1986. Elms *(Ulmus* spp.). In: Y.P.S. Bajaj (Ed.). Biotechnology in Agriculture and Forestry I. Springer-Verlag. New York. Pp. 326340.

Karnosky, D.F., R.A. Mickler and D.D. Lange. 1982. Hormonal control of shoot and root induction in hypocotyl callus cultures of American elm. In Vitro 18:275.

Kay, R., A. Chan, M. Daly and J. McPherson. Duplication of CaMV 35S promoter sequences creates a strong enhancer for plant genes. Science 236: 1299-1302.

Mante, S., P.H. Morgens, R. Scorz, J.M. Cordts and A.M. Callahan. 1991. *Agrobacterium mediated* transformation of plum *(Prunus domestica* L.) hypocotyl slices and regeneration of transgenic plants. Bio/Technology 9:853-857.

McGranahan, G.H., C.A. Leslie, S.L. Uratsu, L.A. Martin and A.M. Candekar. 1988. Agrobacterium-mediated transformation of walnut somatic embryos and regeneration of transgenic plants. Bio/Technology 6:800-804.

Neale, A.D., J.A. Wahleithaer, M. Lund, H.T. Bonnett, A. Kelly, D.R. Meeks-Wagner, W.J. Peacock and E.S. Dennis. Chitinase, 13-1,3 glucanase, osmotin and extensin are expressed in tobacco explants during flower formation. Plant Cell 2:673-684.

Peach, C. and J. Velten. 1991. Transgene expression variability (position effect) of CAT and GUS reporter genes driven by linked divergent T-DNA promoters. Plant Mol. Biol. 17:49-60.

Pijut, P.M., R.D. Lineberger, S.C. Domir and L.R. Schreiber. 1988. Response of elm callus to culture filtrate of *Ceratocystis ulmi* and correlation with whole-plant disease reaction. HortScience 23:787 (Abstract).

Pijut, P.M., S.C. Domir, R.D. Lineberger and L.R. Schrieber. 1990. Use of culture filtrates of *Ceratocystis ulmi* as a bioassay to screen for disease-tolerant *Ulmus americana*. Plant Sci. 70:fl91-196.

Redenbaugh, M.K., D.F. Karnosky and R.D. Westfall. 1981. Protoplast isolation and fusion in three *Ulmus* species. Can. J. Bot. 59: 1436-1443.

Smigocki, A.C. and F.A. Hammerschlag. 1991. Regeneration of plants from peach embryo cells infected with a shooty mutant strain of *Agrobacterium*. J. Amer. Soc. Hort. Sci. 116: 1092-1097.

Sticklen, M.B., S.C. Domir and R.D. Lineberger. 1986. Shoot regeneration from protoplasts of *Ulmus* x *'Pioneer'*. Plant Science 47:29-34.

Sticklen, M.B, M.G. Bolyard, R.K. Hajela and L.C. Duchesne. 1991. Molecular and cellular aspects of Dutch elm disease. Phytoprotection 72: 1-13.

Szczegola, M., R. Jeng and M. Hubbes. 1987. Mansonone induction by *Ophiostoma ulmi* in callus cultures of elm species. Can. J. Plant Path. 9:286 (Abstract).

Umbeck, P., W. Swain and N.S. Yang. 1989. Inheritance and expression of genes for kanamycin and chloromphenical resistance in transgenic cotton plants. Crop Science 29:196-201.

10
Micropropagation and Regeneration of DED-Resistant Elms Using Protoplast and Tissue Culture Techniques

Subhash C. Domir, M. B. Sticklen, P. M. Pijut,
M. G. Bolyard and R. D. Lineberger

Abstract

The American elm has been devastated by Dutch elm disease, caused by the fungus *Ophiostoma ulmi* (Buisman) Nannf. Attempts to introduce disease resistance into *Ulmus americana* have failed because of sexual incompatibility between the tetrapioid *U. americana* and the dipioid resistant species. Two disease free hybrid elm selections, 'Homestead' and 'Pioneer', became available during the last few years. We isolated and cultured protoplasts from *U. americana, Ulmus* x 'Pioneer', and *Ulmus* x 'Homestead'; developed protocols to obtain somatic hybridization between *U. americana* and disease resistant hybrids; evaluated somaclonal variation in plants regenerated from protoplasts; and carried out *in vitro* organogenesis of 'Pioneer' and *U. americana* from stem tip leaf tissue, leaf explants and sections, and callus tissue. Plants were obtained from the protoplast cultures of *Ulmus* x *'Pioneer,"* but not from protoplasts from *U. americana* or *Ulmus* x 'Homestead'. Somaclonal variation occurred in plants regenerated from 'Pioneer' protoplasts and in ramets propagated from them. Shoot tip explants of 'Pioneer', cultured weekly on MS medium containing 2μM benzyladenine (BA), produced shoot proliferating cultures sufficient for micropropagation. Callus, derived from internodes of actively elongating 'Pioneer' shoots, produced shoot organogenesis when transferred from a primary medium to a secondary medium containing 10 μM BA. On transfer to a medium containing 2.5 μM BA these shoots elongated. Shoot differentiation was also obtained from margins of leaf tissue maintained on MS medium containing 2 μM BA. No intervening callus formation was observed in this case. These tissue cultured shoots were successfully rooted, acclimated, and transplanted to greenhouse and field. Shoots were also regenerated from greenhouse grown *U. americana* and *U. parvifolia* leaf sections cultured on a modified MS medium containing 0.1μM Thidiazuron. No variability was detected among seedlings in their ability to regenerate shoots. Attempts to induce shoot organogenesis from callus cultures of *U. americana* were successful; however, attempts to elongate these shoots and subsequently produce plants did not succeed.

Introduction

A significant amount of research has focused on Dutch elm disease (DED) for more than 50 years. Much of the DED literature involves concerted efforts to develop control methods and breeding programs aimed at improving disease resistance. Hybrid elms with ornamental traits characteristic of the North American and European elms and disease resistance characteristic of the Asian elms have been developed and released (Schreiber and Main 1976; Smalley and Lester 1973; Townsend and Masters 1984a; Townsend and Masters 1984b). Individual trees of *Ulmus americana* L. with high levels of resistance or tolerance to DED were identified and clones of these trees subjected to additional testing (Towsend 1982). To date, it has not been possible to introduce disease resistance into *U. americana* because of sexual incompatibility between the tetraploid *U. americana* and diploid resistant European and Asian species (Dermen and May 1966). Several researchers have attempted to apply new approaches using modern tissue culture techniques to overcome this barrier.

The results described in this report are the outcome of research on: (a) micropropagation and regeneration of resistant hybrid elms; (b) *in vitro* organogenesis; (c) propagation of American elm from shoot tip and callus tissue; (d) development of protoplast regeneration protocols to attempt somatic hybridization between *U americana* and the hybrid elms 'Pioneer' or 'Homestead'; and (e) evaluation of the extent of somaclonal variation in plants regenerated from protoplasts.

Micropropaqation and regeneration of resistant hybrid elms, 'Pioneer' and 'Homestead'.

Softwood stem cuttings taken from mature trees were rooted and subsequently used as stock plants. Actively elongating shoot tips from stock plants were sterilized and trimmed 5 to 10 mm in size and cultured either on modified Murashige and Skoog (MS) or woody plant medium (WPM). The procedures for these studies have been reported earlier (Fink et al. 1986). For 'Pioneer' elm, MS medium was far superior to WPM. Shoot proliferation was greatest on MS medium containing 4.4 μM benzylaminopurine (BA). The addition of 1-Napthaleneacetic acid (NAA) did not enhance shoot proliferation. Most cultures exhibited the formation of a white "foamy" callus at the base of the shoots and these cultures eventually died after five weeks. These initial studies indicating low numbers of shoot initiation, inability of cultures to maintain vigor, and presence of "foamy" callus led to additional studies involving cytokinin concentration, explant size, frequency of transfer to fresh medium, medium constituents, and light intensity.

Results indicated that weekly transfer of shoot tips to fresh media reduced production of "foamy" callus and shoot proliferation was considerably enhanced. Weekly transfer and addition of either 2.2 or 4.4 µM BA to the media resulted in increased shoot multiplication. Also, 10 mm explant size resulted in the most consistent shoot initiation. Attempts to further enhance the rate of shoot multiplication by changing the salt content of the MS medium were not successful. Doubling the concentration of NH_4NO_3 was phytotoxic, and increasing the concentration of KNO_3, $CaCl_2$, and $MgSO_4$ seemed to depress shoot growth, as did using half-strength MS medium.

Studies were also conducted to determine if light intensity affected the rate of shoot multiplication. Cultures were either maintained in darkness or under light intensity varying between 45 to 170 µmol/m^2/sec. Light increased the development of axillary as well as adventitious shoots. Light levels higher than 65 µmol/m^2/sec appeared inhibitory to development since shoot tips appeared light green and less vigorous.

Studies using explants from 'Homestead' elm were not as productive. Various combinations and changes in growth regulators, explant size, medium composition, etc. did not enhance shoot proliferation. After several attempts, further investigations with this hybrid were terminated.

When 'Pioneer' leaf tissue was used as an explant source, adventitious shoot development was observed. The leaf explant, cultured on MS medium containing between 2 and 4 µM BA, produced an organogenic callus from which shoots developed. In other instances, primordial shoots developed from the leaf margin without callus formation. The shoot primordia, when detached from the leaf, developed into shoot proliferating cultures. These shoots, although stouter, appeared similar to those generated from axillary buds.

Internodal explants were used to generate callus cultures. Callus growth was enhanced in the presence of light and growth regulators 2iP and zeatin. Calli grown on MS medium containing 10 µM BA developed shoot meristems. However, no differentiation was observed in the presence of various cytokinins or in the absence of light. If the small calli with meristems were transferred to medium with 2.5 µM BA, shoot elongation and leaf development were observed.

Microcuttings larger than 2 cm in length with well developed leaves obtained from shoot tip proliferating cultures, internodal callus cultures, and leaf cultures were rooted in soilless medium under high humidity. Subsequently, these rooted microcuttings were acclimated to greenhouse conditions and planted in the field.

In vitro organoqenesis of American and Chinese elms from leaf explants

Genetic engineering is an exciting new field that may help produce tree species resistant to various natural predators by introducing resistance-inducing genes into susceptible tree species. Progress in this technology may protect American elm against the spread of DED via the elm bark beetles, (*Scolytus scolytus* fabri and *S. multistriatus* Marsh), or Chinese elm (*Ulmus parvifolia* Jacq.) from elm leaf beetle (*Pyrrhalta luteola* Muller). *In vitro* regeneration of elms is essential if genetic engineering technology is to be utilized. Regeneration of elms from leaf tissue may provide such a system.

Thidiazuron (TDZ) (Arndt et al. 1976), has been useful for the regeneration of several species using shoot tip culture or shoot regeneration from leaf explants. Studies were conducted to determine if TDZ could be useful in generating shoots from leaf explants of American and Chinese elms (Bolyard et al. 1991). Fully expanded leaves of American or Chinese elm obtained from 6-month-old greenhouse grown seedlings were sterilized. These leaves were cut into sections and cultured on modified MS media containing 0.1, 0.5, 1, or 5 μM TDZ. Materials and methods for this study have been reported elsewhere (Bolyard et al. 1991). Early indications were that 5 μM TDZ was fatal to explants, and the absence of TDZ in the medium did not produce shoots. Follow-up experiments with 0.1 μM TDZ showed that, after 2 months in culture, more than 70% of the American elm cultures produced between 2 to 9 shoots per leaf section. Variability was not detectable among seedlings in their ability to regenerate shoots. However, variability was dependent upon the section of leaf from which the explant was selected, since most regenerated shoots were produced from the veinal regions. Transfer of these shoots to MS media containing 3 or 5 μM gibberellic acid resulted in their elongation. These shoots developed roots in MS medium containing 0.1 mg/liter lH-indole-3-butyric acid. In studies with Chinese elm, more than 80% of the leaf sections regenerated shoots.

Propagation of American elm from shoot tip and callus tissues

Although American elm can be propagated by rooting shoot cuttings, it is usually difficult to propagate a good percentage of these cuttings into plants. A micropropagation system using tissue culture technology to obtain proliferating shoot cultures would be a significant improvement over present practices. To date, it has not been possible to successfully propagate American elm through tissue culture using material derived from mature trees. However, using cell suspensions or callus derived from hypocotyl or epicotyl tissues (juvenile in nature), researchers were able to propagate American elms (Durzan and Lopushanski 1975; Ho 1985; Karnosky et al. 1982; Ulrich et al. 1984).

Shoot tips obtained from stock plants of mature American elms were used to establish *in vitro* shoot proliferating cultures. Several media formulations, growth regulator combinations, and subculture periods were evaluated. In spite of extensive efforts, these experiments did not succeed in the development of shoot proliferating cultures (unpublished data). Attempts were also made to regenerate shoots from callus derived from mature leaf tissue of American elm. Various timed pulse treatments with growth regulators, subculture periods and light versus dark treatments were evaluated. Callus cultures routinely produced green nodule-like primordia, but further development and differentiation did not consistently occur (unpublished data).

Development of protoplast cultures and somatic hybridization

Since it is not possible to successfully transfer DED resistance into *U. americana* by conventional breeding techniques, other approaches need to be evaluated to attain this objective. This incompatibility may be overcome by: (a) regenerating fertile plants through anther culture of American elm; (b) somatic hybridization of American elm with disease resistant diploid species; and (c) isolation of genes responsible for resistance and their introduction into American elms. Studies with antherderived callus or protoplasts isolated from microspores were not successful in regenerating American elms (Redenbaugh et al. 1980; Redenbaugh et al. 1981). The development of protocols for isolation and culture of elm protoplasts is essential before somatic hybridization can be achieved.

Both 'Homestead' and 'Pioneer' elms have shown a high level of resistance to DED and were selected as logical models for protoplast isolation, culture, plant regeneration, and finally somatic hybridization.

Internodal sections (1-2 mm in length) of 'Homestead' were cultured on a modified MS medium (Sticklen et al. 1985) to obtain friable callus cultures. Callus tissue was digested in solution consisting of appropriate inorganic salts (Power and Davey 1979), cell wall digesting enzymes Cellulysin and Macerase, potassium dextran sulfate, and mannitol. The protoplasts were isolated and then cultured in Kao and Michayluk 8p medium (KM) supplemented with 1 μM NAA, either 0.5 or 1.0 μM 2,4-D, either 0.5 or 1.0 μM zeatin, 0.38 μM glucose, and 0.7 mM sucrose. Three weeks later, colonies developed (8-60 cells per colony). On transfer to semisolid medium containing KM salts, 1 μM NAA, 0.5 μM 2,4-D, 1 μM kinetin, 0.38 M glucose, 0.7 mM sucrose, and 0.5% agar, the colonies developed into callus cultures. However, efforts to regenerate shoots from callus cultures were not successful.

Leaf sections of 'Pioneer' were used to initiate callus cultures on a modified half-strength MS medium containing 2.5 μM 2,4-D and 1.0 μM kinetin (Sticklen et al. 1986). Protoplasts were isolated from these callus cultures using Cellulysin and Macerase enzymes as described above. Callus age at the time of protoplast isolation affected the yield, which progressively decreased over the 80-week course of experimentation. The protoplasts were cultured in KM 8p medium supplemented with growth regulators and sugars (Sticklen et al. 1986). Colonies developed 2 weeks after culture with a plating efficiency of approximately 5%. Colonies developed into friable callus 3 weeks after subculturing on semisolid KM 8p medium containing 1.0 μM 2,4-D, 2.5 μM NAA, and 1.0 μM zeatin.

Protoplast derived calli were cultured on various media to obtain a morphogenic response. Approximately 40% of the calli cultured on MS media containing 10, 20, or 30 μM BA became dark green. About 10% of these calli developed into shoot bud primordia. Some of these buds developed into shoots with well expanded leaves when subcultured on MS medium containing 2 μM BA. In this system, in order to obtain shoot generation, it was important to transfer early stage microcalli from liquid medium into shoot regeneration medium. Microcuttings from these cultures were rooted under nonsterile conditions in a soilless medium and grown in the greenhouse.

Protoplasts from American elm cotyledons, leaf, and callus tissue were also isolated. These protoplasts further subdivided only up to the 8-cell colony stage and did not survive afterward.

Fusion experiments were carried out to combine protoplasts from leaf or cotyledons of American elm seedlings and protoplasts from callus tissue of 'Pioneer' or 'Homestead' (unpublished data). One ml protoplast suspensions obtained from American elm cotyledons and 'Pioneer' callus (density, $1x10^5$ /ml) were mixed in a centrifuge tube. Two-tenths ml of this combined suspension was dispensed into a Petri plate. After this suspension was incubated for 5 min, 2 drops (0.2 ml each) of 50% Polyethylene glycol 8000 (PEG) solution was added to the plate. The resulting mixture was incubated at room temperature for 15 min. Fusion was observed using inverted microscope. Of all the protoplasts observed, 15% were fused products of American elm and 'Pioneer' elm.

Somaclonal variation studies

Plants regenerated through different types of cell culture may exhibit phenotypic variability. Initially, this variation was considered epigenetic, as a tissue culture artifact, and was disregarded. However, it was later discovered that some variation exhibited by regenerants was stable over several clonal propagation cycles. This variation was considered to have genetic basis and was termed somaclonal variation. Success in protoplast regeneration of *Ulmus* x 'Pioneer' provided the opportunity to determine the extent and nature of somaclonal variation in this woody hybrid.

One-hundred sixty-eight *Ulmus* x 'Pioneer' plantlets regenerated from protoplast were grown in growth chambers according to accepted horticultural practices (Kroggel 1988). These plants were characterized by various features, such as: (a) rate of shoot elongation from the base of the most recent flush following plants cut back to 7-8 nodes, (b) increase in stem caliper at the 5th node from the shoot tip of actively growing plants, (c) leaf length-to-width ratio for five leaves, (d) number of leaf serrations, and (e) length of guard cells.

Two-node cuttings from most of these regenerants were rooted, and the 5 most heavily rooted cuttings in each clone were grown into plants under conditions described earlier (Kroggel 1988). These plants were termed ramets and were evaluated using above mentioned quantitative procedures.

A comparison of phenotypic characters was made for the following populations: (a) regenerants which included a random selection of 40 of the 168 plants regenerated from protoplasts, (b) ramets, a group of 5 rooted cuttings derived from each of the 37 of 40 regenerants, and (c) a group of 5 'Pioneer' elm plants derived from rooted cuttings.

Regenerants exhibited noticeable variation in growth rate, occasional occurrence of lobed leaves and leaf variegation. A significant variation was noted in morphological traits related to number of leaf serrations, leaf length by width ratio, caliper size, and growth rate. Growth rate was extremely variable and measured between 0.2843 mm/day to 1.5748 mm/day. To determine whether this variation was somaclonal in nature, and not epigenetic, variation among ramets was compared against variation of the regenerants and to rooted cuttings from the 'Pioneer' stock plants. The range of variation in shoot length and caliper increase was lower in the ramet population. The small population of 'Pioneer' rooted cuttings served as a relative base for comparison. Since only few plants were used, this study did not allow for an accurate comparison of growth rate distribution.

A better way to determine somaclonal variation can be obtained by studying the correlation of the relative rank of the regenerants and the ramets in regards to morphological features, i.e. do the slowest or fastest growing regenerants produce the slowest or fastest growing ramets. In this study, all correlation coefficients were positive, indicating a tendency to clonal fidelity of the growth parameters measured (Lineberger et al. 1990). An extremely high correlation (0.89) was noted for guard cell length. This would indicate that the phenotypes are stable and therefore potentially stable somaclonal variants.

Total protein content and peroxidase isozyme analyses were conducted using polyacrylamide gel and starch gel electrophoresis, respectively, to determine whether any gross differences existed among regenerated plants. These analyses showed no reproducible difference in banding patterns (unpublished data). That gel electrophoresis showed no reproducible difference is not unusual. Other than growth rate differences, there were no exceptional differences. At any given time, there were populations of different sized, but otherwise similar looking plants (Kroggel 1988).

This report suggests avenues which may lead to the development of Dutch elm disease resistant *U. americana.* These approaches would include the possibility of somatic hybridization, use of callus tissue from mature American elms to regenerate somaclonal variants resistant to the disease, and isolation and introduction of gene(s) to impart resistance into American elms.

References

Arndt, F., R. Rusch, and H.V. Stifried. 1976. SN49537, a new cotton defoliant. Plant Physiol. 57:599.

Bolyard, M.G., C. Srinivasan, J. Cheng, and M. Sticklen. 1991. Shoot regeneration from leaf explants of American and Chinese elm. HortScience 26:1554-1555.

Dermen, H. and C. May. 1966. Colchiploidy of *Ulmus pumila* and its possible use in hybridization with *U. americana.* Forest Sci. 12:140-146.

Durzan, D.J. and S.M. Lopushanski. 1975. Propagation of American elm via cell suspension cultures. Can. J. For. Res. 5:273-277.

Fink, C.V.M., M.B. Sticklen, R.D. Lineberger, and S.C. Domir. 1986. *In vitro* organogenesis from shoot tip, internode, and leaf explants of *Ulmus* x 'Pioneer'. Plant Cell, Tissue, and Organ Culture 7:237-245.

Ho, R.H.. 1985. Micropropagation of American elm. IN: New ways in forest genetics. F. Caron, A.G. Corriveau, and T.J.B. Boyle (Eds.). Proc. 20th Canadian Tree Improvement Assoc., 9-22 August, pp. 78-82. Quebec City, Quebec.

Karnosky, D.F., R.A. Mickler, and D.D. Lange. 1982. Hormonal control of shoot and root induction in hypocotyl callus cultures of American elm. *In Vitro* 18(3):275.

Kroggel, M.A. 1988. Characterization of somaclonal variation in *Ulmus* x 'Pioneer' regenerated from protoplasts. M.S. Thesis. The Ohio State University. 71 pp.

Lineberger, R.D., M.B. Sticklen, P.M. Pijut, M.A. Kroggel, C.V.M. Fink, and S.C. Domir. 1990. Use of protoplast, cell, and shoot tip culture in elm germplasm improvement program. Acta Horticulturae 280:247-253.

Power, J.B. and M. R. Davey 1979. Laboratory Manual: Plant protoplasts (isolation, fusion, culture, genetic transformation). University of Nottingham.

Redenbaugh, M.K., R.D. Westfall, and D.F. Karnosky. 1980. Protoplast isolation from *Ulmus americana* L., pollen mother cells, tetrads, and microspores. Can. J. For. Res. 10:284-289.

Redenbaugh, M.K., D.F. Karnosky, and R.D. Westfall. 1981. Protoplast isolation and fusion in three *Ulmus* species. Can. J. Bot. 59:1436-1443.

Schreiber, L.R. and H. V. Main. 1976. 'Urban" elm. HortScience 11:517-518.

Smalley, E.B. and D.T. Lester. 1973. 'Sapparo Autumn Gold' elm. HortScience 8:514-515.

Sticklen, M.B., R.D. Lineberger, and S.C. Domir. 1985. Isolation and culture of protoplasts of *Ulmus* x 'Homestead'. Plant Science 41:117-120.

Sticklen, M.B., S.C. Domir, and R.D. Lineberger. 1986. Shoot regeneration from protoplasts of *Ulmus* x 'Pioneer'. Plant Science 47:29-34.

Townsend, A.M. 1982. Genetic improvement of elms and other genera. J. Arboric. 8:13-17.

Townsend, A.M. and W. O. Masters. 1984a. 'Homestead' elm. HortScience 19:897-898.

Townsend, A.M. and W.O. Masters. 1984b. 'Pioneer' elm. HortScience 19:900.

Ulrich, J.M., R.A. Mickler, B.J. Finkle, and D.F. Karnosky. 1984. Survival and regeneration of American elm callus cultures after being frozen in liquid nitrogen. Can. J. For. Res. 14:750-753.

11
Cerato-ulmin: A Unique Wilt Toxin of Instrumental Significance in the Development of Dutch Elm Disease

Wayne C. Richards

Abstract

The unique wilt toxin cerato-ulmin has been documented, covering its initial discovery and sequential experimental data to support its instrumental significance in the development of Dutch elm disease. Characterization of cerato-ulmin in view of its pathological implications for host specificity has brought up points that strongly suggest that cerato-ulmin is involved in *Ophiostoma ulmi* pathogenesis: 1) cerato-ulmin is produced only by *O. ulmi,* the Dutch elm disease pathogen, 2) cerato-ulmin causes symptoms of toxicity both external and internal that are similar to those of Dutch elm disease, 3) the toxicity of cerato-ulmin is selective and is highest with hosts of the pathogen, 4) high levels of cerato-ulmin apparently enhance the aggressiveness of *O. ulmi,* 5) cerato-ulmin affects host physiology, largely in a host specific manner, by increasing respiration rates and plasma membrane permeability; similar symptoms are observed in hosts affected by Dutch elm disease, in accordance with host susceptibility to Dutch elm disease, 6) *in vitro* production of cerato-ulmin is correlated with pathogenicities of *O. ulmi* wild-type isolates and, 7) cerato-ulmin is produced *in vivo* at relatively early stages in the development of external symptoms of Dutch elm disease.

Overview of Reported Dutch Elm Disease Toxins

In the late forties, a trend developed towards a toxin theory of Dutch elm disease causation. Much evidence in support of this theory is derived from the injection of cell-free filtrates of *Ophiostoma (Ceratocystis) ulmi* (Buisman) Nannf. cultures into healthy plants This was initially demonstrated by Zentmyer (1942a; 1942b) whereby he obtained a wilting of test plant cuttings, discoloration of cell walls, and formation of gum in vessels following treatment with culture filtrates. Since Zentmyer (1942a; 1942b) demonstrated that such production is possible in liquid culture of the fungus the search of toxins produced by *O. ulmi* has been carried out by a number of workers (Claydon et al. 1974, 1980; Dimond 1947; Dimond et al. 1949; Feldman et al. 1949, 1950; Frederick and Howard 1951; Nordin and Strobel 1981; Rebel 1969; Salemink et al. 1965; Scheffer and Elgersma 1981; Strobel et al. 1978; Van Alfen and Turner 1975). The involvement of toxins in Dutch elm disease has been the subject of heated debates decennia.

At this time attention will be focused mainly on non-enzymatic toxins of which four groups are known: phenolics, polysaccharides, glyco-peptides and cerato-ulmin.

Phenolics

The phenolic metabolites (C_{10} phenolic carboxylic acids) of *O. ulmi* investigated by Claydon et al. (1974) expressed phytotoxicity in elm shoot cuttings but attempts to produce phytotoxic effects by injecting these compounds into stems of four-year-old rooted elm saplings were unsuccessful and no evidence pertaining to their *in vivo* production was obtained (Claydon et al. 1980).

Polysaccharides

A polysaccharide fraction can be precipitated from culture filtrates with 70% ethanol. Dimond (1947) and Dimond et al. (1949) isolated a polysaccharide as a prime toxin from the ethanol-insoluble fraction in their screening of toxic substances in the culture filtrate of this fungus and observed some effects of this material on tomato and elm cuttings. The importance of the polysaccharides *in vivo is* questionable; apparently the viscosity of the xylem fluid is not affected after infection by *O. ulmi* in which case the production of polysaccharides should be low in comparison to the production in shake cultures where the viscosity rises rapidly due to the production of polysaccharides; both Feldman et al. (1950) and Rebel (1969) considered this fraction to be of minor or no importance as concentrations necessary to induce wilting were even higher than that occurring in shake cultures. Feldman et al. (1949; 1950) and Frederick and Howard (1951) claimed that the prime toxin component was in the ethanol-soluble fraction but not in the ethanol-insoluble fraction. Their studies on this toxic fraction were discontinued, so that no information is available on its chemical character.

Glycopeptides

A partially purified glycopeptide fraction was obtained by Salemink et al. (1965) and later by Rebel (1969). Phytoxicity of the glycopeptide fraction in elm shoots has been reported by Salemink et al. (1965), Rebel (1969), Strobel et al. (1978) and in more detail by Van Alfen and Turner (1975). Using affinity chromatography, Strobel et al. (1978) obtained a purified polydisperse glycopeptide fraction and Nordin and Strobel (1981) revealed that it was a polydisperse peptidorhamnomannan with a molecular weight range of approximately 105,200 to 120,000. The protein backbone is rich in theronine, alanine, serine, valine, proline and glycine but represents only 6.9% of the mass of the compound with the polysaccharide (consisting of predominantly rhamnose and mannose) comprising the rest of the molecule. The assignment of toxicity to the peptidorhamnomannan (400 μg/ml) was based on a bioassay utilizing fluid conductance in the elm stems and observing the wilting of leaves (Van Alfen and Turner 1975). Reduction in conductance caused by the peptidorhamnomannan was the basis of the assignment of toxicity and there was no evidence for biochemical effects on the host tissue such as electrolyte loss from leaf cells (Van Alfen and Turner 1975). Toxicity of the peptidorhamnomannan is also not host specific and caused wilting in a wide variety of plants (Strobel et al. 1978). The characterization of the peptidorhamnomannan has been extensive (Nordin and Strobel 1981; Strobel et al. 1978). However, the lack of biological selectivity of this compound and the ability of dextrans of the same size to mimic its effects does question the assignment of the peptidorhamnomannan as a toxin. Work by Scheffer and Elgersma (1981) has detected a glycoprotein product of *O. ulmi* in xylem elm sap (uncharacterized but likely the peptidorhamnomannan).

Neither the phenolics, polysaccharides nor the glycopeptides (Scheffer et al. 1987) have been demonstrated as a toxin responsible for the development of Dutch elm disease to settle the dispute concerning their importance in pathogenesis.

Cerato-ulmin

Cerato-ulmin (Richards and Takai 1973; Takai and Richards 1978) is a unique low molecular weight protein, produced by the Dutch elm disease fungus *O. ulmi,* and has demonstrated features as a pathotoxin (Wheeler and Luke 1963). The chapter herein is a documentation of this unique wilt toxin covering its initial discovery and sequential experimental data to support its instrumental significance in the development of Dutch elm disease.

1 - Discovery of Cerato-ulmin

During the routine observations of shake cultures of *O. ulmi* small particles (microstructures) were periodically noticed. The name cerato-ulmin was given to these microstructures in order to distinguish them from any other biological material. Ouellette and Gagnon (1960) also reported small particles in their strain, which they termed microendospores and microhyphae; Brotzman and Campana (1968) pointed out experimentally that these were not viable. In order to characterize the particles observed in our cultures, an isolation technique was necessary. Techniques involving centrifugation and solvent extraction were unsuccessful. However, the supposition that these particles may be associated with a gaseous condition resulted in the development of a novel technique which was successful and is described in this section.

Materials and Methods

Preparation and conditions of O. ulmi *cell-free culture filtrate*

The isolate CESS-16K of the fungus *O. ulmi* was cultured in Wilson modified liquid medium in which 1% sucrose was used as the carbon source. The cultures (25 ml each) were kept at 23 C on a gyratory shaker (120 rpm). After 10 days, the cultures were removed from the shaker and centrifuged for 30 min at 1,000 *g* to eliminate precipitible solid material including fungal spores and hyphae. After centrifugation the supernatant fluid was slowly decanted. It was then filter-sterilized through a series of Millipore membrane filters with pore diameters of 0.8, .45, and .22 µm respectively, in order to remove any additional small spores and hyphae not precipitated by centrifugation. The 0.22 µm membrane filtrate was then divided into 75 ml samples. The physical properties of cerato-ulmin in liquid medium were investigated by shaking cell-free culture filtrates on a gyratory shaker (120 rpm) for 30 min at 23 C and comparing them to unshaken cell-free culture filtrates that were but centrifuged at 1,465 *g* for 30 min at 23 C.

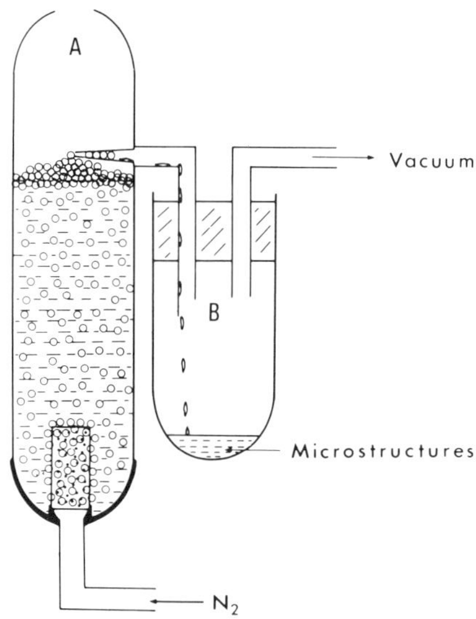

Figure 1. Schematic diagram of the apparatus and the bubbling technique used for the isolation of cerato-ulmin (microstructures)- A) bubbling unit and B) retaining unit.

Bubbling Technique

The apparatus used for the isolation of microstructures is illustrated in Fig. 1. The primary part (A) of the apparatus consists of a capsule constructed of two 50-ml polycarbonate centrifuge tubes, fused together by the solvent, 1,2-dichloroethane. Holes (0.5 inch, approximately 1.25 cm) were drilled through each end. The fritted-glass cylinder portion (medium porosity) of a gas-dispersion tube was inserted into the capsule through the bottom hole and held in a vertical position by a punched rubber serum stopper. The stopper served not only as a holder but also as a seal. A 0.25-inch (approximately 0.63 cm) hole was also drilled through the capsule wall 0.5 inch above that level marked for 75-ml filtrate.

The secondary part (B) of the apparatus consists of a retainer (50-ml polycarbonate centrifuge tube) and a two-hole rubber stopper supporting two tubes, one of which has a tapered end. The tube with the tapered end was used to connect the retainer to the capsule. The tapered portion was pushed through the hole previously mentioned, and the tip was positioned to the center of the capsule. The other tube was used to connect the retainer to the water-aspirator line (vacuum).

The cell-free culture filtrate (75 ml) prepared as described above, was poured slowly into the capsule through the top hole by means of a funnel. N_2 gas was then applied at 6.9 kPa (1 lb/in^2). As the N_2 gas passed through the fritted-glass cylinder, it entered the filtrate as small bubbles. Bubbling resulted in the accumulation of white foamy material on the top of the filtrate. Small bubbles worked better than large ones. A vacuum was then applied to suck this material through the tube with the tapered end and deposit it in the retainer. The procedure above was continued until the foamy material could no longer be collected. This material consisted of the microstructures originally set out to be isolated.

Results and Discussion

Characteristics which were observed for cerato-ulmin are shown in Fig. 2. Either shaking or bubbling caused the "unit" structures first seen in shake cultures of *O. ulmi* to assemble gradually into "rods" and "fibrils" and accumulate on the filtrate surface. As this occurs, the clear filtrate becomes milky or turbid in appearance (Fig. 3C.) and very similar in appearance to the total culture containing spores (Fig. 3A). By centrifugation, the assembled structures disassemble and revert back to "unit" structures, and the filtrate becomes clear once more (Fig. 3B). Reassembling can be achieved by shaking or bubbling. However, when detergent was added (sodium deoxycholate at a final concentration of 0.5%), the "rod" and "fibril" formations were disassembled into "unit" structures. The fact that bubbling at this time had no effect suggests the possible elimination of the means by which bubbling caused the "unit" structures to assemble into "rods" and "fibrils".

From each 75 ml of filtrate, a sample of 1 to 2 ml could be obtained by the bubbling technique. Microscopic examination of the isolated sample revealed numerous "rod" and "fibril" formations and having the same characteristics as that seen in the liquid shake culture. Cerato-ulmin was not viable.

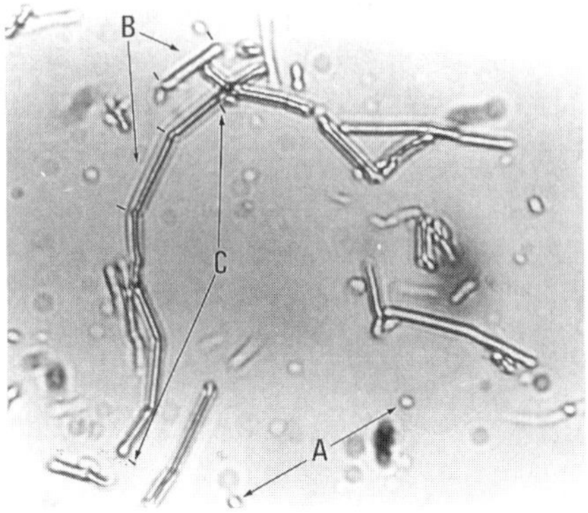

Figure 2. Three unstable forms of cerato-ulmin in cell-free culture filtrates of *O. ulmi* as observed under the microscope. A) "unit" (a primary form), B) "rod" and, C) "fibril" (X256). "Rod" and "fibril" are assemblages of "units".

Figure 3. Transformation of cerato-ulmin in cell-free culture filtrates of *O. ulmi*. A) total *O. ulmi* shake culture without removal of spores, B) cell-free culture filtrate of *O. ulmi* subjected to centrifugation at 1,465 *g* for 30 min at 23 C and C) cell-free culture filtrate placed on a gyratory shaker (120 rpm) for 30 min at 23 C.

2 - Isolation and Characterization of Cerato-ulmin

While studying toxic metabolites of *O. ulmi*, the author observed the abundant formation of microstructures (cerato-ulmin) in liquid shake cultures of this fungus, and developed a bubbling technique (Richards and Takai 1973) for their isolation, as described in section 1. This technique however could not handle large quantities of the culture filtrate for mass isolation. The author subsequently developed a new technique that includes vacuum effervescence whereby cerato-ulmin could be obtained as freeze-dried samples in much greater quantity than by the bubbling technique.

This section deals with the new isolation technique for cerato-ulmin, partial purification of cerato-ulmin, structural elucidation studies on the partially purified cerato-ulmin and the toxicity of purified cerato-ulmin to both host and non-host plants.

Materials and Methods

Culture of the fungus

The fungal strain CESS-16K of *O. ulmi* was used in isolating cerato-ulmin. For plate cultures, a 3% malt agar medium was used and for shake cultures a liquid medium containing the following ingredients: sucrose (10 g), L-asparagine (2 g), yeast extract (2 g), KH_2PO_4 (1 g), $MgSO_4 \cdot 7H_2O$ (0.1 g), $ZnSO_4 \cdot 7H_2O$ (0.44 mg), $FeCl_3 \cdot 6H_2O$ (0.48 mg), and $MnCl_2 \cdot 4H_2O$ (0.36 mg) in 1,000 ml of distilled water. All the media were sterilized by autoclaving (1.1 kg/cm^2 or 15 lb/in^2) for 15 min. Cultures were incubated at 23 C in the dark. One-week-old liquid shake cultures which had been inoculated with an agar block from the plate culture was used as the inoculum. One ml of the inoculum culture was transferred to a 1 L Erlenmeyer flask containing 300 ml of the liquid medium. The inoculated flask was shaken on a gyratory shaker (120 rpm) for 1 week. The uninoculated medium was incubated under the same conditions as the inoculated medium and both were checked for the presence of cerato-ulmin.

Isolation technique for cerato-ulmin

Figure 4 is a flow chart outlining the procedure for the isolation of cerato-ulmin from the shake culture of *O. ulmi*. Fungal spores and hyphae were removed from the culture filtrate by centrifugation and filter-sterilization.

Cerato-ulmin was separated initially from the culture filtrate by vacuum effervescence as follows: the culture filtrate was poured into a 4 L globular separatory funnel (Fig. 5A) which was shaken vigorously by hand for 2 min to force the cerato-ulmin originally in "unit" form to assemble into "rods" and "fibrils" (Richards and Takai 1973). This change in form was accompanied by the development of cloudiness in the filtrate (Fig. 5B). Then a weak vacuum was applied by means of a water aspirator to generate bubbles within the filtrate (Fig. 5C). When the bubbles accumulating on the liquid surface nearly filled the remaining space in the funnel, the vacuum was released; when the bubbles broke, the membranous material remained on the surface of the liquid layer (Fig. 5D, E). At the same time the liquid layer of the filtrate became as clear as it was initially (cf Fig. 5A). The liquid layer was then drained through the bottom outlet of the funnel and only the membranous material remained (Fig. 5F). This material was washed out of the funnel with 80% aqueous ethanol solution which is the highest concentration within the best solubility range (60-80%) so that the cerato-ulmin would dissolve easily and vacuum evaporation would be as efficient as possible. Vacuum effervescence was repeated until further extraction of the membranous material from the filtrate was impossible.

To check the efficiency of vacuum effervescence, the filtrate which showed no membrane formation after the first three attempts, was concentrated to one quarter of the original volume of the sample by molecular filtration with Amicon ultrafiltration membrane UM2. The concentrated sample was subjected to vacuum effervescence three times. The products obtained from the original and the concentrated samples were measured separately.

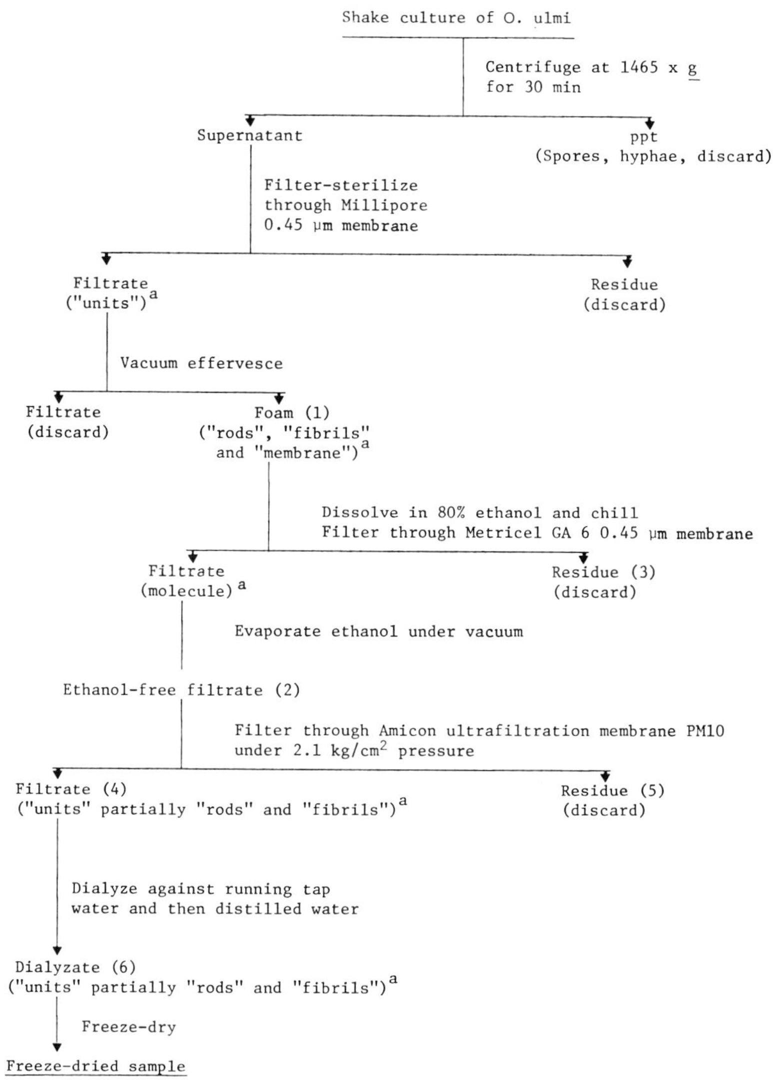

[a] The dominant form of cerato-ulmin

Figure 4. Procedure for isolating cerato-ulmin from shake cultures of *O. ulmi*.

Figure 5. Isolation of cerato-ulmin from the cell-free culture filtrate of *O. ulmi*, by vacuum effervescing in a separatory funnel. A) the culture filtrate free of fungal cells, prepared by centrifugation and filtration (cerato-ulmin is in "molecular" and "unit" form.). B) the sample turns milky by shaking the funnel. (cerato-ulmin is predominantly in "rod" and "fibril" form). C) a mild vacuum is applied by means of a water aspirator and bubbling is amplified [(vacuum effervescence) at which time membranous material accumulates on the surface of the filtrate and cerato-ulmin is in "rod", "fibril" and "membrane" form]. D) separation of cerato-ulmin from the liquid layer by slowly releasing the vacuum; at this time the liquid layer becomes clear again and cerato-ulmin is in "membrane" form out of the liquid, and in molecular and "unit" form in the liquid. E) cerato-ulmin in "membrane" form floating on the liquid layer. F) remaining cerato-ulmin in "membrane" form in the funnel after the liquid layer has been drained out through the bottom outlet; this "membrane" fraction was collected as the cerato-ulmin sample.

To localize cerato-ulmin, each fraction in the isolation process was monitored in the following manner: about 3 ml of the sample from each fraction were placed in a test tube and vigorously bubbled using a Pasteur pipet. If the sample contained cerato-ulmin, it turned milky. This change in appearance can be a very sensitive means of detecting the presence of cerato-ulmin in an aqueous solution. Even when the result was negative, the fraction was concentrated by molecular filtration as described above and the test was repeated to confirm the absence of cerato-ulmin in that fraction.

Since cerato-ulmin displayed a weakly positive color reaction to ninhydrin, a sample for the ninhydrin test was taken from each fraction in the isolation procedure after vacuum effervescence. Each sample was prepared so as to stand on an equal basis throughout the isolation procedure. An aliquot from each fraction was placed on filter paper and the color was developed with ninhydrin after the method of Payne and Kieber (1954). The coloration intensity of each spot was compared visually.

Partial purification of cerato-ulmin

Column chromatography. A loading sample of 50 mg of crude cerato-ulmin dissolved in 2 ml of 60% aqueous ethanol solution was applied to a 1.6 cm x 32 cm Sephadex LH-20 column. The sample was eluted downward with the same solution at a flow rate of 14 ml/h and 1.2 ml fractions were collected. Each collection was checked for protein and carbohydrate as noted in the figure caption for Figure 6, since cerato-ulmin appeared to possess characteristics of both.

Thin layer chromatography (TLC). The crude and chromatographed cerato-ulmin samples were developed on precoated Polygram Sil G plates using the solvent system of *n*-butanol-acetic acid-water (4:1:5). Chromatograms were stained with the reagents listed in Table 1.

High-voltage electrophoresis. The procedure of Stevenson and Smillie (1968) was employed.

Phytotoxicity test

Bioassay of the crude cerato-ulmin and two fractions (peaks I and II) derived from column chromatography of Sephadex LH-20 was done using new developing shoots of white elm 15-20 cm long with 4-5 opened leaves attached. Freeze-dried samples of cerato-ulmin were dissolved in distilled water to make up assay solutions. The concentration of the crude cerato-ulmin and the sample from peak I (cerato-ulmin fraction) were both adjusted to 120 μg/ml, the same as previously reported (Takai 1974). However, the concentration of the sample from peak II was adjusted to 4.3 μg/ml to keep the original yield ratio of 27.9:1 between two chromatographic fractions from peaks I and II. Four elm cuttings were placed in each assay solution and allowed to take up the solution under Agro-lites at room temperature, without application of pressure. The occurrence of symptoms on cuttings was carefully observed.

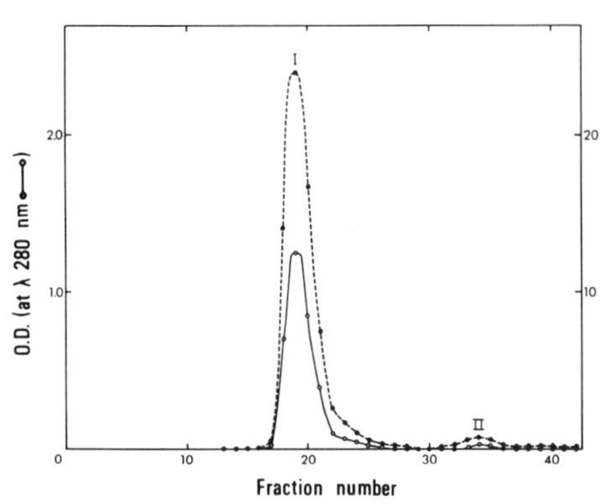

Figure 6. Chromatography of the crude cerato-ulmin on Sephadex LH-20. O.D. at λ 280 nm representing protein concentration O.D. at λ 480 nm for the colored product reacted with the phenol-sulfuric acid reagent representing carbohydrate concentration.

The partially purified cerato-ulmin was characterized according to standard methods unless otherwise stated. The U.V. spectrum of cerato-ulmin and the elution profile of protein from the Sephadex LH-20 column were obtained on a Unicam SP 800 B Ultraviolet Spectrophotometer. The hydrolysis of cerato-ulmin for amino acid analysis was carried out in 6N HCl at 110 C for 24 h. The hydrolysates were analyzed on a Beckman 121 automatic amino acid analyzer. Measurement of the turbidity and carbohydrate content of the cerato-ulmin sample was made on a B&L Spectronic 20 as described below.

Table 1. Color reactions for the characterization of cerato-ulmin

Reagent	Reaction	Compound	Reference
Tollens	negative	Reducing	Tollens 1882
Benedict	negative	group	Koch and Hanke 1948
$CuSO_4$-phosphomolybdate	negative		French et al. 1950
Molish	pinkish purple	Carbohydrate	Koch and Hanke 1948
Anthrone	greyish blue		Morris 1948
Phenol-sulphuric acid	yellowish brown		Dubois et al. 1951
Aniline-diphenylamine	brown		Moczar & Moczar 1970
Bromophenol blue	blue	Sugar acid	Demain and Phaff 1954
o-Phenylendiamine	negative		Macek and Tadra 1952
Ninhydrin	pink (weak)	Amino acid and amino sugar	Payne and Kieber 1954
Folin phenol	blue	Peptide&protein	Lowry et al. 1951
Biuret	purple		Hawk et al. 1947

Estimation of the molecular size range of cerato-ulmin

Three aqueous solutions of the freeze-dried cerato-ulmin were prepared and handled as follows: 1) with no molecular filtration, 2) with the recovery of the filtrate through Amicon ultrafiltration membrane PM10, 3) with the recovery of the residue on Amicon ultrafiltration membrane UM2. The molecular filtration was conducted under 207 kPa (30 lb/in^2) pressure. All samples were subjected to dialysis against water. The location of cerato-ulmin in each resulting fraction was carefully checked by bubbling it, using a Pasteur pipet as previously mentioned. Another tentative estimation of the minimum molecular weight of cerato-ulmin was made on the basis of the results of amino acid analysis.

Quantitative determination of cerato-ulmin

Turbidity measurement. A simple measurement of the cerato-ulmin in culture filtrate and of the isolated cerato-ulmin dissolved in water was used. The culture filtrate free of the fungal cell was prepared according to the procedure outlined in Figure 4. An aqueous solution of partially purified cerato-ulmin was made up at a concentration of 50 µg/ml. Twenty-five ml of the sample were poured into a 125-ml Erlenmeyer flask. The flask was shaken vigorously by wrist action for 1 min and then shaken continuously on a gyratory shaker (120 rpm) for 30 min to obtain maximum assembling (i.e., the highest turbidity). The turbid sample was serially diluted with water when necessary, and was kept for continuous shaking until it was placed in the measuring machine (B&L Spectronic 20). When the culture filtrate was measured the reference was prepared with the same culture filtrate as was used for the sample and at the same dilution, and was handled carefully to prevent the transformation of "units" into "rods" and "fibrils". If the reference became turbid, it was immediately centrifuged at 1,465 *g* for 30 min. so that it would regain its clear appearance. When the partially purified sample was measured distilled water was used as the reference. Turbidity of the sample was read for O.D. at λ 400 nm.

Colorimetry of cerato-ulmin. A) Protein measurement: the improved Lowry method (Oyama and Eagle 1956) was employed for determination of relative protein levels to estimate the relative purity of cerato-ulmin samples. B) Carbohydrate measurement: the phenol-sulfuric acid method of Dubois et al. (1951) was applied for all the measurements of carbohydrate. The O.D. of the colored product was λ 480 nm. Calculation of carbohydrate content was based on glucose value.

Plant materials

Seed. Seeds of white elm *(Ulmus americana* L.), radish *(Raphanus sativus* L.), snapdragon *(Antirrhinum orontium* L.) and tomato *(Licopersicum esculentum* Mill.) were used. Prior to treatment with cerato-ulmin, seeds were washed in running tap water for 3 h.

Shoot cuttings. Two- or three-year-old dormant seedlings of white elm, Siberian pea shrub *(Caragana arborescens* Lam), dogwood *(Cornus stolonifera* Michx.), green ash *(Fraxinus pennsylvanica* Marsh.), Manitoba maple *(Acer saccharum* Marsh.), and red elderberry *(Sambucus racemosa A.* Gray) were grown in pots under controlled greenhouse conditions. Shoot cuttings of white elm were obtained by methods described elsewhere (Richards et al. 1982) and three to four leaves were left at the end of each cutting. Shoot cuttings of other woody plants were obtained when new shoots were 60% mature; the number of leaves left on each cutting depended on the developmental stage of the foliage. Tomato plants were grown from seeds, and stem cuttings instead of shoot cuttings were prepared when seedlings were approximately 15 cm high. As soon as the cuttings were removed from the seedlings a second cut was made within distilled water de-gassed by autoclaving, 2 cm above the first cut, in order to avoid the possibility of air-locking in the water-conducting vessels. The cuttings were kept in tall beakers with the cut ends immersed in distilled water.

Elm leaf discs. Leaf discs were prepared from leaves removed from 2- or 3-year-old seedlings of white elm and Siberian elm (*U. pumila* L.) grown in pots in the greenhouse as described above. The third leaf from the top of the shoot was harvested from seedlings. Leaves were then thoroughly washed in distilled water to eliminate any dirt from the leaf surface, and with a paper punch discs 6 mm in diameter were punched from the leaves.

Bioassay solution

Aqueous solutions of purified cerato-ulmin. Crude cerato-ulmin isolated from the *O. ulmi* culture filtrate as described above, was purified (Fig. 7) by gel filtration on Sephadex LH 60 (Pharmacia Fine Chemicals, Uppsala, Sweden) as described elsewhere (Stevenson et al. 1979). The purity of the product was tested by electrophoresis on SDS polyacrylamide gel (20%); heavy sample loading presented no bands other than the cerato-ulmin band. Purified cerato-ulmin was dissolved in either de-gassed or de-ionized distilled water at various concentrations (10^{-3} to 120 µg/ml). Immediately before the experiments were set up, the solutions were centrifuged at 1,465 g for 30 min at 4 C to keep cerato-ulmin in the form of "units" and free molecules (Richards and Takai 1973; Takai et al. 1979). Purified cerato-ulmin was also dissolved in distilled water at a concentration of 120 µg/ml and the solution was divided into two equal volumes; one was autoclaved at 103.5 kPa (15 lbs/in^2) for 15 min and the other was boiled at 100 Cfor 15 min. Each solution of cerato-ulmin was then diluted to a final concentration of 10 µg/ml with distilled water heat-treated in the same manner.

Aqueous solutions of dextran. Blue dextran 2000 (2 x 10^6 MW), dextran T40 (4 x 10^4 MW) and dextran T10 (1 x 10^4 MW) were dissolved in de-gassed distilled water at a concentration of 10 µg/ml. (All of these dextrans were supplied by Pharmacia Fine Chemicals.)

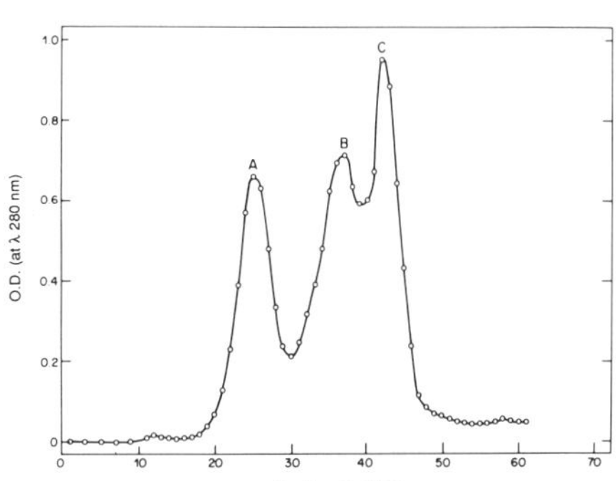

Figure 7. The purification of cerato-ulmin on Sephadex LH60. Cerato-ulmin (40 mg), prepared as described by Takai and Richards (1972), was chromatographed on a 1.6 cm x 25 cm column equilibrated and eluted with 60% ethanol at 23 C. The flow rate was 10 ml/h with 1.0 ml fractions being collected. Peak A is cerato-ulmin and Peaks B and C are non-protein material, likely phenolic resins.

Method of bioassay

Germination of plant seeds. Fifty seeds of each plant, described above, were placed on sterilized filter paper saturated with an aqueous solution of cerato-ulmin (120 μg/ml) in a petri-dish at 25 C. For each seed source four petri-dish replicates were made. The germination of seeds treated with cerato-ulmin was compared with that of seeds treated with distilled water.

External symptom expression on cuttings. Prior to use, all cuttings prepared as described above were incubated for 1 h in the greenhouse and those cuttings showing no effects from excision were transferred to the growth chamber conditioned at 25 C, 50% R.H. and 600 f.c. light where experiments were conducted. A cutting was placed in a vial containing one of the bioassay solutions described above, by the method reported by Van Alfen and Turner (1975). External symptom expression was periodically observed and recorded at each reading of transpiration. In most cases drooping was the initial state in which the leaves and the top of the shoot began to hang down slightly as a result of a decrease in normal turgor of tissues. Wilting, on the other hand, was considered an advanced stage of drooping; with this advanced stage the rigidity of the leaves and shoot tissue was completely lost. Since some cuttings exhibiting drooping did not proceed to the wilting state, the symptom of drooping was not considered a reliable parameter for the toxicity of cerato-ulmin.

Transpiration of white elm cuttings. Elm cuttings were prepared as described above. Cuttings were placed in the various bioassay solutions described above for various incubation time periods and measurements of transpiration were conducted by the method described by Richards et al. (1982). The relative transpiration rate was defined as the ratio between the transpiration rate measured as ml per g fresh weight in 1 h (ml/g fr.wt/h) for cuttings placed in the bioassay solutions and cuttings placed in de-gassed distilled water. To determine the critical transpiration rate for wilting, white elm cuttings were exposed to purified cerato-ulmin and measurements of transpiration were made; the critical transpiration rate for wilting development was calculated by averaging the transpiration rate measurements obtained from elm cuttings for which wilting was initiated.

Recovery of wilted white elm cuttings. Cuttings were incubated with various concentrations of cerato-ulmin (1, 15, 30, 60 and 120 μg/ml) and transferred to de-gassed distilled water as soon as wilting was initiated. Observations of external symptom expression and measurements of transpiration were continued up to 48 h.

Measurement of respiration of elm leaf tissue. Prior to measurement of respiration of the leaf discs, prepared as described above, were floated for 1 h in 10^{-4} M phosphate buffer adjusted to pH 5.9, to coincide with the pH of the sap. The leaf discs were then placed in a solution of 1% glucose in 10^{-4} M phosphate buffer, pH 5.9, to which purified cerato-ulmin was added to make up the following concentrations: 20, 30, 40, 60, 80 and 120 μg/ml. Controls contained no cerato-ulmin. To determine the oxygen uptake (QO_2), a Gilson Differential Respirometer, single valve style installed with all-glass volumometers, was used. Thirty leaf discs were floated in the main chamber (15 ml volume) of each reaction flask containing 3 ml of the test solution. Flasks containing the sample were vacuumed at 85 kPa (25" Hg) for 10 min to infiltrate the test solution into leaf tissue and then 2 ml of 20% KOH solution were delivered to the center well of the reaction flask. The reaction flasks were covered with aluminum foil to shut out light and were then shaken at 120 fpm at 2.75 cm amplitude in a water bath set at 25 C. Manometer readings were recorded every 10 min within a 2-h period, at the end of which the discs were taken out of the flask, washed with distilled water and oven-dried at 95 C until the weight became constant. The QO_2 was expressed as µL of O_2 absorbed by 1 mg dry weight of leaf tissue in 1 h (µl of O_2/mg dry wt/h). The relative respiration rate represents a ratio of the QO_2 obtained between leaf tissue treated with cerato-ulmin and leaf tissue treated with distilled water.

Measurement of electrolyte loss from elm leaf tissue. The electrolyte loss from elm leaf tissue was determined according to the method of Damann et al. (1974) and Van Alfen and Turner (1975). Two grams of leaf discs prepared as described above were vacuum infiltrated with a de-ionized distilled water solution of cerato-ulmin (120 µg/ml) and another 2 g with only de-ionized distilled water (control), in the manner described above for measurement of respiration of elm leaf tissue. After treatment elm discs were washed thoroughly with de-ionized distilled water and incubated at 25 C in 50 ml of de-ionized distilled water. Readings of conductance were made with a Radiometer CD2e Conductivity Bridge and Cell after 2, 4 and 6 h incubation. An increased conductance represents an electrolyte loss from leaf tissue (Damann et al. 1974). Relative conductance represents a ratio of the conductance obtained between leaf tissue treated with cerato-ulmin and leaf tissue treated with de-ionized distilled water.

Results

Isolation of cerato-ulmin

Cerato-ulmin was recovered from the liquid shake culture of *O. ulmi* according to the procedure outlined in Figure 4.

Three consecutive applications of vacuum effervescence resulted in an 85% recovery of extractable cerato-ulmin. Cerato-ulmin was not detected either in the fraction precipitated with 80% ethanol (Fig. 4), residue 3) or in the residue on the Amicon ultrafiltration membrane PM10 (Fig. 4), residue 5). It was found to be nondialyzable. As shown in Table 2, the sample at each stage of the procedure after vacuum effervescence was checked for ninhydrin reactivity since the final product is weakly positive to ninhydrin (Table 1). Ninhydrin positivity of the sample was substantially reduced by dialysis (Table 2).

Table 2. Changes in ninhydrin reactivity of fractions during extraction of cerato-ulmin from the shake culture filtrate of *O. ulmi.*

	Fraction*					
Intensity of coloration with ninhydrin	1	2	3	4	5	6
	++	++	+	++	±	±~+

Coloration: pink
Intensity of coloration: ++, distinct, +, weak, ±, very weak.
*) Fraction number: according to Figure 4

An aqueous solution of this product assembled into "units", "rods" and "fibrils" which are characteristic assemblages for cerato-ulmin (Fig. 2). The U.V. absorption spectrum of the product exhibited the following maxima: $\lambda_{max}^{60\%EtOH}$ nm: 269, 278 and 285 (Fig. 8).

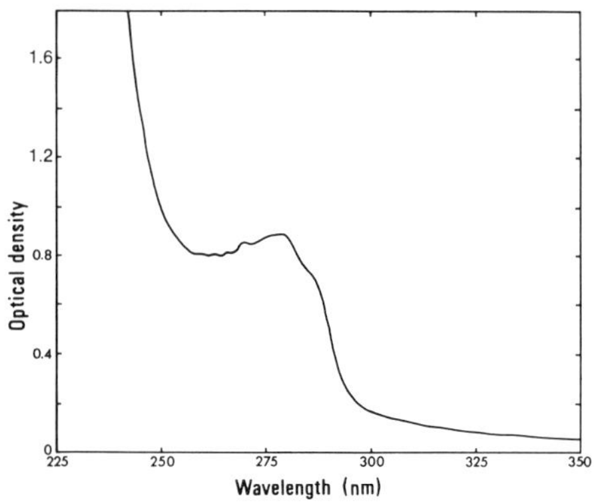

Figure 8. U.V. absorption spectra of cerato-ulmin in 60% aqueous ethanol. Sample concentration (2 mg/ml).

After the sample was freeze-dried, cerato-ulmin was harvested as a fluffy white powder at the rate of 25 mg/L from the culture filtrate; this was obtained from a culture with a carbon source of 1% sucrose. No cerato-ulmin could be detected in the medium without the fungus.

Partial purification of the crude cerato-ulmin

The crude cerato-ulmin was chromatographed on Sephadex LH-20 (Fig. 6). Two separated peaks were obtained.

The sample of the major peak (I). The solution was pale yellow and presented weak fluorescence at λ 285 nm. The freeze-dried eluate was whiter than the crude cerato-ulmin. Its aqueous solution was able to assemble into "units", "rods," "fibrils" and "membrane". The U.V. absorption of this sample in 60%

ethanol exhibited the following maxima: λ nm: 269, 278 and 285 (Fig. 8), which corresponded exactly with those of the crude cerato-ulmin. The yield of the sample was 36.3% of the crude product (Table 3). Typical toxicity symptoms (Fig. 9A) as described previously (Takai 1974) developed on white elm cuttings when these were tested. Symptoms appeared at the times indicated: an initiation of a substantial reduction in uptake after 3 h; obvious wilting after 19 h; irreversible wilting and chlorosis on living leaves after 43 h; and necrosis on living leaves after 91 h. When the xylem was checked for brown discoloration at the end of the incubation (168 h), it was detected in all cuttings. No abnormal symptom was found, however, in the control cuttings (Fig. 9B) incubated with de-gassed distilled water. Hence this product was identified as cerato-ulmin.

Figure 9. External symptoms expressed on white elm cuttings incubated in A) aqueous cerato-ulmin solution (120 μg/ml) for 48 h and B) de-gassed distilled water (control) for 48 h.

The sample of the minor peak (II). The eluate appeared yellowish brown, presented strong fluorescence at λ 285 nm, and reacted positively to the ferric chloride-potassium ferricyanide reagent. It was slightly contaminated with cerato-ulmin by tailing from peak I. The U.V. absorption spectrum of this sample was λ_{max} nm: 299, λ_{min} nm: 274, while the yield was only 1.3% of the crude product. This sample presented very weak toxic symptoms on white elm cuttings. Therefore, it was concluded that the sample was not cerato-ulmin.

Reactivity to ninhydrin and phenol-sulfuric acid reagents, fluorescence, and visible color were absent with eluates before peak I. However, these characteristics tended to become apparent as elution proceeded beyond peak I.

The contribution of chromatography on Sephadex LH-20 to the purification of cerato-ulmin is summarized in Table 3. As mentioned above chromatography separated the cerato-ulmin fraction from the substance of peak II as confirmed on the elution profile (Fig. 6). When the chromatographed cerato-ulmin sample was submitted to TLC and high-voltage electrophoresis no spots were observed other than cerato-ulmin. Checks were made for carbohydrates and peptides. As a result of chromatography of the crude cerato-ulmin on Sephadex LH-20, protein content increased 40% while carbohydrate content decreased 30%. Phytotoxicity of the chromatographed sample was distinct; however, a slight increase in toxicity was observed with the chromatographed cerato-ulmin.

Table 3. Contribution of chromatography on Sephadex LH-20 to purification of cerato-ulmin

	Sample chromatographed	
	Before	After
No. peaks by Sephadex LH-20 chromatography	2	1
No. of spots by TLC	4	1
Stained with ninhydrin	3 (R_F 0.08, 0.14, 0.28)	1 (R_F 0.08)
$2N\text{-}H_2SO_4$*	3 (R_F 0.08, 0.28, 0.36)	1 (R_F 0.08)
aniline-diphenylamine	3 (R_F 0.08, 0.28, 0.36)	1 (R_F 0.08)
by H-V electrophoresis**		
stained with cadmium-ninhydrin	-	1 (R_F 0)
Erhlich's reagent	-	0
Pauly reagent	-	1 (R_F 0)
Relative protein value	100	140
Carbohydrate content	14.7%	10.3%
(Relative value)	100	70
Relative yield	100	36.3
Toxicity	Positive	Positive (slightly increased)

* For carbohydrate detection

** High-voltage electrophoresis at pH 2

General properties of cerato-ulmin

Cerato-ulmin, a non-hygroscopic, white, fluffy powder, is soluble in water but physically unstable; in molecular form it changes to "units", "rods" or "fibrils" by assembling (Richards and Takai 1973; Takai 1973; Takai 1974; Takai and Richards 1972). It is slightly soluble in ethanol, methanol and acetone. However, its solubility increases in an aqueous solution of these solvents. The extent of this increase in solubility varies from one solution to another: the increase is greater in an aqueous solution of ethanol than in one of acetone, and least of all in an aqueous solution of methanol. The best range of solubility among aqueous ethanol solutions appeared to be between 60% and 80%. However, *n*-butanol, *n*-amyl alcohol, iso-propanol, chloroform, DNFA, *n*-hexane, ether, benzene, ethylacetate and some other water-immiscible solvents do not dissolve cerato-ulmin.

Cerato-ulmin in 80% aqueous ethanol solution (5 mg/ml) had a pH of 5.8 Under normal pressure cerato-ulmin dissolved in water was not dialyzable while that in 60% aqueous ethanol solution was dialyzable. "Units" were observed in the aqueous solution of cerato-ulmin under the microscope during dialysis, while nothing was observed in the aqueous ethanol solution (60%) of cerato-ulmin even when it was vigorously shaken.

Cerato-ulmin aqueous solution passed through Amicon ultrafiltration membrane PM10 but not through Amicon ultrafiltration membrane UM2. These filtrations were conducted under pressure. Neither cerato-ulmin in the filtrate through Amicon ultrafiltration membrane PM10 nor that in the residue on Amicon ultrafiltration membrane UM2 was dialyzable.

From the cutoff points of these membranes the molecular weight of cerato-ulmin was assumed at the time to be between 1,000 and 10,000.

Cerato-ulmin showed carbohydrate characteristics when subjected to Molish, anthrone, phenol-sulfuric acid and aniline-diphenylamine tests. The positive result from bromophenol blue suggests that cerato-ulmin contains sugar acid characters, but this is uncertain (Table 1). The carbohydrate content appeared to be 10.3% (Table 3).

It is weakly positive to ninhydrin, and positive to the Folin phenol and biuret reagents (Table 1). Cerato-ulmin was precipitated when its aqueous solution was 60% saturated with ammonium sulfate. The presence of a least 16 amino acids was tentatively confirmed in the cerato-ulmin molecule. Neither tryptophan or methionine was detected in the sample. A very high value was obtained for proline. Assuming a histidine value of one, a total of 126.4 residues were calculated (Table 4). Cerato-ulmin in the water layer, both heated and unheated, did not move into the chloroform/methanol layer (1:1), indicating a lack of fatty acid character.

The U.V. absorption spectrum of the partially purified cerato-ulmin exhibited maxima as follows (Fig. 8): λ nm: 269, 278 and 285.

Quantitative determination of cerato-ulmin

Turbidity measurement. The cell-free culture filtrate at different dilutions and the aqueous solution of the partially purified cerato-ulmin at different concentrations showed linear relationship between concentration and turbidity (O.D. at λ 400 nm in the range of 0.19-1.2 in O.D., or 7-30 μg/ml of cerato-ulmin concentration (Fig. 10A,B). Thus Beer's law is applicable to this system. However, turbidity of the sample at O.D. lower than 0.6 was not stable enough for constant reading.

Colorimetric measurement. The O.D. of the reaction product obtained when the partially purified cerato-ulmin and the phenol-sulfuric acid reagent were combined was measured at λ 480 nm. Within the range of 46.7-280 μg of cerato-ulmin there was a linear relationship with O.D. (Fig. 11), indicating that Beer's law is also applicable to this system.

Table 4. Summary of amino acid analysis of partially purified cerato-ulmin

Amino acid	Amount* (nmole)	Ratio (Let H's = 1)
Lysine	13.1	1.4
Histidine	9.7	1.0
Arginine	26.5	2.7
Tryptophan	0	0
Aspartic acid and asparagine	136.7	14.1
Theronine	73.6	7.6
Serine	139.7	14.4
Glutamic acid and glutamine	71.6	7.4
Proline	164.0	16.9
Glycine	115.5	11.9
Half cystine	111.1	11.5
Valine	62.0	6.4
Methionine	0	0
Isoleucine	30.9	3.2
Leucine	155.1	16.0
Tyrosine	11.9	1.2
Phenylalanine	17.3	1.8
Total number of residues		126.4

* Average of two repetitions

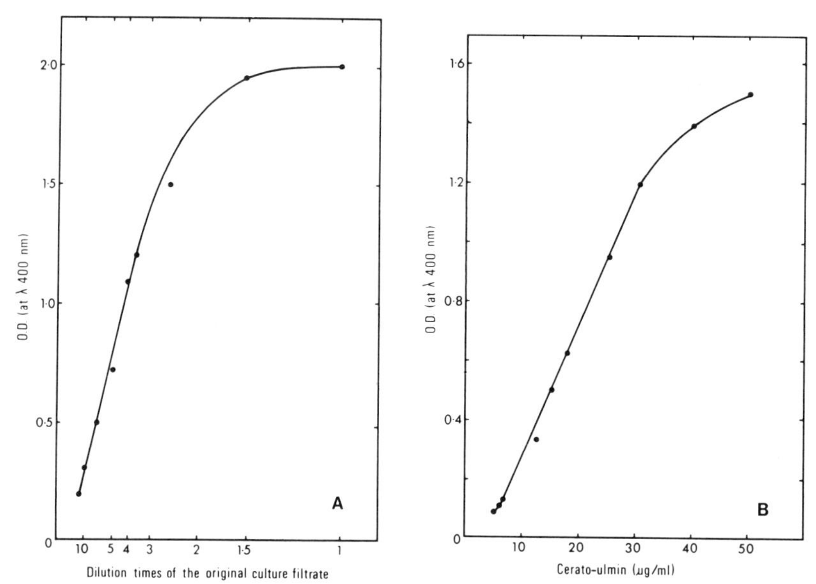

Figures 10A and 10B. Relationship between turbidity and concentration of cerato-ulmin. A) the culture filtrate free of fungal cells (a dilution value of 1 equals the original culture filtrate) and B) partially purified cerato-ulmin in water.

Effect of cerato-ulmin on seed germination

When seeds of white elm, radish, snapdragon and tomato were germinated in a solution of cerato-ulmin at a concentration of 120 µg/ml the germination rates did not differ significantly from those obtained in distilled water.

Screening of plants against cerato-ulmin

External symptom expression for plant cuttings incubated with cerato-ulmin (120 µg/ml) were as follows: white elm was the most sensitive in response to cerato-ulmin among those plants tested. White elm cuttings developed drooping within 1-2 h after cerato-ulmin was taken up by transpiration, and within 5 h 50% of the elm cuttings wilted. Green ash developed wilting infrequently and much later than did white elm. Manitoba maple, sugar maple and red elderberry developed drooping after 8 h and this condition remained the same even when the incubation time was extended to 48 h. Dogwood, Siberian pea shrub and tomato expressed no visible effect even after the incubation time was extended to 48 h.

Effect of cerato-ulmin on external symptom expression and transpiration of white elm cuttings

Because fluctuating results were obtained for both external symptom expression and transpiration, a considerable number of replicates had to be taken. The only external symptom expression observed during the observation period were drooping and wilting. In most cases drooping of the leaves in the top portion of the shoot appeared at the initial stage of wilting. Wilting was observed in 50% or more of the white elm cuttings when these were incubated in cerato-ulmin solutions at concentrations of 30 µg/ml in the 0-2 h time range, 15 µg/ml in the 2-8 h time range and 10^{-1} µg/ml in the 8-24 h time range. However, even concentrations of cerato-ulmin as low as 10^{-3} µg/ml caused infrequent wilting.

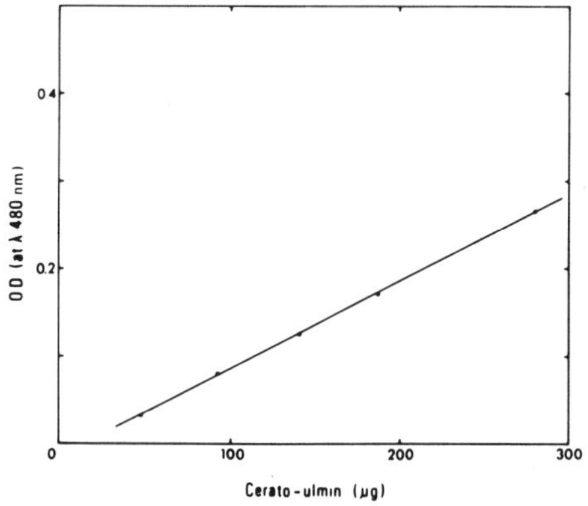

Figure 11. Relationship between cerato-ulmin concentration and color density of the reaction product of the partially purified cerato-ulmin with the phenol-sulfuric reagent.

A significant reduction (P=0.05) in the transpiration rate of elm cuttings treated with cerato-ulmin from that of the de-gassed distilled water control was also observed for the cerato-ulmin concentrations of 30 μg/ml in the 0-2 h incubation time range, 15 μg/ml in the 2-8 h incubation time range and 10^{-1} μg/ml in the 8-24 h incubation time range. (Table 5). The critical transpiration rate for wilting development was calculated to be 0.19 ml/g fr.wt/h; this figure was calculated on the basis of the average of the transpiration rate measurements obtained from 52 samples from 12 different experiments in which wilting was initiated in white elm cuttings exposed to purified cerato-ulmin.

This figure is less than one-third the hourly transpiration rate of white elm cuttings in de-gassed distilled water which was calculated to be 0.65 ml/g fr. wt/h. There was no difference in the transpiration rate for elm cuttings incubated for 0-2 h and 8-24 h in de-gassed distilled water (control). However, a lower transpiration rate was observed for elm cuttings incubated for 2-8 h in de-gassed distilled water (control).

Table 5. Transpiration of white elm cuttings treated with cerato-ulmin at different incubation time ranges.

Time range (h)	Transpiration rate of control[a] (ml/g fr.wt/h)	RT[b] (%) Concentration of cerato-ulmin (μg/ml)							
		10^{-3}	10^{-2}	10^{-1}	1	15	30	60	120
0-2	0.65 +/-0.07 (46)	126	97	83	86	61	53*	66	36*
2-8	0.40 +/-0.05 (41)	103	92	76	65	43*	32*	32*	9*
8-24	0.65 +/-0.05 (24)	84	77	61*	43*	16*	18*	17*	8*

[a] Mean +/- standard error (number of replicates).
[b] Transpiration rate relative to that of the de-gassed distilled water control, average of 22 replicates.
* Significant difference (P = 0.05) from that of the de-gassed distilled water control (Tukey's test).

Cuttings which began to wilt following the uptake of cerato-ulmin at concentrations of 30, 60 and 120 μg/ml showed no significant recovery from either wilting or transpiration rate when transferred to de-gassed distilled water.

No difference in the degree of wilting was observed among cuttings treated with unheated, autoclaved or boiled cerato-ulmin at the concentration of 10 μg/ml.

Effect of molecular weight simulated by the use of dextrans on the transpiration rate of white elm cuttings

To determine whether or not the molecular weight of cerato-ulmin (Feng et al. 1991; Yaguchi et al. 1992) was a factor mechanically contributing to its toxic effect on white elm cuttings, Blue dextran 2000 (2×10^6 MW), dextran T40 (4×10^4 MW) and dextran T10 (1×10^4 MW) were tested against elm cuttings by measuring their transpiration (Table 6). Dextran T10, comparable with cerato-ulmin in molecular weight, and dextran T40, greater in molecular weight than cerato-ulmin caused no significant difference in transpiration rate. Blue dextran 2000, however, showed a reduction in transpiration rate to some extent but was not significantly different from the de-gassed distilled water control. Cerato-ulmin, however, caused a significant reduction in the transpiration rate after 2 h incubation.

Table 6. Comparison of transpiration of white elm cuttings treated with dextrans of various molecular weights with that of cuttings treated with cerato-ulmin (concentration of assay solutions, 10 μg/ml; average of 15 replicates).

	RT[a] (%)			
Time range (h)	Blue dextran 2000 (2×10^6 MW)	Dextran T40 (4×10^4 MW)	Dextran T10 (1×10^4 MW)	Cerato-ulmin (7.6×10^3 MW)
0-2	78	124	94	53
2-8	50	93	100	31*
8-24	48	83	90	28*

[a] Transpiration rate relative to that of the de-gassed distilled water control.
* Significant difference ($P = 0.05$) from that of the de-gassed distilled water control (Tukey's test).

Effect of cerato-ulmin on the respiration of elm leaf tissue

The oxygen uptake (QO_2) measured as μl of O_2/mg dry wt/h for elm leaf tissue in distilled water (control) was 0.585 for white elm and 0.361 for Siberian elm (Table 7). Cerato-ulmin treatment resulted in an increase in respiration rate for both elms; however, the rate for each elm started to increase at different concentrations of cerato-ulmin and the pattern of the increase was also different. The respiration rate of white elm started to increase in response to cerato-ulmin at a concentration between 20 and 30 μg/ml whereas that of Siberian elm started to increase at a concentration between 30 and 40 μg/ml. The respiration rate of white elm continued to increase at every concentration tested beyond 20 μg/ml, whereas that of Siberian elm peaked at 60 μg/ml, and beyond this concentration remained unchanged. At the concentration of 120 μg/ml the percentage increase in the respiration rate of white elm was 2.3 times greater than that of Siberian elm.

Effect of cerato-ulmin on electrolyte loss from elm leaf tissue

Electrolyte loss from elm leaf tissue was measured as an increase in conductivity in de-ionized distilled water. As shown in Table 8 there was a basic electrolyte loss from leaf tissue for both elms incubated in only de-ionized distilled water (control) and this loss increased as the incubation was extended. The increase was accelerated at 6 h incubation, an indication of some changes in membrane permeability as a result of prolonged incubation. The pattern of electrolyte loss from leaf tissue treated with cerato-ulmin varied in the two elms. White elm, susceptible to Dutch elm disease, showed a pronounced increase in electrolyte loss when compared with the control, whereas Siberian elm, resistant to Dutch elm disease, showed a negligible increase when compared with the control. The electrolyte loss over a 4-h incubation period was seven times greater for white elm than for Siberian elm. The accelerated increase in electrolyte loss from the control leaf tissue after a 6-h incubation could explain the decline in relative conductance for both elms.

Table 7. Effect of cerato-ulmin at various concentrations on respiration of leaf tissue from white elm and Siberian elm.

Elms tested	QO_2 of control*	Relative QO_2[a] (%) Cerato-ulmin concentration (μg/ml) 20	30	40	60	80	120
White elm	0.585	98	123	173	220	232	248
Siberian elm	0.361	100	100	120	166	163	165

Average of 6 replicates
[a] Oxygen uptake relative to that of the control
* (μl of O_2/mg dry wt/h)

Table 8. Effect of cerato-ulmin on electrolyte loss from leaf tissue of white elm and Siberian elm (concentration of cerato-ulmin, 120 μg/ml; average of three replicates).

Elms tested	Treatment	Conductance μmhos/dry wt (g) Time after treatment (h) 2	4	6
White elm	Cerato-ulmin (T)	39.31	64.33	75.06
	Control (C)	30.63	37.44	51.06
		128*	172*	147*
Siberian elm	Cerato-ulmin (T)	52.26	91.03	123.58
	Control (C)	50.04	82.24	122.93
		104*	111*	101*

* Conductance relative to that of the de-ionized distilled water control (%).

Discussion

Vacuum effervescence is an obvious improvement over the initial bubbling technique (Richards and Takai 1973; Takai and Richards 1972) for isolating cerato-ulmin from the culture filtrate. It has resulted in greater efficiency, in that assembling of cerato-ulmin seems to be affected by the gaseous content of the liquid (Richards and Takai 1973); shaking before vacuum effervescing enables the liquid to become saturated with gas, while vacuum effervescing increases this gaseous condition, and results in maximum assemblage of cerato-ulmin. The bubbles generated by vacuum effervescing are distributed evenly throughout the liquid and efficiently transport the assembled cerato-ulmin to the liquid surface, whereas the bubbles generated by the initial technique were insufficient for this purpose. In the second place, it has enabled us to handle samples on a larger scale than was possible with the initial bubbling technique. Finally, it has the virtue of simplicity: no special apparatus is required.

The principal procedure for isolating cerato-ulmin is that of applying selective pressure to the sample to keep cerato-ulmin in the form desired (Fig. 12). Centrifugation as a form of centrifugal force applies positive pressure to the sample. This procedure was designed to limit cerato-ulmin to "unit" size by disassembling "rods" and/or "fibrils". A vacuum (negative pressure) was applied for vacuum effervescence to extract cerato-ulmin after it had been grouped into larger assemblages. The effect of the procedure is the opposite of that of centrifugation. Positive pressure was applied to the molecular filtration system with Amicon ultrafiltration membrane PM10 to maintain cerato-ulmin in molecular form, while under normal pressure dialysis was performed to maintain it in assemblages only. Thus pressure has a distinct effect on the form of cerato-ulmin in water.

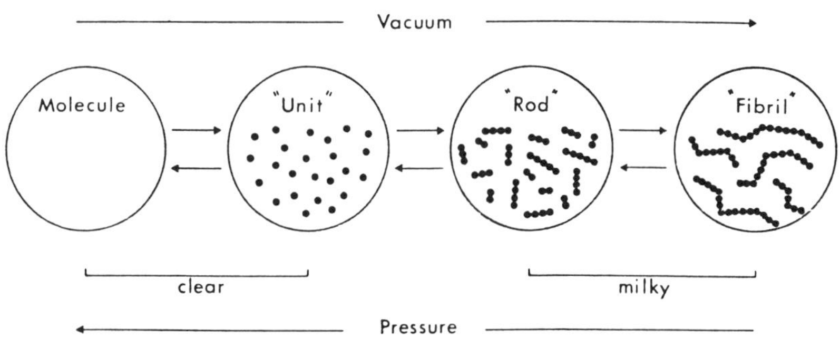

Figure 12. Schematic diagram illustrating characteristic changes in cerato-ulmin structure as observed under the microscope. Application of a vacuum to the aqueous cerato-ulmin solution initiates assembling (shown by the arrow direction). However, application of pressure to the assembled cerato-ulmin causes disassembling (shown by the arrow direction). These changes are reversible. When free molecules and/or "units" are present, the solution remains clear; however, when "rods" and/or "fibrils" are formed the system turns milky.

Centrifugation made the milky filtrate clear in appearance and prepared it for subsequent filtration. Filtration of the supernatant resulting from centrifugation through Millipore membrane (0.45 μm) did not cause any appreciable loss of cerato-ulmin because the membrane pores were large enough to let "units" pass through. In vacuum effervescence designed to pick up cerato-ulmin in larger assemblages such as "rods" and/or "fibrils" and "membranes", some cerato-ulmin which remain in molecular and/or "unit form are inevitably missed by this new technique.

However, the recovery of the extractable cerato-ulmin indicates that this technique is satisfactory in terms of efficiency. In dialysis cerato-ulmin seems to stay in "units" (i.e., assemblages of cerato-ulmin molecules) as evidenced by microscope observations. However, some molecular cerato-ulmin must remain unassembled: this would be dialyzed out automatically as evidenced by the fact that cerato-ulmin in 30% aqueous ethanol solution was partially dialyzed out along with ethanol. This loss, however, is so small that it can be ignored. It is unlikely that many constituents of the sample in the range below the cutoff point of Amicon ultrafiltration membrane PM10 change their structure according to pressure change, as does cerato-ulmin: hence dialysis of the sample at this stage, combined with molecular filtration, is considered to be selective for cerato-ulmin. In fact the substantial amount of ninhydrin positive impurities removed by dialysis from the cerato-ulmin fraction was evidenced by an obvious reduction in color intensity (Table 2).

Formation of such multiple assemblages of molecules has led to the erroneous conclusion that it is insoluble in water (Takai 1973). When cerato-ulmin is assembled as "units" in water it has a clear appearance, but it is still questionable whether or not this form of cerato-ulmin can be described as dissolved. However, in molecular form, as a true solution without cerato-ulmin assemblages, it readily dissolves in 60-80% aqueous ethanol, to give a clear appearance. Russo et al. (1981;1982) have demonstrated on the basis of optical, surface tension, X-ray, light scattering and nuclear magnetic resonance that all aqueous solutions of cerato-ulmin are in fact dispersions and not true molecular solutions. These authors have characterized cerato-ulmin as a very compact, hydrophobic molecule that is very surface active, has immeasurably low solubility in water and the unusual ability to coat and stabilize air bubbles in water, even at nanogram per milliliter concentrations. These findings by Russo et al. (1981;1982) have confirmed the author's hypothesis that the physical properties of cerato-ulmin described in Section 1 are in fact the result of an association with a gaseous condition (Richards and Takai 1973).

Because cerato-ulmin seemed to retain only "units" and larger assemblages during dialysis, but molecular forms and/or "units" during molecular filtration, the author assumed that the molecular weight range is between 1,000 and 10,000 (based on the cutoff point of Amicon ultrafiltration membrane PM10 and UM2), or not larger than 12,000 (based on the cutoff point of the cellophane dialyzing membrane.

Cerato-ulmin has been quantitatively isolated from the culture filtrate by the new method consisting of centrifugation, vacuum effervescence, ethanol fractionation, molecular filtration, and dialysis. However, no recovery of cerato-ulmin was made from the medium itself, and this indicates that cerato-ulmin is certainly of fungal origin. Most of these techniques were adopted to extract cerato-ulmin selectively on the basis of its liquid crystal-like characteristics (Brown et al. 1971). As already noted, cerato-ulmin takes different forms in an aqueous solution: free molecular form, "unit", "rod" and/or "fibril", and "membrane" when forced onto the surface of the solution (Richards and Takai 1973; Takai 1973; Takai 1974; Takai and Richards 1972).

The 25 mg/L yield of crude cerato-ulmin obtained in the present studies is not the maximum. If the level of sucrose in the medium is doubled the yield is doubled (Takai 1978).

Owing to the structural diversity of cerato-ulmin in water which caused non-uniform chromatographic migration, any chromatography using a mobile phase of water resulted in unsatisfactory separation. Since cerato-ulmin remains in molecular form in 60% aqueous ethanol solution, a better result was expected from chromatography using this elution system. Sephadex LH-20 was chosen for this reason.

In fact, cerato-ulmin was separated satisfactorily from other fractions (Fig. 6). a) Cerato-ulmin was recovered from the peak I fractions and was proved completely different from the major component of the eluate from peak II; U.V. absorption spectrum of cerato-ulmin was undoubtedly different from the sample from peak II. b) TLC and high-voltage electrophoresis did not reveal the presence of any impurities with carbohydrate, amino acid or peptide characteristics. This indicates that the eluate from peak I may contain just one substance. c) Phytotoxicity, which was characteristic of the original (crude) cerato-ulmin, increased slightly, indicating that the final product was that at which we were aiming in our studies.

Cerato-ulmin displayed carbohydrate characters. This is supported by positive results of reagent tests with Molish, anthrone, phenol-sulfuric acid, aniline-diphenylamine reagents (Table 1). In fact, molecular carbohydrate content appeared to be about 10.3%. However, it does not seem to possess reducing groups (Table 1). The failure to extract cerato-ulmin with a methanol-chloroform mixture suggests a negative character for fatty acid.

Cerato-ulmin appears to have proteinous characters. This is supported by the following interrelated facts: cerato-ulmin is positive to ninhydrin, Folin and biuret reagents (Table l); it is precipitated when saturated with ammonium-sulfate, and it very likely makes up a peptide complex with at least 16 amino acids, resulting in about 126 residues. The high value of proline is of special interest, as is the fact that L-proline contributed substantially to cerato-ulmin production by *O. ulmi* (Takai 1978).

Two methods, one physical and the other chemical, for the quantitative determination of cerato-ulmin have been developed. The physical method is based on the turbidity of a cerato-ulmin aqueous solution. Turbidity and concentration of cerato-ulmin in culture filtrate are found to be in linear relationship (Fig. 10A). This relationship was further confirmed with the partially purified cerato-ulmin in distilled water (Fig. 10B). It is highly unlikely that materials other than cerato-ulmin stable turbidity in the culture filtrate when the filtrate is shaken. Thus this method has the advantage of being very specific for cerato-ulmin. It has been found to be simple, practical and sensitive, and therefore has been used routinely to measure cerato-ulmin production by the fungus in the author's laboratory. However, it is applicable only to the relative determination of cerato-ulmin in culture filtrate since turbidity expression depends on the assembling of "units" into "rods" and/or "fibrils". These assemblages are not stable at a low concentration; hence, the practical working range is virtually limited to between 0.6 and 1.2 in O.D.

The chemical method is based on measurement of the carbohydrate content in cerato-ulmin by the color reaction with the phenol-sulfuric acid reagent (Dubois et al. 1951). A wide range of linearity as shown in Figure 11 for the relationship between color density (O.D. at λ 480 nm) and the amount of cerato-ulmin (46.7-280 μg/ml) gives a good practical working range for measurement. This chemical method is obviously more accurate than the turbidity method mentioned above, since the stability of the color developed by the chemical method is much greater than that of cerato-ulmin assemblages. On the other hand, the turbidity method appeared to be handier and more sensitive than the chemical method.

Cerato-ulmin has been purified (Fig. 7) by gel chromatography in 60% ethanol on Sephadex LH-60 (Stevenson et al. 1979); the carbohydrate content has been reduced from 10.3% to 4.4%. It was first revealed that cerato-ulmin was a small protein of molecular weight ca. 13,000 (Stevenson et al. 1979) and most recently 7,618 (Feng et al. 1991; Yaguchi et al. 1992). It has also been confirmed that cerato-ulmin is not a glycoprotein but a single polypeptide chain and the previously reported 4.4 % carboyhdrate found in cerato-ulmin preparations must be bound non-covalently (Yaguchi et al. 1992).

Although the toxicity of cerato-ulmin appeared to be not entirely host specific, cuttings from white elm, the host of Dutch elm disease, were affected the most severely among the plants tested. Some plants, including tomato, which are commonly used for toxicity assays, were not affected. This indicates that the toxicity of cerato-ulmin is selective.

The toxicity of cerato-ulmin to white elm was high and the toxic effect remained even after the cutting was removed from the cerato-ulmin solution and transferred to sterilized distilled water. It is notable that the toxicity of cerato-ulmin is highly stable, as indicated by its thermostability.

The toxicity to elm cuttings of dextran T10 which is similar to cerato-ulmin in molecular weight, and dextran T40 which has a molecular weight much higher than that of cerato-ulmin was found to be nil; this suggested that the toxicity of cerato-ulmin does not appear to be attributable to its molecular weight.

Sensitivity of elms to cerato-ulmin appears to parallel their response to *O. ulmi* as indicated by results in measurements of respiration and electrolyte loss with white elm, which is susceptible to Dutch elm disease, and Siberian elm, which is resistant to Dutch elm disease. The plasma membrane potential consists of the passive diffusion potential and active electrogenic potential. Passive diffusion potential is a measure of the degree of electrolyte loss from cells, whereas the active electrogenic potential is a measure of the driving force for active ion transport, which is ATPase dependent (Novacky et al. 1983). Membrane potential analysis with white elm (Okamoto et al. 1985) revealed that cerato-ulmin acts against the host cell to alter the membrane function (Okamoto et al. 1986) in a manner that reduces the passive diffusion potential (Fig. 13); in other words, it acts to increase electrolyte loss from the cells. However, the potential reduction is compensated for by an increase in the active electrogenic potential for as long as the supply of ATP continues to be replenished (from either respiration or photosynthesis).

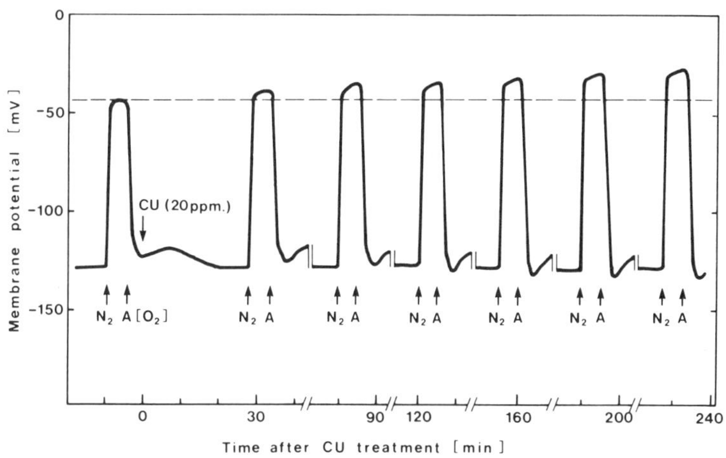

Figure 13. Effect of cerato-ulmin (CU) on membrane potential at the parenchyma symplast / xylem interface of the young shoot segment of susceptible white elm seedlings. The times when air/oxygen (A) is replaced with nitrogen gas (N_2) and vice versa, are indicated with arrows as is cerato-ulmin (CU at time = 0) perfusion into the shoot. The increase in potential after N_2 is added represents the net active electrogenic potential (which is canceled by N_2); the remaining potential before air replaces N_2 represents the passive diffusion potential.

The result of the membrane potential analysis mentioned above suggests that unless the supply of ATP is inhibited, net electrolyte loss from elm tissue treated with cerato-ulmin may not be measurable since the compensating operation of electrogenic ion pumps turns on and continues in response to a reduction in the passive diffusion potential. Therefore, if the activity of electrogenic ion pumps is reduced or arrested it is logical that electrolytes leak from the cell and thus the measurement of a net loss becomes possible. A clear-cut host-specific response to cerato-ulmin is observed in the pattern of alteration of membrane permeability. The study was conducted with leaf discs of plants (Otani et al. 1986; Otani et al. personal communication) that were vacuum infiltrated with cerato-ulmin (50 μg/ml) and vanadate (0.5 mM). The non-host plants tested (apple, *Malus pumila* L. cv. Red Gold; Japanese pear, *Pyrus serotina* Rded. cv. Nijisseiki; Walker poplar; tobacco, *Nicotiana tabacum* L. cv. Burley; and tomato, *Lycopersicon esculentum* Mill. cv. Earlypak 7) were not affected at all. Host plants responded to cerato-ulmin in accordance with the type of response they exhibited to the pathogen itself. The susceptible white elm exhibited the greatest increase in electrolyte loss from leaf cells upon exposure to cerato-ulmin and maintained this high level longer than less susceptible hosts, i.e., for at least 48 h. The increased electrolyte loss for Japanese elm of intermediate resistance was approximately 50% that of white elm, and the rate of loss started declining 2 h after treatment. The highly resistant Siberian elm did not respond to cerato-ulmin by exhibiting increased electrolyte loss (Fig. 14).

A comparative analysis of membrane potential, made between the susceptible white elm and the resistant Chinese elm *(Ulmus parvifolia* Jacq.) with cerato-ulmin perfused into xylem vessels, gave similar results: resistant elm did not exhibit changes in either the passive diffusion potential or the active electrogenic potential, whereas susceptible elm exhibited decreases in the passive diffusion potential, as already described (Okamoto et al., personal communication).

To date, as far as the author can determine, no toxins possessing liquid crystal-like characteristics, similar to cerato-ulmin (Richards and Takai 1973; Takai 1973; Takai 1974; Takai and Richards 1972), have been reported for any other organism except *O. ulmi.* To support this a previous test of five *Ceratocystis* spp. in addition to *O. ulmi* revealed that none of these other than *O. ulmi,* was capable of producing cerato-ulmin *in vitro* (Takai 1974). A recent extensive screening for in *vitro* microbial production of cerato-ulmin led to the conclusion that cerato-ulmin is most likely a metabolite specific to the Dutch elm disease pathogen *O. ulmi* (Nordin et al. 1987). In this screening 12 species in the genus *Ceratocystis,* 4 fungal families (that included 7 genera), 6 bacterial genera (that included 7 species), 29 unidentified bacterial isolates and 19 yeast isolates from elm tissues were tested for cerato-ulmin production by ELISA (enzyme- linked immunosorbent assay) with murine ascitic antibodies against purified cerato-ulmin. Only *O. ulmi* isolates reacted with the antibodies, an indication that *O. ulmi* may be the sole producer of cerato-ulmin. However, tests gave a negative reaction in 28 of 86 isolates of this fungus; 2 isolates gave possible positive results and the remaining 56 isolates tested positive. It was also shown by Miller and Yaguchi (personal communication), through a computer search, that no proteins similar to cerato-ulmin have been reported.

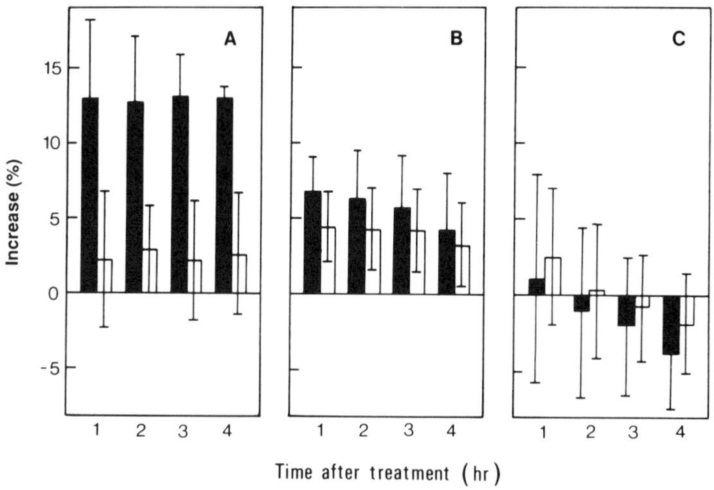

Figure 14. Differential impact of cerato-ulmin on electrolyte loss from leaves of elm with different types of response to *O. ulmi.* A) white elm, (susceptible), B) Japanese elm, (intermediate resistance) and C) Siberian elm,(high resistance). Leaf discs were vacuum infiltrated with 50 µg/ml cerato-ulmin plus 0.5 mM sodium vanadate (black bars) or with 50 µg/ml cerato-ulmin only (white bars).

3 - Criterion for *O. ulmi* Pathogenicity

The occurrence of variants or sectors in *O. ulmi* led early workers to investigate their pathogenicity. Although some of the variants proved to differ from their parent cultures for pathogenicity, no single character could be identified that could be correlated with pathogenicity. Holmes (1965;1968) reported that the variations in pathogenicity of progeny cultures could not be correlated with the compatibility type or with the ability to form concentric zones of sparse and dense mycelial growth.

A few years later, however, Gibbs and Brasier (1973) reported a correlation between cultural characters and pathogenicity of wild-type isolates. These authors described aggressive isolates as "fast-growing and fluffy", due to the formation of aerial mycelium aggregated into radial strands, which gave the cultures a fibrous striate appearance. The non-aggressive isolates grew more slowly and generally exhibited a waxy, yeast-like appearance with sparse aerial mycelium and little zonation. Moreover, the non-aggressive isolates were observed to be more variable in culture morphology than the aggressive ones.

Other cultural characters were eventually associated with pathogenicity. Schreiber and Townsend (1976) reported that the aggressive strain produced more synnemata on wood disks than the non-aggressive strain. Hindal and his co-workers (1981) reached similar conclusions when they grew isolates on a defined synnema production medium. They also reported that on this medium, the aggressive isolates produced more mycelial pigment than the non-aggressive isolates. Brasier and his co-workers (1981) reported that the two strains also differed in their *in vitro* temperature-growth responses. The optimum temperature for the aggressive isolates was established to be 20-22 C, whereas the non-aggressive isolates grew best at 30 C. The upper limit for growth was 32-33 C for the aggressive stain, and 35 C for the non-aggressive strain.

Brasier and Gibbs (1975) indicated that in Great Britain, both A- and B-mating types were found in the non-aggressive strain, whereas type B was predominant in the aggressive strain.

Elgersma (1976) failed to correlate polygalacturonase and cellulase production with pathogenicity, when the fungus was grown on pulverized elm wood in shake culture. In a subsequent study, however, Svaldi and Elgersma (1982) reported differences in the activity of glycosidases and exo-glycanases produced *in vitro* by the pathogen. Enzyme preparations from aggressive isolates released more arabinose, xylose and rhamnose from cell walls of elm wood than did preparations from non-aggressive isolates.

Janutolo and Stipes (1979) tested the *in vitro* response of *O. ulmi* to the benzamidazole carbamate fungitoxicants MBC, MBC-HCL and MBC-H_3PO_4. In all cases, aggressive cultures exhibited a more tolerant (or less sensitive) reaction to the fungitoxicants than the non-aggressive cultures.

Hubbes and Jeng (1981) reported that in liquid still culture, aggressive isolates grew on the surface of the medium and produced fluffy aerial mycelium. Non-aggressive isolates grew submerged, without any aerial hyphae. These investigators also determined that aggressive isolates utilized the amino acid valine much faster than the non-aggressive ones.

While most researchers were searching for cultural markers to differentiate the two strains *in vitro,* Brasier (1979) reported the occurrence of two culturally distinct races within the aggressive strain. One group of cultures from the U.S., Canada, Holland, France, Germany and the U.K. appeared regular and striate, whereas cultures collected in Italy and Iran were uneven and less striate. The race distributed in North America and northwest Europe was termed North American (NAN), and the race present in Italy and Iran was termed Eurasian (EAN). According to Brasier (1979; 1982) the NAN race would have been introduced into Britain in the late sixties and would have spread eastwards and southward into the adjacent European countries, whereas the EAN race would spread westwards.

Although most researchers seem to have observed unequivocal differences between the aggressive and non-aggressive strains *in vitro,* some discrepancies and exceptions have been reported. Schreiber and Townsend (1976) indicated that their attempts to correlate pathogenicity and cultural characters were generally inconclusive, even though some trends were observed. Fleming (1980) reported similar results.

It thus appeared that more specific markers might be appropriate and it is the intent of this chapter to introduce the toxin cerato-ulmin as a possible criterion for the pathogenicity of *O. ulmi.*

Materials and Methods

Pathogenicity rating for O. ulmi *isolates*

Sub-groupings of O. ulmi. As previously reported by Gibbs and Brasier (1973) and Brasier (1979; 1982a; 1982b) *O. ulmi* exists as three distinct sub-groupings: a weakly pathogenic non-aggressive strain and a highly pathogenic aggressive strain (Brasier and Gibbs 1975) which itself divides into two distinct races, the North American (NAN) and the Eurasian (EAN) races. As shown in Table 9, Brasier (1982a) categorized *O. ulmi* isolates according to their appropriate sub-group; he also evaluated each isolate's pathogenicity against 5-year-old Commelin elm, an early Dutch selection having little resistance (Gibbs et al. 1977). This elm was chosen since it has more resistance to *O. ulmi* and can be used to assess variation within or between the races of the aggressive strain (Brasier 1982a). The chief inoculation technique was the same as that used in the breeding program in The Netherlands and employed in the earlier experiments of Gibbs et al. (1972) and Gibbs and Brasier (1973). A spore suspension was prepared for each isolate in mid-June by shaking pieces of mycelium in 10 ml of Tchernoff medium for 2 days. An incision was made in the main stem of each elm with a knife to expose the wood vessels and 3 drops of the spore suspension were allowed to enter the cut. The inoculum averaged 5×10^5 spores per tree. Disease was assessed as percentage defoliation after an incubation period of 12 weeks. Control elms were inoculated with sterilized distilled water. Three replicates were taken for each isolate and control.

O. ulmi *non-sporulating isolate WRB2-1.* The non-sporulating isolate WRB2-1 that had originally sectored out of the *O. ulmi* isolate Al aggressive in pathogenicity was confirmed *O. ulmi* as a result of fertility tests (Richards et al. 1982).

Pathogenicity evaluation was carried out as follows: Petri dish cultures of *O. ulmi* isolates WRB2-1 and Al were prepared in 3% Malt agar and after 5 days incubation at 23 C each isolate was liquid shake cultured in Wilson medium (Takai and Richards 1972). Spores of Al were suspended in distilled water and the density of the spore suspension was adjusted for host inoculation as previously described (Takai and Kondo 1979). Since the growth of isolate WRB2-1 in liquid shake cultures was mycelium (Richards et al. 1982), the sample taken for host inoculation had to consist of mycelial segments (Tyler and Parker 1945). To obtain a sample of this nature the mycelium was removed from the culture medium and washed three times with distilled water. The washed mycelium were then placed in a Virtis "23" homogenizer containing distilled water, and mildly cut into small segments, followed by filtration through two layers of cheesecloth. By mild centrifugation at 164 g the small segments of mycelium were collected for the inoculum and the inoculum density adjusted to that of the Al spore inoculum.

Three-year-old nursery grown white elm seedlings were transplanted into pots in the greenhouse. After the leaves attained almost full expansion following the breaking of dormancy, 12 seedlings were inoculated by means of the slit method (Takai and Kondo 1979) with a drop of one of the above prepared fungal suspensions. Three control seedlings were similarly inoculated, but with sterilized distilled water only.

Table 9. Relationship between production of cerato-ulmin *in vitro* and pathogenicity of *O. ulmi* strains classified as aggressive (NAN), aggressive (EAN) and the non-aggressive.

O. ulmi strain and isolate No.*	CPI[a]	Commelin elm pathogenicity % defoliation 12 wk*	Origin of *O. ulmi* isolate*
Aggressive(NAN)			
H623	585	41.8	Jersey (Channel Islands)
H340	2336	63.3	Denmark
H279	842	65.8	West Germany
H248	458	67.7	Holland
H175	1920	74.0	U.S.A.
H220	2669	78.8	Ireland
H618	989	79.3	Sweden
H646	922	87.6	Spain
H106	986	88.7	France
H589	696	89.9	West Germany
Aggressive(EAN)			
AST-27	1018	14.8	Iran
H590	643	42.9	East Germany
T88	320	45.1	Turkey
CKT-11	470	52.9	Iran
H648	528	63.4	Tashkent (USSR)
H219	498	69.0	Ireland
H346	1283	69.7	Turkey
H323	1830	80.2	Czechoslovakia
H513	931	82.0	West Germany
H328	1139	86.8	Chaikov (USSR)
Non-aggressive			
YU99	0	0	Yugoslavia
R21	0	0	Romania
H503	9	0	Holland
H173	7	0	U.S.A.

Summary of data in Table 9

O. ulmi strain and race	CPI[a]		Commelin elm pathogenicity % defoliation 12wk*	
	Range	Average	Range	Average
Aggressive-NAN (10)	458-2669	1240	41.8-89.9	73.7
Aggressive-EAN (9)	320-1830	764	42.9-86.8	65.8
Aggressive-EAN (1)	1018	1018	14.8	14.8
Non-aggressive (4)	0-8	2	0	0

* This data was established by Brasier (1982a).
North American race (NAN), Eurasian race (EAN)

[a] Cerato-ulmin production index = turbidity x 100 x dilution factor of the culture filtrate (Takai 1978).

Twelve 10-year-old field grown white elm saplings were inoculated by means of the drill method (Takai and Kondo 1979) with two drops of the above prepared inoculum suspensions, at the time the leaves attained almost full expansion following the breaking of dormancy. Three control saplings were similarly inoculated, but with sterilized distilled water only. Incubation time for both greenhouse and field grown inoculated elms was 5 weeks, during which time external symptoms were monitored daily. At the end of the 5-week period the elms were removed by cutting at the base (ground level) and internal symptoms of brown staining and fungus distribution were checked (Takai and Kondo 1979). Inoculation trials were repeated three times for both elm seedlings and elm saplings.

Upon recovery of the *O. ulmi* isolate WRB2-1 from inoculated elms, host inoculation was repeated on elm seedlings as described above, to check for any changes in pathogenicity that might occur following the previous incubation of the isolate in the host.

Synnemata formation was also investigated. Wood ships were prepared and applied to 3-day-old petri dish cultures of WRB2-1 and A1, prepared as described above. After 6 days incubation at 23 C synnemata formation was evaluated by the stereomicroscope (Takai 1980); synnemata produced by isolate WRB2-1 and isolate A1 were also prepared for observation by scanning electron microscope (SEM) as described by Takai et al. (1980).

O. ulmi *isolate X6*. Pathogenicity evaluation of *O. ulmi* isolate X6 was carried out in potted 3-year-old seedlings of white elm in the greenhouse. At the time the leaves attained almost full expansion following the breaking of dormancy, the seedlings were inoculated with one drop of spore suspension (10^4 spores/ml) by the slit method, described elsewhere (Takai and Kondo 1979). The scale developed by Takai (1980) to evaluate the pathogenicity of *O. ulmi* isolates was applied; an isolate causing wilting symptoms within 3 weeks was said to be causing acute symptom development A, while an isolate which failed to cause wilting within this period was said to be causing chronic symptom development C. At the end of the 5-week incubation period an isolate that caused foliage loss exceeding 50% was rated E while an isolate causing foliage loss of less than 30% was rated L; an isolate rate of AE or Cl, therefore, was considered aggressive or non-aggressive, respectively. Since the pathogenicity of the *O. ulmi* isolate CESS-16K was previously evaluated as aggressive (Takai 1980) it was used as a reference and white elm seedlings were inoculated with it in the same manner as described above for the *O. ulmi* isolate X6. Control elms were inoculated with sterilized distilled water. At each inoculation six seedlings were inoculated. Inoculation trials were repeated three times.

Production of cerato-ulmin in vitro *by* O. ulmi *isolates*

The *O. ulmi* isolates listed in Table 9, the *O. ulmi* isolates CESS-16K and A1 previously classified as aggressive in pathogenicity (Takai 1980) and the *O. ulmi* isolates X6 and WRB2-1 classified as non-aggressive in pathogenicity (data to be described later in this section) were shake-cultured for 7 days at 23 C on the standard medium; methods of culturing *O. ulmi* and preparing cell-free filtrates were described previously in sections 1 and 2. For each isolate 50 ml of cell-free culture filtrate were taken and the cerato-ulmin production index (CPI) was determined on the basis of the turbidity method (Takai 1978; Section 2); this method utilized the physical properties of cerato-ulmin which caused the medium to become turbid (assembling of cerato-ulmin into "rods" and "fibrils") when shaken (Section 1, 2).

Bioassay solution

Cell-free culture filtrates of O. ulmi *isolates.* Cell-free filtrates of the *O. ulmi* isolates CESS-16K, X6 and WRB2-1 were obtained from the same preparation as was used for the determination of cerato-ulmin production.

Aqueous solutions of fractions obtained from cell-free culture filtrates of O. ulmi *isolates.* For *O. ulmi* isolates CESS-16K, X6 and WRB2-1, 200 ml of the cell-free culture filtrate, prepared as described above, were taken and fractionated by stepwise molecular filtration through Amicon ultrafiltration membranes with different retention capacities as described previously (Richards and Takai 1984). The types of ultrafiltration membranes employed (supplied by Amicon Corporation, U.S.A.) and the order of use were as follows: XM50 (general retention, 5 x 10^4 MW), PM 30 (general retention, 3 x 10^4 MW), PM10 (general retention, 1 x 10^4 MW), UM2 (general retention, 1 x 10^3 MW). At the end of each filtration, the residue on the ultrafiltration membrane was washed three times with 50 ml of de-gassed distilled water. The filtrate which resulted from washing the residue was kept and used only for washing the residue obtained on the subsequent ultrafiltration membrane. The residue obtained on each ultrafiltration membrane was dissolved with de-gassed distilled water and the final volume was adjusted to 200 ml in order to keep each fraction at the same volume as that taken for the cell-free culture filtrate. The filtrate recovered through the UM2 ultrafiltration membrane described as UM2-F fraction was also freeze-dried and dissolved in de-gassed distilled water up to a volume of 200 ml.

Serological detection of cerato-ulmin in fractions of the cell-free culture filtrate of the O. ulmi *isolate CESS-16K*

A 25-ml sample was taken from each fraction obtained from the cell-free culture filtrate, described above. The same volume was taken from the original cell-free culture filtrate. All samples were freeze-dried and dissolved in 0.5 ml HEPES-saline buffer (0.01 M HEPES, pH 7.5, 0.9% NaCl). Sample solutions were assayed against the antiserum developed against purified cerato-ulmin by double diffusion; methods for the antiserum development against purified cerato-ulmin and double diffusion assay were described by Krywienczyk et al. (1979).

Preparation of white elm cuttings

Dormant 3-year-old white elm seedlings were potted and grown under greenhouse conditions until leaves on new shoots had expanded 10-50% (Takai et al. 1979). Timing of cutting preparation is critical to obtain a sensitive response. Three to four leaves at the tip of shoots were left on the cutting. As soon as cuttings were removed from seedlings, the cut ends were immersed in de-gassed water (previously autoclaved), and were cut back further (as much as 2 cm) at the bottom end, within water, to avoid air-locking in xylem vessels. Prepared cuttings were incubated in tall beakers for 1 h in the greenhouse and then transferred to the growth chamber conditioned at 25 C, 50% R.H. and 600 f.c. light, where experiments were conducted.

Transpiration rate of white elm cuttings

A cutting was inserted in a vial containing one of the bioassay test solutions described above according to the method described by Van Alfen and Turner (1975). Five replicates were taken for each bioassay test solution and for the planned incubation periods. The weight of each cutting was predetermined at 0 h incubation. Transpiration was measured as the weight difference between two weighings of the total apparatus (vial + elm cutting + bioassay test solution). The transpiration rate was defined as milliliters per gram of fresh weight per hour (ml/g fr.wt/h). Measurements of transpiration were made at 0 h, 2 h and/or 8 h incubation times, giving two incubation ranges, 0-2 h and 2-8 h. External symptoms were monitored throughout the incubation periods. The above transpiration rate trials were repeated three times and data were analyzed by Tukey's test.

Results

Pathogenicity and production of cerato-ulmin in vitro *for* O. ulmi *isolates from sub-groupings*

As shown in Table 9. *O. ulmi* isolates were placed into three sub-groups: the NAN race of the aggressive strain, the EAN race of the aggressive strain and the non-aggressive strain. *O. ulmi* isolates from the NAN sub-group caused the highest percentage defoliation of Commelin elm after an incubation period of 12 weeks, ranging from 41.8-89.9% and averaging 73.7%; the EAN sub-group was slightly different than the NAN sub-group in that the range was similar (42.9-86.8%), but the average was lower at 65.8%; in contrast to the NAN and EAN races of the aggressive strain the percentage defoliation on Commelin elms by *O. ulmi* isolates from the non-aggressive strain was 0%.

O. ulmi isolates in the NAN race of the aggressive strain also produced the highest levels of cerato-ulmin *in vitro* ranging from 458-2669 as CPI and averaging 1240 as CPI (Table 9); *O. ulmi* isolates from the EAN race of the aggressive strain produced slightly lower levels of cerato-ulmin *in vitro* than the

NAN sub-group ranging from 320-1830 as CPI and averaging 764 as CPI; in contrast to the NAN and EAN races of the aggressive strain the levels of cerato-ulmin production *in vitro* by the non-aggressive strain was measurably low ranging from 0-8 as CPI and averaging 2 as CPI.

One isolate in the EAN race of the aggressive strain was an exception in that after incubation in Commelin elm for 12 weeks it caused only 14.8% defoliation but produced cerato-ulmin *in vitro* up to a rather high level which was indicated by a CPI of 1018 (Table 9).

Pathogenicity of the non-sporulating O. ulmi *isolate WRB2-l*

When elm seedlings in the greenhouse were inoculated with isolate WRB2-1 no external symptoms were noticed; both leaf expansion and shoot development were very vigorous and comparable with those of control seedlings throughout the total incubation period. However, seedlings inoculated with isolate A1 displayed typical Dutch elm disease symptoms; wilting and curling of leaves began within a week of inoculation and at the end of 5 weeks' incubation the majority of leaves were curled and brown, and some had necrotic and chlorotic symptoms. The seedlings inoculated with isolate A1 were also severely stunted in comparison with the control elms and those inoculated with WRB2-1. Light microscope observations of elms inoculated with isolate WRB2-1 (Fig. 15B) clearly show that the internal symptom of browning development was limited to a few xylem vessels, and the new growth ring was actively developing in a manner similar to that of the control (Fig. 15A).

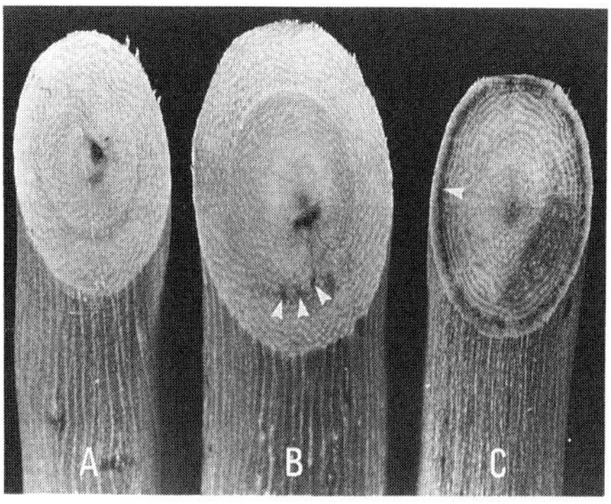

Figure 15. Stereoscope micrographs comparing disease staining symptoms in white elm seedlings inoculated with A) a distilled water control (no stain in outer sapwood), B) the non-sporulating isolate WRB2-l (very limited brown strain, and C) the sporulating isolate A1 (dense brown ring typical of Dutch elm disease. Arrow designates brown stain area.

As shown in Figure 15C the internal browning symptoms of seedlings inoculated with isolate A1 were characteristic of Dutch elm disease in that browning was very extensive and was distributed throughout the entire annual ring. Disease symptoms caused by isolate A1 spread throughout the entire seedling whereas the browning in seedlings inoculated with isolate WRB2-1 was associated only with limited xylem vessels below actively growing new wood, and the browning appeared to diminish gradually within the range of 15-18 cm above and below the inoculation point in the seedling. Where either isolate caused browning the fungus could always be re-isolated. When the isolate WRB2-1 was re-isolated from elm seedlings and re-inoculated into elm seedlings the same results as described above were obtained, which suggests that isolate WRB2-1 was stable. Details of internal symptoms caused by these isolates on elm seedlings are shown by SEM in Figure 16. Where browning occurred, the xylem vessels in seedlings inoculated with either isolate WRB2-1 or isolate Al were very similar in appearance: a coating was distributed over the vessel wall and bubbles were formed (Fig. 16A and B, respectively). The only internal difference observed by SEM was that isolate WRB2-1 (Fig. 16A) had mycelium but no spores, whereas isolate A1 (Fig. 16B) had spores and mycelium within the xylem vessel.

Light microscopic observations of slides prepared for *O. ulmi* isolate WRB2-1 and A1 indicated that isolate WRB2-1 produced mycelium only and was incapable of producing any form of spores, whereas isolate A1 was normal; the same lack of sporulation was noticed when isolate WRB2-1 was cultured in liquid medium as shake cultures, a procedure normally used to promote rapid budding of conidia with minimum mycelial growth. It should be mentioned that the growth of isolate WRB2-1 developed as compact round white balls of mycelium under shake culture conditions. Synnemata formation was also investigated, and stereomicroscope and light microscope observations indicated that isolate WRB2-1 produces synnemata with clear heads and no spores. The scanning electron micrograph in Figure 17 definitely show that isolate WRB2-1 (Fig. 17A) remains pointed and abortive whereas strain A1 (Fig. 17B) is characteristic of a normally developing synnema head both in shape and abundant spores.

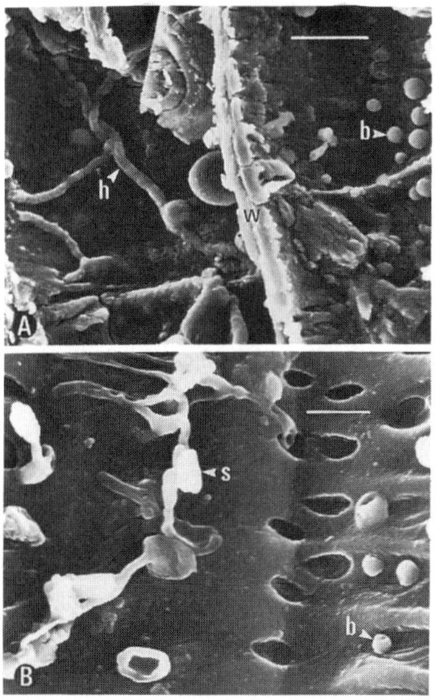

Figure 16. Scanning electron micrographs showing fungal elements and internal disease symptoms in xylem vessels of white elm seedlings inoculated with A) the non-sporulating isolate WRB2-1 and B) the sporulating isolate A1. s = spore, h = hyphae or mycelial growth, b = coating and bubbles in xylem vessel, w = xylem vessel wall. Bar represents 20 μm (Fig. 16A) and 5 μm (Fig. 16B).

Pathogenicity of the O. ulmi *isolate X6.*

Seven of the 18 white elms inoculated with the *O. ulmi* isolate X6 exhibited only slight symptoms of wilting between 3 and 5 weeks after inoculation (rated at C) and showed an average defoliation of 22% at 5 weeks after inoculation (rated as L). The overall rating for the *O. ulmi* isolate X6 appeared to be CL according to the scale proposed by Takai (1980) therefore its pathogenicity was determined as non-aggressive. Control seedlings were normal. Eleven of the 18 white elm seedlings inoculated with the *O. ulmi* isolate CESS-16K classified as aggressive in pathogenicity (Takai 1980) exhibited distinct wilting symptoms between 2 and 3 weeks after inoculation which gives a rating of acute symptom development A. The defoliation rate 5 weeks after inoculation was averaged at 83% (rated as E). The overall rating for *O. ulmi* isolate CESS-16K was AE (aggressive) which was the same as that previously reported by Takai (1980). Control seedlings were normal.

Toxicity distribution in fractions from cell-free culture filtrate of the O. ulmi *isolate CESS-16K*

To find out what the molecular weight range was for the components in the culture filtrate showing toxicity to white elm cuttings, the cell-free culture filtrate was fractionated as shown in Table 10. Cerato-ulmin could be detected by double diffusion assays in only the cell-free culture filtrate and the UM2-R fraction. Significant toxicity in terms or reduction in transpiration was caused only in the cell-free culture filtrate, the XM50-R fraction and the UM2-R fraction. Similarly, wilting incidence was high with the cell-free culture filtrate, the XM50-R fraction and the UM2-R fraction. In the case of XM50-R fraction, however, wilting developed later (24 h or more) than that in the case of the cell-free culture filtrate (2-8 h). The PM30-R fraction, the PM10-R fraction and UM2-F fraction caused no significant toxicity expressed as reduction in transpiration; wilting incidence was low for the PM10-R fraction (after 24 h or more) and nil for the PM30-R fraction and the UM2-F fraction.

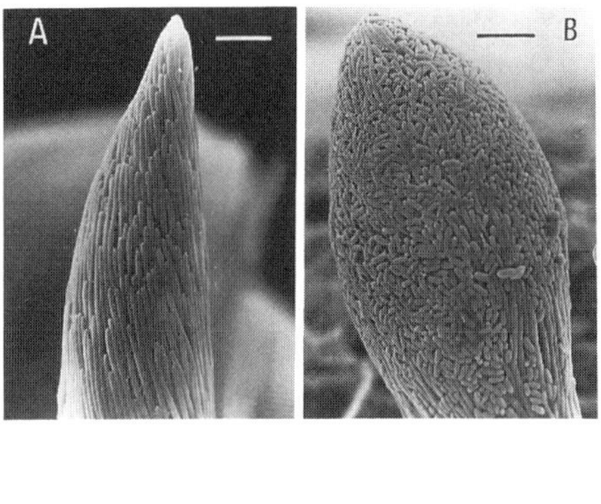

Figure 17. Scanning electron micrographs of synnema formed on white elm wood chips. A) the non-sporulating isolate WRB2-1 and B) the sporulating isolate A1. For Fig. 17A and B the bar represents 10 μm.

The UM2-F fraction which contains low molecular weight components, < 1 X 10^3 MW, could include phenolic carboxylic acids (metabolites also produced *in vitro* by *O. ulmi)* as reported by Claydon et al. (1974, 1980).

Table 10. Presence of cerato-ulmin in the various molecular weight culture filtrate fractions of the *O. ulmi* isolate CESS-16K and the effect of these fractions on transpiration and wilting of white elm cuttings (incubation time, 8 h; average of 15 replicates).

Cell-free culture filtrate fraction	Molecular weight range[a]	Presence of cerato-ulmin[b]	RT[c] (%)	WI[d] (%)
Cell-free culture filtrate	All fractions	+	19*	70
XM50-R	>5 x 10^4	-	22*	58[e]
PM30-R	>3x 10^4 <5x 10^4	-	73	0[e]
PM10-R	>1 x 10^4 <3x 10^4	-	58	25[e]
UM2-R	>1 x 10^3 <1 x 10^4	+	18*	75
UM2-F	<1 x 10^4	-	76	0[e]

[a] Based on Publication 553 Amicon, Products of Separation Technology, Amicon Co.
[b] Detected by double diffusion assay with antiserum developed against purified cerato-ulmin (Krywienczyk et al. 1979).
[c] Transpiration rate relative to that of the de-gassed distilled water control.
[d] Wilting incidence.
[e] Incubation period extended beyond 24 h.
* Significant difference (P = 0.05) from that of the de-gassed distilled water control (Tukey's test).
Note: Although the actual level of cerato-ulmin produced in the culture filtrates was not determined it is assumed to be close to 25 µg/ml on the basis of a previous study (Takai and Richards 1978).

Toxicity of the UM2-R fraction produced by the O. ulmi *isolates WRB2-1, X6 and CESS-16K*

As previously documented by Takai (1980) the production of cerato-ulmin *in vitro* by WRB2-1 was positive and the CPI as a relative number was 175 and for the parent isolate A1 it was 431. Data for the transpiration rate of white elm cuttings placed in cell-free culture filtrate and the UM2-R fraction for the *O. ulmi* isolate WRB2-1 were analyzed and the results are shown in Table 11. These trials show clearly that timing of the cutting is very critical in order to obtain a sensitive host response. A transpiration rate decrease in elm cuttings appears to begin as early as 2 h when the cuttings are incubated in either the WRB2-1 cell-free culture filtrate or the WRB2-1 culture filtrate residue obtained on the UM2 membrane (membrane used to retain the cerato-ulmin) bioassay solution, whereas a more obvious transpiration rate decrease is observed after 8-

h incubation. There is no significant difference in the transpiration rate decrease between elm cuttings placed in the cell-free culture filtrate or in the UM2 residue bioassay solution. These results indicate a definite production of toxic metabolites by isolate WRB2-1, among which cerato-ulmin is included.

The *O. ulmi* isolates CESS-16K, aggressive in pathogenicity, and X6, non-aggressive in pathogenicity as described above, were compared for cerato-ulmin production in the cell-free filtrate, measured as CPI, and for the toxicity to elm cuttings of the cell-free culture filtrate and the UM2-R fraction. As shown in Table 12, the isolate CESS-16K had a CPI of 278 while that of X6 was 0. In terms of wilting incidence, the toxicity of both the cell-free culture filtrate and the UM2-R fraction was high with CESS-16K and nil for X6. Reduction in transpiration caused by both the cell-free culture filtrate and the UM2-R fraction was approximately twice as high for CESS-16K as for X6. Cerato-ulmin contributed more to wilting incidence than to reduction in transpiration.

Table 11. Effect of the cell-free culture filtrate (b), and the UM2-R fraction (c) from the *O. ulmi* isolate WRB2- 1 on the transpiration of white elm cuttings.

Range of incubation period (h)	Bioassay solution	Trial No. 1 Mean (ml/g fr.wt/h)	Trial No. 2 Mean (ml/g fr.wt/h)	Trial No. 3 Mean (ml/g fr.wt/h)
0-2	(a)	0.4086*	0.9140	*0.7048
	(b)	0.3282*	0.3900*	*0.5056*
	(c)	0.2646*	0.2454*	0.6454*
2-8	(a)	0.4830*	0.8114	0.6644
	(b)	0.2424*	0.1806*	0.2714*
	(c)	0.2400*	0.1380*	0.3538*

(a) De-gassed distilled water (control).
Mean is the average of 5 replicates.
* no significant difference (Tukey's test).

Table 12. Cerato-ulmin production in the culture filtrate of the *O. ulmi* isolates CESS-16K and X6 and the effect of the cell-free culture filtrate and the UM2-R fraction from each isolate on the transpiration and wilting of white elm cuttings (incubation time, 8 h; average of 15 replicates).

		RT[b] (%)		WI[c] (%)	
O. ulmi isolate	CPI[a]	Cell-free culture filtrate	UM2-R	Cell-free culture filtrate	UM2-R
CESS-16K (aggressive)[d]	278	14	31	75	75
X6 (non-aggressive)[d]	0	23	60	0	0

[a] Cerato-ulmin production index = turbidity x 100 x dilution factor of the culture filtrate (Takai 1978).
[b] Transpiration rate relative to that of the de-gassed distilled water control.
[c] Wilting incidence.
[d] Pathogenicity rating.

Discussion

It has been clearly shown in this section that the production of cerato-ulmin *in vitro* is correlated with the pathogenicity of *O. ulmi* wild type isolates; this supports previous claims made by Takai (1980), Hubbes and Jeng (1981) and Pusey and Wilson (1981).

In a recent study 10 *O. ulmi* isolates from the three wild-type sub-groups (the EAN [Eurasian] aggressive strain, the NAN [North American] aggressive strain and the non-aggressive strain [Brasier et al. 1990]) were carefully tested for cerato-ulmin production and pathogenicity under strict and uniform conditions, in order to reach a conclusive viewpoint about this issue. Dual assays for cerato-ulmin production, by turbidimetry and enzyme-linked immunosorbent assay (ELISA), and dual host elms, less susceptible clonal Commelin elm and English elm (*U. procera* Savlisb.) were employed for evaluation of pathogenicity. Results also demonstrated a clear correlation between in *vitro* levels of cerato-ulmin production and pathogenicities of *O. ulmi* wild-type isolates. The average pathogenicity of the non-aggressive group was extremely low in comparison with that of the two aggressive groups. Cerato-ulmin production by the non-aggressive group was also extremely low. Only three isolates in this group tested positive for cerato-ulmin production in the ELISA test; the highest reading was 56 ng cerato-ulmin/ml culture filtrate. Four isolates had a cerato-ulmin production index (CPI) value greater than 1.0, with a high of 24.8 in the turbidimetric assessment. In contrast, one isolate from these two aggressive groups gave the lowest ELISA reading (2.0×10^3 ng/ml), which amounted to 36 times the lowest reading among non-aggressive isolates. Whether or not isolates without a positive ELISA reading produce cerato-ulmin, or some form which is slightly different structurally, remains to be determined. Nevertheless, cerato-ulmin seems to be an essential factor in *O. ulmi* pathogenesis.

A complex of reproductive barriers operates between the three sub-groups of *O. ulmi* to prevent free hybridization between them (Brasier 1977). In nature, hybridization between the subgroups is probably effectively limited by such reproductive barriers (Brasier 1978); however, hybridization of the subgroups can be found in the laboratory. The behavior of F_1 progeny from crosses between the aggressive and non-aggressive strains has been analyzed (Brasier 1977; 1982a; Brasier and Gibbs 1976). Our latest analysis of the correlation between cerato-ulmin production and pathogenicity of the F_1 progeny has revealed that although there is a strong correlation for wild-type isolates, this correlation is absent in some progenies (Richards et al. 1983). It appears that cerato-ulmin production may be one of the most important factors, however, it may not be the sole factor that regulates *O. ulmi* pathogenicity; in nature cerato-ulmin production and the additional factor(s) may operate together; in hybridization these factors somehow separate and eventually cause a conflict with the pattern observed for wild-type isolates.

The results obtained from this section clearly show that the isolate WRB2-1 is an abnormal stain of the Dutch elm disease fungus *O. ulmi;* it is unable to sporulate and is non-aggressive in pathogenicity. At the same time, however, it is able to produce cerato-ulmin, the toxin correlated with those *O. ulmi* stains rated as aggressive in pathogenicity. The mechanism related to the lack of sporulation is at present unknown, however, through fertility tests that were recently conducted, the data obtained from the segregation patterns of F_1 progeny as well as through backcrossings clearly provided evidence for the involvement of a single nuclear gene (Wayne C. Richards, personal communication).

When the cell-free culture filtrate from the aggressive isolate of *O. ulmi*, CESS-16K, was fractionated and tested against white elm cuttings, the toxicity of the XM50-R fraction was fairly high although it was lower than that of the UM2-R fraction or the cell-free culture filtrate. The XM50-R fraction is a complex containing metabolites in molecular weights of 5×10^4 and over and might contain phytotoxic glycopeptides in molecular weights of 5×10^5 - 2×10^6 (Van Alfen and Turner 1975) and 1×10^6 (Rebel 1969; Salemink et al. 1965), although identity of these compounds at the present time remains entirely unclear. Information of the actual toxic agent in this fraction, therefore is still not available. However, as far as the glycopeptides reported by Van Alfen and Turner (1975) are concerned which may contain the peptidorhamnomannan purified by Nordin and Strobel (1981) the action was speculated as physical plugging of pit membranes since no evidence for the biochemical effects on the host tissue such as electrolyte loss from leaf cells was observed. Biochemical effects on the host tissue such as electrolyte loss and respiration increase was observed, however, for cerato-ulmin as reported in Section 2.

The present findings indicate that the UM2-R fraction was the only fraction from the fractionated cell-free culture filtrate to contain cerato-ulmin. The UM2-R fraction also presented the highest toxicity among the cell-free culture filtrate fractions obtained from the *O. ulmi* isolate CESS-16K, aggressive in pathogenicity. The amount of cerato-ulmin in the corresponding fraction of the *O. ulmi* isolate X6, non-aggressive in pathogenicity, was neither measurable or detectable. This suggests that the toxicity of the cell-free culture filtrate of the *O. ulmi* isolate CESS-16K may result from the action of cerato-ulmin. Furthermore, it indicates that cerato-ulmin may be the prime toxin produced by *O. ulmi in vitro* as previously suggested by Takai and Hiratsuka (1984) and supports the claim that the production of cerato-ulmin *in vitro* is correlated with the pathogenicity of *O. ulmi* wild type isolates (Takai 1980) and that cerato-ulmin might be used as a criterion for the pathogenicity of *O. ulmi* isolates (Takai 1974).

4 - Cerato-ulmin at the Diseased Site

Zentmyer (1942a; 1942b) demonstrated the presence of a toxin in the culture filtrate of *O. ulmi* the pathogen of Dutch elm disease. Since then several metabolites have been reported as toxins (Claydon et al. 1980; Dimond et al. 1949; Nordin and Strobel 1981; Rebel 1969; Salemink et al. 1965; Scheffer and

Elgersma 1981; Strobel et al. 1978; Van Alfen and Turner 1975). However, there is no sound evidence that these metabolites are pathotoxins of Dutch elm disease. Since the discovery of cerato-ulmin (Section 1), which is produced *in vitro* by *O. ulmi* isolates (Section 2) that are classified as aggressive in pathogenicity (Section 3), attempts have been made to elucidate further the pathotoxic features of cerato-ulmin. In this respect it has been shown that white elm develops both external (Section 2) and internal (Takai and Hiratsuka 1981) symptoms similar to those of Dutch elm disease when treated with cerato-ulmin. Cerato-ulmin also causes a reduction in transpiration and an increase in respiration of elm and damage to membrane permeability measured as electrolyte loss from elm leaf tissue; impacts measured by respiration and damage to membrane permeability are pronounced in white elm which is susceptible to Dutch elm disease and insignificant in Siberian elm which is resistant to Dutch elm disease (Section 2). The reaction of the elm to *O. ulmi is* parallel to the sensitivity of the elm to cerato-ulmin (Section 2). Cerato-ulmin is the prime toxin in the culture filtrate (Section 3) and cerato-ulmin production is correlated with the pathogenicity of *O. ulmi* isolates (Section 3). In addition to these documented pathotoxic features of cerato-ulmin the key pathotoxic feature for establishing the significance of cerato-ulmin may lie in whether or not cerato-ulmin can be produced *in vivo* by *O. ulmi.*

This section will provide evidence for the production of cerato-ulmin in white elm naturally infected with *O. ulmi* and artificially inoculated with *O. ulmi,* which strengthens the claim of cerato-ulmin as a pathotoxin of Dutch elm disease.

Materials and Methods

Elm material naturally infected with O. ulmi

Nine diseased and three healthy white elms grown at different sites (average 10 years old, 10 cm DBH) were cut during June and July.

Of the diseased elms three were dying and showed no indication of new shoot development; three exhibited dieback and had branches bearing dead shoots which had developed in the spring and three had crowns which were 50% wilted. All samples from these diseased trees showed extensive discoloration of xylem wood, typical of Dutch elm disease, while samples from the healthy elms had no discoloration. Diseased and control wood samples were considered to be incubated under "dynamic" conditions.

Bolts from sampled elms were debarked; discolored xylem wood from the diseased elm bolts and wood from the comparable xylem area of unaffected (control) elms were stripped off (Fig. 18). The wood samples were pulverized to a powder in a Wiley mill. Wood powder (400 g) was suspended in 5 L of distilled water containing 0.05% Tween 20, and the suspension was stirred overnight at 4 C. Solid materials were removed by centrifugation at 1,500 g for 30 min. The supernatant was 60% saturated with $(NH_4)_2SO_4$ and held at 4 C overnight. Centrifugation was then repeated and the resulting precipitate was dissolved in 0.01 M HEPES-saline buffer, pH 7.5. The ample was concentrated by dialyzing down to 1 ml with Aquacide II-A (Calbiochem-Behring Corp., California, U.S.A.) and tested by double diffusion assay as described later in this section. The supernatant resulting from salting-out with $(NH_4)_2SO_4$, as described above, was discarded because it did not react with the cerato-ulmin antiserum.

Elm material artificially inoculated with O. ulmi.

Preparation of fungus. The *O. ulmi* isolate chosen for this study was RR-2; this isolate was previously evaluated as aggressive in pathogenicity and according to the rate established for the cerato-ulmin production index it was the highest of the O. *ulmi* isolates tested (Gibbs et al. 1975; Takai 1980). To initiate fungal growth on synthetic medium wood chips from the diseased area of a white elm freshly infected with RR-2 were removed and transferred to petri-dishes containing 3% Bacto malt agar. After 5 days incubation at 23 C in the dark a 2 mm^2 agar block of the actively growing fungal colony was transferred to a 125-ml Erlenmeyer flask containing 25 ml of our standard liquid medium (Takai and Richards 1978), and the fungus was cultured by shaking (120 rpm) at 23 C for 7 days. In addition to the normal culturing period of 7 days two other culturing periods were taken; a very short culturing period of 3 days and a longer culturing period of 13 days. At the end of each culturing period the culture was filtered through eight layers of cheese cloth to remove mycelium and the filtrate centrifuged at 1,085 g for 15 min at 4 C. The spores (ppt) were washed by resuspending in sterilized distilled water and re-centrifuging. This was repeated four times and the final spore pellet resuspended in sterilized distilled water (6.0 x 10^6 spores/ml).

Preparation of white elm. Eleven field grown white elms were selected. All elms were actively growing in the same geographical location. The area had also been assessed as free from Dutch elm disease.

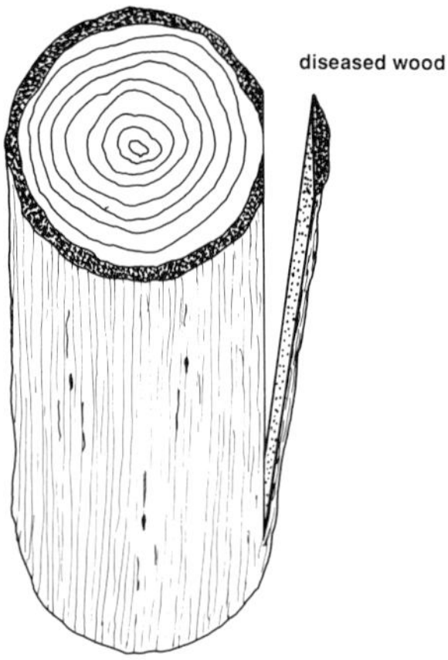

Figure 18. Schematic diagram of the removal of diseased wood, expressed as brown-streaking from white elm infected with *O. ulmi.*

One of the elms (50 cm DBH) selected above was strictly used as the source from which branch sections would be prepared on a weekly basis. Branches were first removed on approximately mid May at which time the buds were breaking; removal of additional branches were on a weekly basis up to the end of June. The last sampling date was considered to be approximately the climax of seasonal susceptibility of white elm to *O. ulmi* in central Ontario (Takai and Kondo 1979).

On the same date that the branches were removed from the elm they were immediately cut into smaller branch sections (15-20 cm long and 1.5-2.0 cm wide) and prepared for *O. ulmi* spore inoculation.

Ten of the elms (10-15 cm DBH) selected above were monitored on a daily basis until leaf expansion and fruit maturation had progressed about 10 and 25%, respectively; at this developmental stage, the elms were immediately prepared for O. *ulmi* spore inoculation since it has been previously shown to be the most susceptible growing stage for seasonal development of Dutch elm disease (Takai and Kondo 1979).

Inoculation of white elm branch section. To inoculate *O. ulmi* spores into a branch section, prepared as described above, sterilized tygon tubing of the appropriate inside diameter was first placed over the upper end of the branch section (Fig. 19) and firmly held by a stainless steel adjustable hose clamp; the other end of the tygon tubing was connected to a peristaltic pump. By placing the lower end of the branch section in one of the three *O. ulmi* spore suspensions prepared as described above, a volume of 25 ml was flushed through each branch section by operating the pump at 25 rpm. The pump was then stopped and the tygon tubing removed from the branch section. Sterilized distilled water without *O. ulmi* spores was used for the control. The tips of the branch section were immediately coated with paraffin wax to prevent moisture loss. For each inoculation 15 replicates were taken.

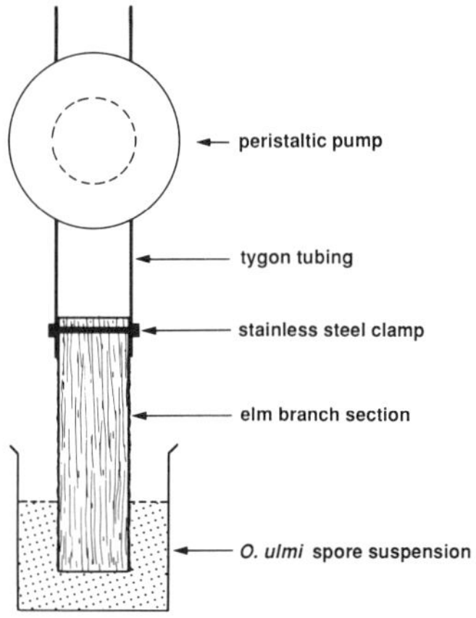

Figure 19. Schematic diagram of the apparatus implemented to both administer an *O. ulmi* spore suspension into a white elm branch section and to water-extract sample from the infected branch section.

Inoculation of intact white elm. To inoculate *O. ulmi* spores into an intact elm, selected as described above, the root injection method was used (Kondo and Jorgensen 1973). A portion of the root system was first exposed and in this experiment only one major root tip, about 2 cm in diameter was severed. Sterilized tygon tubing containing sterilized distilled water was immediately attached to the severed root tip by clamping with a stainless steel adjustable hose clamp. The tygon tubing was also connected to a reservoir containing a sterilized distilled water dispenser of *O. ulmi* spores (6.0 x 10^6 spores/ml) that were cultured for the normal culturing period of 7 days, as described above. Another length of tygon tubing was used to connect the reservoir to a nitrogen gas cylinder, and through a regulator a pressure of 34.5 kPa (5 lb/in^2) was introduced to the system. All air between the reservoir and the severed root tip was removed by use of a bleeder valve. A total of 400 ml of the prepared spore suspension was pressure injected into the elm. The tygon tubing was then removed, and the severed root tip sealed with grafting wax, and the soil replaced. Sterilized distilled water without *O. ulmi* spores was used for the control. Seven intact elms were inoculated with *O. ulmi* spores and three were inoculated with sterilized distilled water (control).

Processing of inoculated elm branch sections. Branch sections inoculated with *O. ulmi* spores, as described above, were incubated at 23 C for 4 and 10 days. At the end of each incubation period both ends of the branch section were cut back 1 cm and 50 ml of sterilized distilled water flushed through by the method previously described in this report (Fig. 19). The water extract obtained was centrifuged at 1,500 g for 30 min at 4 C and the supernatant filtered through a 0.45 µm pore size Millipore membrane to remove any remaining fungal elements. Water extracts were freeze-dried and dissolved in 1.0 ml of 0.01 M HEPES-saline buffer (0.01 M HEPES, pH 7.5, 0.9% NaCl) in preparation for double diffusion assays. Water extracts from control branch sections incubated at 23 C for 4 and 10 days were also obtained and prepared for double diffusion assay in the same manner. Diseased and control branch sections, as described above, are considered to be incubated under "static" conditions.

Processing of inoculated intact elms. The intact elms inoculated with *O. ulmi* spores, as described above, were monitored on a daily basis until wilting developed in 50% of the crown. At this time the elms were felled and elm bolts, free from bark and 2 m in length, were stripped of the brown streaked outer xylem wood (Fig. 18). The strips were cut into small pieces (1 cm in length, 1 cm wide and 0.5 cm thick) then ground with dry ice in a Wiley mill and obtained as wood powder through a 2-mm sieve. Diseased wood powder (700 g) was suspended in 5 liters of distilled water containing 0.05% Tween 20. The suspension was stirred overnight at 4 C and then vacuum filtered through Whatman No. 1 filter paper, on a Buchner funnel, and the filtrate collected. The wood powder residue was pressed to near dryness, using a Carver press. The liquid exudate was combined with the Whatman No. 1 filtrate (total water extract), and then centrifuged at 1,500 g for 30 min at 4 C. The supernatant was filtered through a 0.45 µm pore size Millipore membrane to remove remaining fungal elements. The filtrate was ultrafiltered under a pressure of 103.5 kPa (15 lb/in2) through the PM10 and the UM2 Amicon ultrafiltration membrane. The concentrate obtained on the UM2 membrane was freeze-dried and prepared for double-diffusion assay by dissolving in 3.0 ml of 0.01 M HEPES-saline buffer, pH 7.5; it has been previously shown that when cerato-ulmin is isolated from the cell-free culture filtrate cerato-ulmin is retained on the UM2 membrane (Section 2).

For the control intact elms the outer xylem wood comparative to that of the diseased elms was stripped off, processed, and water extracts obtained and prepared for double diffusion assay, as directed above for diseased wood powder. At the time the intact elms were felled branches were also removed. In the diseased elms branches were removed from the crown that expressed wilting whereas in the control elms branches were removed from the crown area at random. Branch sections were immediately prepared as described above and 50 ml of sterilized distilled water flushed through by the method previously described in this report. Water extracts obtained were freeze-dried and prepared for double-diffusion assay by dissolving in 1.0 ml of 0.01 M HEPES-saline buffer, pH 7.5.

For each elm 15 branch sections were taken. Diseased and control branch sections and elm bolts, as described above, are considered to be incubated under "dynamic" conditions.

O. ulmi *isolation and production of cerato-ulmin* in vitro

Wood chips were removed from the outer xylem wood of all diseased and control elm samples immediately following the removal of the bark and determining the presence of brown-streaking. The wood chips were transferred to 3% Malt agar. All cultures were incubated in the dark at 23 C for 10 days. Fungal colonies were identified as *O. ulmi* by their morphological appearance and by the formation of synnemata which developed on sterilized wood chips applied to actively growing fungal colonies (Takai 1980). *O. ulmi* isolates recovered from diseased elm samples were rated for production of cerato-ulmin *in vitro*. Since we showed previously that the production of cerato-ulmin *in vitro* is correlated with pathogenicity of *O. ulmi* isolates (Section 3) we evaluated the pathogenicity of the recovered isolates by calculating their cerato-ulmin production index (CPI) (Takai 1978). This was done as follows: each isolate was shake-cultured on our standard medium and the cell-free culture filtrate was obtained by the method described in Section 1. The culture filtrate was shaken on a gyratory shaker at 120 rpm for 30 min in order to achieve maximum turbidity which was the result of cerato-ulmin assembling into "rods" and "fibrils" (Section 1, 2); the turbidity was then measured as the optical density (O.D.) at λ 400 nm (Section 2); the CPI was calculated as O.D. x 100 x the dilution factor of the sample measured (Takai 1978).

Purification of cerato-ulmin

Crude cerato-ulmin isolated from the *O. ulmi* culture filtrate was purified by gel filtration on Sephadex LH 60 (Pharmacia Fine Chemicals, Uppsala, Sweden) as described in Section 2. The purity of the product was tested by electrophoresis on SDS polyacrylamide gel (20%); heavy sample loading presented no bands other than the cerato-ulmin band.

Antiserum against purified cerato-ulmin

We developed the antiserum against purified cerato-ulmin (Stevenson et al. 1979) as described elsewhere (Krywienczyk et al. 1979). Rabbit and murine ascitic antibodies against purified cerato-ulmin have also been developed and used extensively by others for detection of cerato-ulmin (Ipsen and Abul-Hajj 1982; Nordin et al. 1987; Svircev et al. 1988).

Purification of the antibody specific to purified cerato-ulmin (AB)

To eliminate the possibility of a non-specific reaction that could occur with the use of the antiserum, the antibody specific for purified cerato-ulmin was purified by means of affinity chromatography. Purified cerato-ulmin was first coupled with CNBr-activated Sepharose 4B (Pharmacia Fine Chemicals); the coupling mixture (10 ml) consisted of 20 mg of the purified cerato-ulmin and 1.5 g of CNBr-activated Sepharose 4B suspended in the coupling buffer (1 M $NaHCO_3$-saline buffer, pH 8.3). The mixture was transferred and kept in a Standard Amicon Stirred Cell No. 52 in which the bottom disc normally used to support the membrane filter was sealed with Parafilm. The mixture was stirred mildly at 4 C overnight under 207 kPa (30 lbs/in^2) pressure; the pressure was maintained to keep cerato-ulmin in molecular and "unit" forms only and to prevent further assembling in to "rods" and "fibrils" (Section 1), a requisite for maximum coupling. The uncoupled groups on the gel were blocked with 1 M ethanolamine at pH 8.0. The gel was then alternately washed first by filtration on a sintered glass filter and then by centrifugation at 12,000 *g* for 20 min, with 50 ml of 0.1 M acetate-saline buffer, pH 4.0, and with the coupling buffer, pH 8.3, respectively. Alternate washing procedures consisting of five washing cycles of low and high pH were needed to remove completely all excess cerato-ulmin that did not couple to the gel. The gel was then packed in a column (1.0 x 5.5 cm) with the coupling buffer. The antiserum against purified cerato-ulmin (7 ml) was passed through the column at a flow rate of 5 drops/min. and the gel was washed with 0.01 M HEPES-saline buffer, pH 7.5, until O.D. at λ 280 nm of the eluate reached 0. The absorbed antibody was then eluted with 0.5 M acetate-saline buffer, pH 2.4 at a flow rate of 10 drops/min. Collection of the eluate (1 ml/tube) was monitored for protein by O.D. at λ 280 nm. The protein eluate was immediately neutralized with 1N NaOH and dialyzed against 0.01 M HEPES-saline buffer, pH 7.5 at 4 C overnight. The dialyzate was concentrated to 0.5 ml (AB) with Aquacide II-A and was used in the double diffusion assays.

Double diffusion assay

Antiserum development against purified cerato-ulmin and purification of the antibody specific to purified cerato-ulmin (Ab) were obtained by the methods described above. The water extracts obtained from each diseased and control wood sample as well as the normal rabbit serum (control), and purified cerato-ulmin (Stevenson et al. 1979) were challenged against AB. Assays were done in petri-dishes containing 1.6% agarose and 2% polyethylene glycol 6000 dissolved in 0.01 M HEPES-saline buffer pH 7.5. Detection of cerato-ulmin was confirmed by the formation of a precipitin line and sample homogenicity confirmed by the formation of a line of identity (the joining of adjacent ends of two precipitin lines).

Results

Natural infection

O. ulmi was recovered from every diseased wood sample, but none from the healthy wood sample. Each isolate from its respective diseased tree was rated in the range of 400 to 600 CPI, an indication that it was aggressive in pathogenicity (Section 3; Takai 1980). The indirect pathogenicity rating of each isolate was further supported by abundant synnemata formation (Takai 1980), fluffy mycelial habit and a fast-growing rate of the colony (54 mm or more after 6 days on malt-agar plates, as proposed by others (Gibbs and Brasier 1973).

In double diffusion assays the antiserum developed against purified cerato-ulmin did not react with any water extracts from healthy wood or the normal rabbit serum; however, it did react with each water extract from diseased wood. Also, another antiserum against purified cerato-ulmin developed by Ipsen and Abul-Hajj (1982), when compared with our antiserum reacted in the same manner. When normal rabbit serum was used as the control and challenged with the purified cerato-ulmin and each water extract from diseased and healthy wood, no precipitin line occurred. These results suggested the presence of cerato-ulmin in diseased wood samples, this was confirmed when the samples were tested against Ab. Each water extract from diseased wood, regardless of the severity of Dutch elm disease symptoms, reacted with Ab but did not react with the water extract from healthy wood. A line of identity was also formed when Ab was simultaneously challenged with each water extract from diseased wood and the purified cerato-ulmin, indicating sample homogenicity (Fig. 20).

Artificial inoculation

Diseased and control wood samples were incubated under "dynamic" and "static" conditions, in this report; samples considered to be incubated under "dynamic" conditions are composed of branch sections and elm bolts harvested either from diseased intact elms at the time wilting developed in 50% of the crown or from healthy intact elms (control) which were normal; on the other hand, samples considered to be incubated under "static" conditions are comprised of branch sections that were removed from a healthy intact elm and incubated at 23 C for periods of 4 and 10 days immediately following inoculation with *O. ulmi* spores or sterilized-distilled water (control).

Typical internal symptoms of Dutch elm disease expressed as brown-streaking in the outer most xylem wood was observed in all diseased wood samples incubated under "dynamic" conditions but was not observed in any diseased wood sample incubated under "static" conditions or in any control wood sample incubated under either condition.

O. ulmi could be re-isolated from all diseased wood samples but from no control wood samples; re-isolation of *O. ulmi* in diseased wood incubated under "dynamic" conditions was limited to the brown-streaked area only, whereas for diseased wood samples incubated under "static" conditions in which case typical brown-streaking was not observed, re-isolation of *O. ulmi* was successful irrespective of brown-streaking.

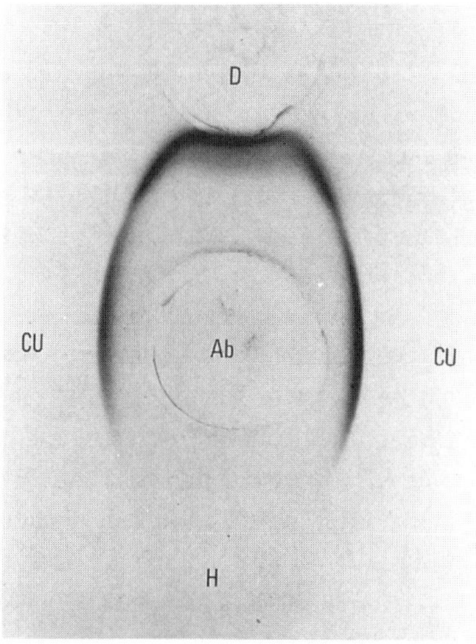

Figure 20. Double diffusion assay in agarose. Ab) purified antibody specific to purified cerato-ulmin, D) water extracts from diseased elm wood samples, H), water extracts from healthy elm wood samples and CU) the purified cerato-ulmin. The plate was stained with 1% Buffalo Black.

The cerato-ulmin production index (CPI) established for each re-isolate was in agreement with that previously reported for the *O. ulmi* isolate RR-2 (Takai 1980).

Water extracts were obtained from diseased and control wood samples as described above; at the time of extraction water extracts obtained from diseased branch sections incubated under "static" conditions for 10 days rapidly changed from amber color to dark brown within seconds of the extraction.

A precipitin line was formed in double diffusion assays when the purified antibody specific to purified cerato-ulmin (Ab) was challenged against water extracts obtained from diseased wood samples incubated under "dynamic" conditions and water extracts obtained from diseased branch sections incubated under "static" conditions for 10 days. No precipitin line, however, was formed in double diffusion assays when Ab was challenged against water extracts obtained from diseased branch sections incubated under "static" conditions for 4 days or from control wood samples incubated under "dynamic" and "static" conditions; a precipitin line did also not develop when normal rabbit serum was used as the control and challenged against the purified cerato-ulmin (Stevenson et al. 1979) and each water extract from diseased and control wood samples. When Ab was challenged with the purified cerato-ulmin (Stevenson et al. 1979) and water extracts from diseased wood samples a line of identity was formed between the purified cerato-ulmin and those water extracts from which cerato-ulmin was detectable, indicating sample homogenicity (Fig. 20).

Discussion

Cerato-ulmin was detected in water extracts obtained from diseased wood of white elm naturally infected with *O. ulmi* as well as from diseased wood artificially inoculated with *O. ulmi;* at the time the infected elms were sampled external symptoms of wilting in 50% of the crown were observed; although different methods of inoculation were undertaken the results were the same which strengthens the evidence for production of cerato-ulmin *in vivo.*

Since cerato-ulmin was detectable by double diffusion assay in water extracts obtained from diseased branch sections incubated under "static" conditions for 10 days and not in water extracts for the 4-day incubation period it would appear that an incubation period of 4-10 days is required for *O. ulmi* to produce cerato-ulmin *in vivo* up to a micro-detectable level.

There also appears to be a difference in the oxidation that takes place within diseased wood samples incubated under "dynamic" conditions and diseased wood samples incubated under "static" conditions. This is indicated by the fact that diseased branch sections incubated under "dynamic" conditions expressed brown-streaking in the outermost xylem wood, but the water extracts collected were amber in color, whereas diseased branch sections incubated under "static" conditions for 10 days did not express typical brown-streaking in the outermost xylem wood but when the water extracts were collected, they rapidly changed from amber to dark brown. This brown oxidized material that was collected in water extracts might be of similar nature to the brown material expressed as brown-streaking.

The pattern of chronological production of cerato-ulmin *in* vivo (between 4-10 days) was very similar to that reported for production of cerato-ulmin *in vitro* in which cerato-ulmin was first observed in the cell-free culture filtrate by the turbidity method on day 3 after inoculation and increased steadily until 13 days after inoculation (Takai 1978).

Neither the age of spore inoculum (3-, 7- or 13-day-old) nor the developmental stage of the elm at the time the branch sections were inoculated with *O. ulmi* spores appeared to be significantly different in their contribution to the production of cerato-ulmin *in vivo.*

The present results clearly indicate that cerato-ulmin is indisputably involved *in vivo* during the early stages of disease development.

General Discussion

The idea of the presence of a toxin in the culture filtrate of *O. ulmi,* the pathogen of Dutch elm disease, was initially demonstrated by Zentmyer (1942a; 1942b). Since then other researchers have isolated several metabolites from the culture filtrate of *O. ulmi* that have been claimed as toxins (Claydon et al. 1980; Dimond et al. 1949; Nordin and Strobel 1981; Rebel 1969; Salemink et al. 1965; Scheffer and Elgersma 1981; Strobel et al. 1978; Van Alfen and Turner 1975); however, none has been established as the toxin responsible for the development of Dutch elm disease (pathotoxin). Owing to a lack of detailed characterization of such toxins, it is difficult to compare one with another. However, most were classified as polysaccharides or were related to carbohydrate. In fact, Dimond (1947) and Dimond et al. (1949) isolated a polysaccharide as a prime toxin from the ethanol-insoluble fraction in their screening of toxic substances in the culture filtrate of this fungus. However, Feldman et al. (1949; 1950) and Frederick and Howard (1951) claimed that the prime toxin component was in the ethanol-soluble fraction and not in the ethanol-insoluble fraction. Studies on this toxin fraction were discontinued, so that no information is available on its chemical character. If the culture filtrate is partitioned with ethanol (70-80% as the final ethanol concentration) as was done by previous workers, cerato-ulmin will never precipitate but will remain in the solution. In this sense, the toxic components claimed by Feldman et al. (1949; 1950) could include cerato-ulmin, if their strain of *O. ulmi* produced it. It has been shown in this paper that cerato-ulmin is not a pure carbohydrate. Furthermore, it is not precipitated by ethanol; hence, similarity to the polysaccharides isolated by Dimond (1947) and Dimond et al. (1949) can be discounted. The fact that cerato-ulmin possesses both carbohydrate and proteinous characters suggests that it is somewhat similar to toxins isolated by Salemink et al. (1965), Rebel (1969), Van Alfen and Turner (1975), Scheffer and Elgersma (1981) and Strobel et al. (1978). The glycopeptide, peptidorhamnomannan has been structurally analyzed by Nordin and Strobel (1981) and has been shown to contain protein of MW 25,000 and a carbohydrate moiety ranging from a MW of 105,000-120,000. Due to the lack of clear descriptions it is difficult to determine whether or not the same toxins were studied by these laboratories. The peptidorhamnomannan is similar to that obtained by Van Alfen and Turner (1975) and Scheffer and Elgersma (1981). However, the lack of biological selectivity of peptidorhamnomannan and the ability of dextrans of the same size to mimic its effects does question the assignment of peptidorhamnomannan as a toxin. The glycopeptides described above are in fact quite different from cerato-ulmin. The purified cerato-ulmin (Stevenson et al. 1979) contains 95.6% protein and 4.4% carbohydrate which is bound non-covalently (Yaguchi et al. 1992), and the molecular weight calculated from the complete amino acid sequence is 7,618.7 and the molecular weight determined by Ion-Spray Mass Spectrometry using Sciex API III is reported to be 7,618.4 (Feng et al. 1991, Yaguchi et al. 1992) whereas the MW's from the glycopeptide toxins are much greater 2.5 x 10^4 and 10^6 (Salemink et al. 1965), 3 x 10^4 and 10^6 (Rebel 1969), 5 x 10^5 and 2 x 10^6 (Van Alfen and Turner 1975), 7 x 10^4 and 2.7 x 10^5 (Strobel et al. 1978) and 10.5-12.0 x 10^4 (Nordin and Strobel 1981). In marked contrast the toxins studied in other laboratories have the opposite compositions: 90% carbohydrate (Rebel 1969), 95 carbohydrate, 5% protein (Van Alfen and Turner 1975) and 93.1 % carbohydrate, 6.9% protein (Nordin and Strobel 1981; Strobel et al. 1978). It is clear that cerato-ulmin is different from all other toxins produced by *O. ulmi* that have thus far been reported. No toxins possessing liquid crystal-like characteristics, similar to cerato-ulmin (Richards and Takai 1973; Takai 1973; Takai 1974; Takai and Richards 1972) have been reported for any other organism except *O. ulmi* (Nordin et al. 1987; Miller and Yaguchi, personal commun-ication).

Cerato-ulmin had no inhibitory effects on the seed germination of any plant tested and the toxicity of cerato-ulmin to the plant shoot cuttings tested was selective and expressed as wilting. White elm, the most susceptible host of Dutch elm disease, was most severely affected; in addition to wilting, the toxic symptoms were expressed as reduction in transpiration, increase in respiration of leaf tissue and electrolyte loss from leaf tissue. Siberian elm, highly resistant to Dutch elm disease when tested against cerato-ulmin, showed only half the increase in respiration obtained from white elm and no significant increase in electrolyte loss. The toxicity to elm cuttings of dextran T10, which is similar to cerato-ulmin in molecular weight much higher than that of cerato-ulmin, was found to be nil. The present results appear to contradict the claim by Van Alfen and Turner (1975) that dextrans in the molecular weight range similar to those of dextran T10 and T40 caused a slowdown in elm stem conductance. However, the concentration used by these authors was extremely high (400 μg/ml) in comparison with that used in the present study (10 μg/ml). The theory that the toxicity of cerato-ulmin may be derived from its "high" molecular weight (Van Alfen and MacHardy 1978), and that it may plug pit membranes in this molecular weight class, seems unjustified. When the cell-free culture filtrate from the aggressive isolate of *O. ulmi,* CESS-16K, was fractionated and tested against white elm cuttings, the toxicity of the XM50-R fraction to white elms was fairly high although it was lower than that of the UM2-R fraction or the cell-free culture filtrate. The XM50-R fraction is a complex containing metabolites in molecular weights of 5 x 10^4 and over and might contain phytotoxic glycopeptides in molecular weights of 5 x 10^5 - 2 x 10^6 (Van Alfen and Turner 1975) and 1 x 10^6 (Rebel 1969; Salemink et al. 1965) although identity of these compounds at the present time remains entirely unclear. Information of the actual toxic agent is this fraction therefore is still not available. However, as far as the glycopeptides reported by Van Alfen and Turner (1975) are concerned which may contain the peptidorhamnomannan purified by Nordin and Strobel (1981) the action was speculated as physical plugging of pit membranes since no evidence for the biochemical effects on the host tissue such as electrolyte loss from leaf cells was observed. Biochemical effects on the host tissue such as electrolyte loss and respiration increase was observed, however, for cerato-ulmin as reported in Section 2. The present findings indicate that the UM2-R fraction was the only fraction from the fractionated cell-free culture filtrate to contain cerato-ulmin. The UM2-R fraction also presented the highest toxicity among the cell-free culture filtrate fractions obtained from the *O. ulmi* isolate CESS-16K, aggressive in pathogenicity. The amount of cerato-ulmin in the corresponding fraction of the *O. ulmi* isolate X6, non-aggressive in pathogenicity, was neither measurable nor detectable. This suggests that the toxicity of the cell-free culture filtrate of the *O. ulmi* isolate CESS-16K may result from the action of cerato-ulmin. Furthermore it indicates that cerato-ulmin may be the prime toxin produced *in vitro* by *O. ulmi* (Section 3).

The surfactant nature of cerato-ulmin, however, enables the toxin to envelop fine air bubbles in the xylem sap. This has been recently suggested as a possible mechanism for the transport of cerato-ulmin in a concentrated form to the site of pit membranes, where the inevitable plugging results (Ipsen and Abul-Hajj 1982; Russo et al. 1981). However, even if plugging of pit membranes by cerato-ulmin were to occur in this manner, it might cause reduction in the transpiration rate but would not cause or explain the biochemical responses of white elm tissue, such as enhanced respiration or electrolyte loss. Cerato-ulmin as free molecules (Takai et al. 1979), however, might pass through the pit membranes and reach the parenchyma cells where most of the physiological activities in woody tissue of elm are maintained (Wilson 1970). If this occurs than it is quite possible that the parenchyma cells in the xylem vessels might response to cerato-ulmin in a fashion similar to that reported herein for elm leaf tissue (Section 2; Richards and Takai 1984).

Sensitivity of elms to cerato-ulmin appears to parallel their response to *O. ulmi* as indicated by results in measurements of respiration and electrolyte loss with white elm, which is susceptible to Dutch elm disease, and Siberian elm, which is resistant to Dutch elm disease. A reduction in transpiration was observed for elms infected with *O. ulmi* (Roberts 1966); it is very possible that cerato-ulmin could share in the contribution to this reduction in transpiration. The present findings of reduction in transpiration, and especially the increase in respiration and electrolyte loss from elm leaf tissue, indicate overall interactions between cerato-ulmin and elm tissue at the cellular level. As discussed by Landis and Hart (1972), physiological changes as measured by an increase in respiration and electrolyte loss from pathogen-free tissue of white elm induced by *O. ulmi* corresponds to the seasonal development of the disease; subsequent to their findings they introduced the hypothesis of a translocatable toxin; they also suggested that the hypothesis can be established only by the isolation and purification of the substance, and demonstrate that it has such a role in disease development. Cerato-ulmin could, in fact, be this substance. Internal symptoms of Dutch elm disease in both *O. ulmi*-infected samples and cerato-ulmin treated samples were investigated by Takai and Hiratsuka (1984); all internal xylem vessel symptoms observed in *O. ulmi*-infected samples were observed in the cerato-ulmin treated samples. Internal symptoms could be detected earlier in the cerato-ulmin treated samples (as early as 2 h) than in the *O. ulmi*-infected samples (about 10 h). Internal vessel symptoms that were observed in both treatments were 1) edema-like wall alteration, 2) granular deposit, 3) pit membrane heaving, 4) smooth coating, 5) bubble and/or tylosis, 6) muddy coating and 7) droplet and callus-like development. Obviously much more work is needed to find out the true nature of these structures and the meaning of these xylem vessel symptoms in relation to stoppage of water transported through xylem vessels; it is believed that these results support the contention that cerato-ulmin which is produced by *O. ulmi* is involved in the disease syndrome expression of Dutch elm disease.

It is clearly shown in Section 3 that the pathogenicity of *O. ulmi* is correlated with the production of cerato-ulmin *in vitro* as suggested by Takai (1980); Hubbes and Jeng (1981); Pusey and Wilson (1981) and Brasier et al. (1990). Cerato-ulmin has also been detected in the synnemata head fluid (Takai et al. 1980) and synnemata formation correlated with the pathogenicity of *O. ulmi* (Hindal et al. 1981; Schreiber and Townsend 1976; Takai 1980). For *O. ulmi* to be transferred by beetles to healthy elms the beetle must first come in contact with the fungus; this occurs primarily upon beetle contact with synnemata that develop in the beetle galleries (Sherald 1976); the disease cycle is completed when the contaminated beetles fly to healthy elms and as a result of feeding in the outer sapwood spores and cerato-ulmin are deposited in the exposed xylem vessels. The fact that cerato-ulmin is found in the synnemata head fluid indicates that the spores that are transferred to the healthy elm have the ability to produce the toxin and the ability to be aggressive in pathogenicity. It has also been shown by Takai (1974) that there seems to be no correlation between spore production and either cerato-ulmin production or pathogenicity since the aggressive strain of *O. ulmi* produces a similar number of spores *in vitro* to that of the non-aggressive strain; however production of cerato-ulmin *in vitro* by the aggressive strain is high and for the non-aggressive is low. When these results are combined with that reported herein for the non-sporulating *O. ulmi* isolate, WRB2-1 (Richards et al. 1982) classified as non-aggressive in pathogenicity but comparable in production of cerato-ulmin *in vitro* to *O. ulmi* isolates classified as aggressive in pathogenicity they support the hypothesis that cerato-ulmin is a contributing factor in the symptom development of Dutch elm disease and the theory that both spore production and quality of spores produced by *O. ulmi* (ability to produce cerato-ulmin) are essential to transmit and deposit the toxin throughout the vascular system of the actively growing elm. As previously discussed in Section 3, experimental fertility tests have shown that factor(s) controlling non-sporulation in *O. ulmi* may be linked to a single nuclear gene (Wayne C. Richards, personal communication). At present, investigations are being initiated to study at the genetic level aspects for this mutagenesis with hope of applying the technology gained for application as a biological control agent or means to arrest asexual sporulation in *O. ulmi*.

Cerato-ulmin was detected in water extracts obtained from diseased wood of white elm naturally infected with *O. ulmi* as well as from diseased wood artificially inoculated with *O. ulmi* (Section 4; Richards and Takai 1988). At the time the infected elms were sampled external symptoms of wilting in 50% of the crown were observed. Although different methods of inoculation were undertaken the results were the same which strengthens the evidence for production of cerato-ulmin *in vivo.*

The pattern of chronological production of cerato-ulmin *in* vivo (between 4-10 days) as reported in Section 4 suggests its involvement in Dutch elm disease possibly prior to the external symptom of wilting.

Cerato-ulmin is indisputably involved in the development of Dutch elm disease at early stages of disease development (Section 4) and findings for pathotoxic features whereby the toxin is likely a metabolite specific to *O. ulmi* (Nordin et al. 1987; Takai 1974); white elm develops both external (Section 2; Takai 1974) and internal (Takai and Hiratsuka 1984) symptoms similar to those of Dutch elm disease when treated with cerato-ulmin; to initiate symptom development *O. ulmi* may need to produce cerato-ulmin *in vivo* up to a certain level (Takai and Hiratsuka 1984); cerato-ulmin also causes a reduction in transpiration and an increase in respiration of elm and damage to membrane permeability measured as electrolyte loss from elm leaf tissue and impacts measured by respiration and damage to membrane permeability are pronounced in white elm which is susceptible to Dutch elm disease and insignificant in Siberian elm which is resistant to Dutch elm disease (Section 2); the reaction of the elm to *O. ulmi is* parallel to the sensitivity of the elm to cerato-ulmin (Section 2); cerato-ulmin is the prime toxin in the culture filtrate and cerato-ulmin production is correlated to the pathogenicity of the *O. ulmi* wild type isolates (Section 3) clearly indicate that cerato-ulmin is a unique wilt toxin of instrumental significance in the development of Dutch elm disease.

Acknowledgements

In memory of a dear friend and working colleague, I dedicate this chapter to Dr. Shozo Takai.

Appreciation is greatly expressed to Edna Morningstar for her excellent typing skills, to Joanne Theriault and Geoffrey Hart for their invaluable layout and editorial consultation and to Ed Rayner for his photographic expertise and professional preparation of final photographic materials.

References

Brasier, C.M. 1977. Inheritance of pathogenicity and cultural characters in *Ceratocystis ulmi.* Hybridization of protoperithecial and non-aggressive strains. Trans. Brit. Mycol. Soc. 68:45-52

Brasier, C.M. 1978. Mites and reproduction in *Ceratocystis ulmi* and other fungi. Trans. Brit. Mycol. Soc. 70:81-89.

Brasier, C.M. 1979. Dual origin of recent Dutch elm disease outbreaks in Europe. Nature 281:78-80.

Brasier, C.M. 1982a. Genetics of pathogenicity in *Ceratocystis ulmi* and its significance for elm breeding. pp.224-235 *in* Resistance to disease and pests in forest trees. Proceedings of the third international workshop on the genetics of host-parasite interactions in forestry, Wageningen, the Netherlands. (eds.) H.M. Heybroek, B.R. Stephan and K. von weissenberg. PUDOC, Wageningen.

Brasier, C.M. 1982b. Occurrence of three sub-groups within *Ceratocystis ulmi.* In: Kondo, E.S., Hiratsuka, Y., Denyer, W.B.G. (eds.) Proc. symp. and workshop on Dutch elm disease. Winnipeg, Manitoba, October 1981. Environment Canada, Province of Manitoba, p.298-321.

Brasier, C.M., and Gibbs, J.N. 1975. Highly fertile form of the aggressive strain of *Ceratocystis ulmi.* Nature 257:128-131.

Brasier, C.M., and Gibbs, J.N. 1976. Inheritance of pathogenicity and cultural characters in *Ceratocystis ulmi:* hybridization of aggressive and non-aggressive strains. Ann. Appl. Biol. 83:31-37.

Brasier, C.M., Lea, J., and Rawlings, M.K. 1981. The aggressive and non-aggressive strains of *Ceratocystis ulmi* have different temperature optima for growth. Trans. Brit. Mycol. Soc. 76:213-218.

Brasier, C.M., Takai, S., Nordin, J.H., and Richards, W.C. 1990. Differences in cerato-ulmin production between the EAN, NAN and non-aggressive sub-groups of *Ophiostoma ulmi.* Plant Pathology 39:231-236.

Brotzman, H.G., and Campana, R.J. 1968. Viability *in vitro* of the smallest spores of *Ceratocystis ulmi* following centrifugation and microfiltration. Phytopathology 58:398.

Brown, G.H., Doan, J.W., and Neff, V.D. 1971. A review of the structure and physical properties of liquid crystals. CRC Press, Cleveland, Ohio.

Claydon, N., Grove, J.F., and Hosken, M. 1974. Phenolic metabolites of *Ceratocystis ulmi.* Phytochemistry 13:2567-2571.

Claydon, N., Elgersma, D.M., and Grove, J.F. 1980. The phototoxicity of some phenolic metabolic products of *Ophiostoma ulmi* to *Ulmus sp.* Neth. J. Plant Pathol. 86:229-237.

Damann, K.E. Jr., Gardner, J.M., and Scheffer, R.P. 1974. An assay for *Helminthosporium victoriae* toxin based on induced leakage of electrolytes from oat tissue. Phytopathology 64:652-654.

Demain, A.L., and Phaff, H.J. 1954. The preparation of tetragalacturonic acid. Arch. Biochem. Biophys. 51:114-121.

Dimond, A.E. 1947. Symptoms of Dutch elm disease reproduced by toxins of *Graphium ulmi* in culture. Phytopathology 37:7 (Abstr.).

Dimond, A.E., Plumb, G.H., Stoddard, E.M., and Horsfall, J.G. 1949. An evaluation of chemotherapy and vector control by insecticides for combating Dutch elm disease. Conn. Agric. Exp. Sta., New Haven, Bull. 531:69 pp.

Dubois, M., Gilles, D.A., Hamilton, J.D., Rebers, P.A., and Smith, F. 1951. Colorimetric method for the determination of sugars. Nature 168:167.

Elgersma, D.M. 1976. Production of pectic and cellulolytic enzymes by aggressive and non-aggressive strains of *Ophiostoma ulmi.* Neth. J. Plant Pathol. 82: 161-172.

Feldman, A.W., Caroselli, N.E., and Howard, F.L. 1949. Physiological and chemotherapeutric investigations of *Ceratostomella ulmi.* Phytopathology 39:8 (Abstr.).

Feldman, A.W., Caroselli, N.E., and Howard, F.L. 1950. Physiology of toxin production by *Ceratocystis ulmi.* Phytopathology 40:341-354.

Feng, R., Konishi, Y., and Bell, A.W. 1991. High accuracy molecular weight determination and variation characterization of protein up to 80 Ku by ion spray mass spectrometry. J. Am. Soc. Mass Spectrom. 2:387-401.

Fleming, M. 1980. Virulence of some strains of *Ceratocystis ulmi* (Buism.) C. Moreau and induced resistance in elm seedlings. M.Sc.F. thesis, University of Toronto, Toronto, Canada.

Frederick, L., and Howard, F.L. 1951. Comparative physiology and pathogenicity of eight isolates of *Ceratocystis ulmi*. Phytopathology 41:12-13 (Abstr.).

French, D., Knapp, D.W., and Pauzu, J.H. 1950. Studies on the Schardinger dextrins. VI. The molecular size and structure of the Y-dextrin. J. Am. Chem. Soc. 72:515~5152.

Gibbs, J.N., and Brasier, C.M. 1973. Correlation between cultural characters and pathogenicity in *Ceratocystis ulmi* from Britain, Europe and America. Nature 241:381-383.

Gibbs, J.N., Heybroek, H.M., and Holmes, F.W. 1972. Aggressive strain of *Ceratocystis ulmi* in Britain. Nature 236:121-122.

Gibbs, J.N., Burdekin, D.A., and Brasier, C.M. 1977. Dutch elm disease. Forestry Commission Forest Record 155:12 pp.

Gibbs, J.N., Brasier, C.M., McNabb, H.S. Jr., and Heybroek, H.M. 1975. Further studies on pathogenicity in *Ceratocystis ulmi*. Eur. J. For. Path. 5(3):161-174.

Hawk, P.B., Oser, B., Bernard, L., and Summerson, W.H. 1947. Practical physiological chemistry. 12th ed. Blakiston. Philadelphia, Pennsylvania.

Hindal, D.F., Harner, E.J., and MacDonald, W.L. 1981. Cultural characteristics, pathogenicity predictions and pathogenicity tests for some isolates of *Ceratocystis ulmi*. Can. J. Bot. 59:2543-2549.

Holmes, F.W. 1965. Virulence in *Ceratocystis ulmi*. Neth. J. Plant Pathol. 71:97-112.

Holmes, F.W. 1968. The causal organism, *Ceratocystis ulmi*. Paper presented at the Int. Symp. on Dutch elm disease, Ames, IA, Feb. 1968.

Hubbes, M., and Jeng, R.S. 1981. Aggressiveness of *Ceratocystis ulmi* and induction of resistance in *Ulmus americana*. Eur. J. Forest Pathol. 11 :257-264.

Ipsen, J.D., and Abul-Hajj, Y.J. 1982. Fluorescent antibody technique as a means of localizing *Ceratocystis ulmi* toxins in elm. Can. J. Bot. 60:724-729.

Janutolo, D.B., and R.J. Stipes.1979. Response of aggressive versus non-aggressive isolates of *Ceratocystis ulmi* to benzamidazole carbamate fungitoxicants. Virginia J. Science 30: 132-134.

Koch, F.C., and Hanke, M.E. 1948. Practical methods in biochemistry. Williams & Wilkinson, Baltimore, Md.

Kondo, E.S., Roy, D.N., and Jorgensen, E. 1973. Salts of methyl-2-benzimidazole carbamate (MBC) and assessment of their potential in Dutch elm disease control. Can. J. For. Res. 3:548-555.

Krywienczyk, J., Takai, S., and Mathieson, B.A. 1979. Cerato-ulmin, a wilting toxin of *Ceratocystis ulmi:* Development of antiserum against the toxin. Ann. Phytopathol. Soc. Japan 45:745-747.

Landis, W.R., and Hart, J.H. 1972. Physiological changes in pathogen-free tissue of *Ulmus americana* induced by *Ceratocystis ulmi*. Phytopathology 62:909-913.

Lowry, O.H., Rosebrough, N.J., Farr, A.L., and Randall, R.J. 1951. Protein measurement with Folin phenol reagent. J. Biol. Chem. 193:265-275.

Macek, K., and Tadra, M. 1952. Rozdelovac 2m chromatografie cukernych ketokyselin na papire. Chem. Listy. 46:450-452.

Moczar, E., and Moczar, M. 1970. Thin layer chromatography in studies of carbohydrate side chains of glycoprotein. pp. 169-214 in Niederwieser, A. and Pataki, G. (eds.), Progress in thin layer chromatography and related methods. I. Ann Arbor-Humphrey Science, Ann Arbor, Michigan.

Morris, D.L. 1948. Quantitative determination of carbohydrates with Dreywood's anthrone reagent. Science 107:254-255.

Nordin, J.H., and Strobel, G.A. 1981. Structural and innunochemical studies on the phytotoxic peptidorhamnomannan of *Ceratocystis ulmi*. Plant Physiol. 67:1208-1213.

Nordin, J.H., Mason, T.L., Smith, L.L., Willmann, P.A., Richards, W.C., and Takai, S. 1987. Use of an enzymelinked immunosorbent assay with murine ascitic antibodies to screen microorganisms for production of cerato-ulmin, a toxin of *Ceratocystis ulmi*. Phytopathology 77:96-100.

Novacky, A. 1983. The effects of disease on the structure and activity of membranes. In: Callow, J.A. (ed.) Biochemical plant pathology, John Wiley & Sons, Chichester, New York, Brisbane, Toronto, Singapore. p. 347-366.

Okamoto, H., Namiki, F., Nishimure, S., and Takai, S. 1985. Application of the xylem perfusion method to plant pathology study. Proc. 50th Ann. Meet. Bot. Soc. Japan 217 (Abstr. in Japanese).

Okamoto, H., Otani, H., and Takai, S . 1986. Effect of cerato-ulmin (CU) on electrogenic pumps of elm cell membrane and ion loss from cells. Phytopathology 76:113 (Abstr.).

Otani, H., Takai, S., Richards, W.C., Kohmoto, K., and Nishimura, S. 1986. Effect of cerato-ulmin (CU) on electrolyte loss from elm leaf cells. Ann. Phytopath. Soc. Japan 52: 133 (Abstr. in Japanese).

Ouellette, G.F., and Gagnon, C. 1960. Formation of microendospores in *Ceratocystis ulmi* (Buism.) C. Moreau. Can. J. Bot. 38:235-241.

Oyama, V.I., and Eagle, H. 1956. Measurement of cell growth in tissue culture with a phenol reagent (Folin-Ciocateau). Proc. Soc. Exp. Biol. Medic 91:305-307.

Payne, W.J., and Kieber, R.J. 1954. The chromatographic determination of glucoseamine with ninhydrin. Arch. Biochem. Biophys. 52:1-4.

Pusey, L., and Wilson, G.L. 1981. Toxin production and pathogenicity of *Ceratocystis ulmi*. J. Arboricult. 7(10):258-260.

Rebel, H. 1969. Phytotoxins of *Ceratocystis ulmi*. Isolation and structure investigation. Ph.D. Thesis, University of Utrecht, The Netherlands. 77 pp.

Richards, W.C., and Takai, S. 1973. Novel technique for isolating microstructures present in shake cultures of the fungus *Ceratocystis ulmi*. Appl. Microbiol. 26:443-444.

Richards, W.C., and Takai, S. 1984. Characterization of the toxicity of cerato-ulmin, the toxin of Dutch elm disease. Can. J. Plant Pathol. 6:291-298.

Richards, W.C., and Takai, S. 1988. Production of cerato-ulmin in white elm following artificial inoculation with *Ceratocystis ulmi*. Physiol. Mol. Plant Pathol. 33:279-285.

Richards, W.C., Takai, S., Lin, D., Hiratsuka, Y., and Asina, S. 1982. An abnormal strain of *Ceratocystis ulmi* incapable of producing external symptoms of Dutch elm disease. Eur. J. For. Pathol. 12:193-202.

Richards, W.C., Brasier, C.M., and Takai, S. 1983. Relationship between cerato-ulmin production and pathogenicity of *Ceratocystis ulmi* strains and F_1 progenies. IIIrd Intern. Mycol. Congr. Tokyo.

Roberts, B.R. 1966. Transpiration of elm seedlings as influenced by inoculation with *Ceratocystis ulmi*. For. Sci. 12:44-47.

Russo, P.S., Blum, F.D., Ipsen, J.D., Abul-Hajj, Y.J., and Miller, W.G.1981. The solubility and surface activity of the Ceratocystis ulmi toxin cerato-ulmin. Physiol.PlantPathol.19:113-126.

Russo, P.S., Blum, F.D., Ipsen, J.D., Abul-Hajj, Y.J., and Miller, W.G. 1982. The surface activity of the phytotoxin cerato-ulmin. Can. J. Bot. 60:1414-1422.

Salemink, C.A., Rebel, H., Kerling, L.C.P., and Tchernoff, V. 1965. Phytotoxin isolated from liquid cultures of *Ceratocystis ulmi*. Science 149:202-203.

Scheffer, R.J., and Elgersma, D.M. 1981. Detection of a phytotoxic glycopeptide produced by *Ophiostoma ulmi* in elm by enzyme linked immuno-specific assay (ELISA). Physiol. Plant Pathol. 18:27-32.

Scheffer, R.J., Liem, J.I., and Elgersma, D.M. 1987. Production in *vitro* of phytotoxic compounds of non-aggressive and aggressive isolates of *Ophiostoma ulmi,* the Dutch elm disease pathogen. Physiol. Mol. Plant Pathol. 30:321-335.

Schreiber, L.R., and Townsend, A.M. 1976. Variability in aggressiveness, recovery and cultural characteristics of isolates of *Ceratocystis ulmi*. Phytopathology 66(3):239-244.

Sherald, J.L. 1976. Monitoring emergence periods of smaller European elm bark-beetle *Scolytus multistriatus* in the National Capital Region with Multilure baited traps. Proc. First Conf. on Sci. Res. in the National Parks, New Orleans, La. Vol. II:1081-1083.

Stevenson, K.J., and Smillie, L.B. 1968. The inhibition of chymotrypsins A4 and B with chloromethyl ketone reagents. Can. J. Biochem. 46:1357-1370.

Stevenson, K.J., Slater, J.A., and Takai, S. 1979. Cerato-ulmin, a wilting toxin of Dutch elm disease fungus. Phytochemistry 18:235-238.

Strobel, G., Van Alfen, N., Hapner, K.D., McNeil, M., and Albersheim, P. 1978. Some phytotoxin glycopeptides from *Ceratocystis ulmi,* the Dutch elm disease pathogen. Biochem. Biophys. Acta. 538:60-75.

Svaldi, R., and Elgersma, D.M. 1982. Further studies on the activity of cell wall degrading enzymes of aggressive and non-aggressive isolates of *Ophiostoma ulmi.* Eur. J. Forest Pathol. 12:29-36.

Svircev, A.M., Jeng, R.S., Hubbes, M. 1988. Detection of cerato-ulmin on aggressive isolates of *Ophiostoma ulmi* by immunocytochemistry and scanning electron microscopy. Phytopathology 73:322-327.

Takai, S. 1973. Cerato-ulmin, a wilting agent of the Dutch elm disease fungus. *In* Abstracts of Papers, 2nd International Congress of Plant Pathology 73:322-327.

Takai, S. 1974. Pathogenicity and cerato-ulmin production in *Ceratocystis ulmi.* Nature 252: 124-126.

Takai, S. 1978. Cerato-ulmin, a wilting toxin of *Ceratocystis ulmi:* Cultural factors affecting cerato-ulmin production by the fungus. Phytopathol. Z. 91:147-158.

Takai, S. 1980. Relationship of the production of the toxin cerato-ulmin to synnemata formation, pathogenicity, mycelial habit and growth of *Ceratocystis ulmi* isolates. Can. J. Bot. 58:658-662.

Takai, S., and Richards,W.C. 1972. Isolation of microstructures present in cultures of the Dutch elm disease fungus, *Ceratocystis ulmi.* Proc. Can. Phytopathol. Soc. 39:42-43 (Abstr.).

Takai, S., and Richards, W.C. 1978. Cerato-ulmin, a wilting toxin of *Ceratocystis ulmi:* Isolation and some properties of cerato-ulmin from the culture of *C. ulmi.* Phytopathol. Z. 91:129-146.

Takai, S., and Hiratsuka, Y. 1981. Toxicity of cerato-ulmin (CU) to white elm. Phytopathology 71:225-226 (Abstr.).

Takai, S., and Hiratsuka, Y. 1984. Scanning electron microscope observations of internal symptoms of white elm following *Ceratocystis ulmi* infection and cerato-ulmin treatment. Can. J. Bot. 62:1365-1371.

Takai, S., and Kondo, E.S. 1979. Seasonal development of Dutch elm diseasee in white elms in central Ontario, Canada. I. Following wound inoculation. Can. J. Bot. 57:341-352.

Takai, S., Richards, W.C., Hiratsuka, Y., and Stevenson, K.J. 1979. Cerato-ulmin, a semipathotoxin of *Ceratocystis ulmi.* pp. 147-151 *in* J.M. Daly and Ikuzo Uritani (eds.), Recognition and Specificity in Plant Host-Parasite Interactions. Japan Sci. Soc. Press, Tokyo and Univ. Park Press, Baltimore.

Takai, S., Hiratsuka, Y., Krywienczyk, J., Richards, W.C., and Davies, Y.P. 1980. Evidence for the presence of the toxin cerato-ulmin in the synnema head fluid of *Ceratocystis ulmi.* Can. J. Bot. 58:669-675.

Tollens, B. 1882. Ueber ammon-alkalische Silberlosung als Reagens auf Aldehyd. Berichte 15: 1635-1639.

Tyler, L.J., and Parker, K.G. 1945. Pathogenicity of the Dutch elm disease fungus. Phytopathology 35:257-261.

Van Alfen, N.K., and Turner, N.C. 1975. Influence of *Ceratocystis ulmi* toxin on water relations of elm *(Ulmus americana).* Plant Physiol. 55:312-316.

Van Alfen, N.K., and MacHardy, W.E. 1978. Symptoms and host-pathogen interactions. pp. 20-25 *in* W.A. Sinclair and R.J. Campana (eds.) Dutch elm disease perspectives after 60 years. Cornell Univ. Agric. Exp. Stn. Search (Agriculture) 8:52 pp.

Wilson, C.L. 1970. What we don't know about some shade tree diseases. Arborists 35:9-14.

Wheeler, H., and Luke, H.H. 1963. Microbial toxins in plant disease. Annual Review of Microbiology 17:223-243.

Yaguchi, M., Pusztai, M., Roy, C., Surewicz, W.K., Carey, P.R., Stevenson, K.J., Richards, W.C., and Takai, S. 1992. Amino acid sequence, infrared and raman spectroscopy of Dutch elm disease toxin cerato-ulmin. International workshop "Dutch Elm Disease: Cellular and Molecular Approaches", Michigan State University, East Lansing, Aug. 1-4, 1992. (in current volume)

Zentmyer, G.A. 1942a. Toxin formation and chemotherapy in relation to Dutch elm disease. Phytopathology 32:20 (Abstr.).

Zentmyer, G.A. 1942b. Toxin formation by *Ceratostomella ulmi*. Science 95:512-513.

12
Amino Acid Sequence and Spectroscopic Studies of Dutch Elm Disease Toxin, Cerato-ulmin

M. Yaguchi, M. Pusztai-Carey, C. Roy, W. K. Surewicz, P. R. Carey, K. J. Stevenson, W. C. Richards and S. Takai.

Abstract

The amino acid sequence of cerato-ulmin (Stevenson et al., 1979), a toxin produced by *Ophiostoma (Ceratocystis) ulmi* (Buisman) Nannf., the causal agent of Dutch elm disease, has been completely determined. The protein has 75 amino acid residues and the following amino acid sequence:

1 5 10 15
Ser-Asp- Ser-Tyr-Asp-Pro-Cys-Thr-Gly-Leu-Leu-Gln-Lys-Ser-Pro-
16 20 25 30
Gln-Cys-Cys-Asn-Thr-Asp-Ile- Leu-Gly-Val-Ala-Asn-Leu-Asp-Cys-
31 35 40 45
His-Gly- Pro- Pro- Ser-Val- Pro-Thr-Ser- Pro- Ser-Gln- Phe- Gln-Ala-
46 50 55 60
Ser-Cys-Val- Ala-Asp-Gly-Gly-Arg-Ser-Ala-Arg-Cys-Cys-Thr-Leu-
61 65 70 75
Ser-Leu-Leu- Gly-Leu-Ala-Leu-Val-Cys-Thr-Asp-Pro-Val-Gly- Ile

Cerato-ulmin consists of two regions (residues 1-38 and 39-75), and each contains two disulfide bonds. The molecular weight calculated from the amino acid sequence is 7618.7 and the molecular weight determined by Ion-Spray Mass Spectrometetry using a Sciex API III mass spectrometer is reported to be 7618.4 (Feng et al. 1991).

The infrared and Raman spectra show that the protein contains about 15% alpha-helix and about 50% beta-sheet. The Raman spectra also revealed no evidence for an S-H stretching vibration near 2580 cm^{-1} but an intense S-S stretching mode is apparent at 514 cm^{-1}. Thus all eight cysteine residues are linked in disulfide bonds.

Introduction

Dutch elm disease (DED) is recognized as one of the more destructive tree diseases and a catastrophic threat to elms. The agent causing DED has been identified as a proteinaceous toxin, cerato-ulmin (CU) (Takai, 1974), which is produced by *Ophiostoma (Ceratocystis) ulmi* (Buisman) Nannf. CU is known to be a small, highly hydrophobic and surface active protein (Stevenson et al. 1979; Russo et al. 1981; 1982; Stevenson and Takai 1982). A correlation has been observed for various wild type isolates of *O. ulmi* between the *in vitro* production of CU and pathogenicity (Takai 1980).

Cerato-ulmin affects the physiology of susceptible elms in a host-specific manner. Exposure to CU causes susceptible white elm to raise cellular respiration dramatically, while respiratory increase by the resistant Siberian elm (*U. pumila* L.) is moderate (Richards and Takai 1984). *In vivo* production of CU in white elms both naturally infected with *O. ulmi* via vector transmission (Takai et al. 1983) and artificially inoculated with *O. ulmi* spores (Richards and Takai 1988) has been observed.

In view of the role of CU as a potent pathogen, there is an obvious interest in elucidating structural and functional properties of the protein. Understanding of the structure-function relationships in CU should provide basis for developing strategies for molecular genetic control of DED. Unfortunately, data pertaining to the structure of the protein is very scarce. In previous studies (Stevenson et al. 1979; Stevenson and Takai 1982), the N-terminal sequence and the sequence of several peptide fragments of CU have been determined. This paper reports the complete amino acid sequence of the protein. Furthermore, we present infrared, Raman and NMR spectroscopic data on the secondary structure of CU and the conformation of S-S linkages.

Materials and Methods

Preparation of Cerato-ulmin (CU)

Cerato-ulmin was produced by the RR-2 isolate of *Ophyostoma (Ceratocystis) ulmi* (Buisman) Nannf. (Brasier and Gibbs 1973) by the methods described previously (Takai and Richards 1978) except that the medium was replaced with the modified Salemink's medium (Takai 1978). The reduced and carboxymethylated CU was prepared as previously described (Stevenson and Takai 1982).

Isolation of Peptides

Peptides for amino acid sequence and composition analysis were obtained by digestion with trypsin, chymotrypsin, Lys/C enzyme, and mild acid hydrolysis (Asp/C).

The large peptides were isolated using four tandem TSK SW columns (Spherogel TSK 2000, Beckman, 7.5 X 600 mm) in 6M guanidine HCl, and the smaller peptides were separated by reverse-phase HPLC column and elution with an acetonitrile gradient (0-70% at 1 %/min) in 0.01 M trifluoroacetic acid at a flow rate of 1.0 ml/min from an Altex Ultrapore RPSC C3 column (Beckman Instruments) previously equilibrated in 0.05% trifluoroacetic acid.

Amino Acid Sequence Analysis

Automated gas-phase (Applied Biosystems 475A) and liquid-phase (Beckman 890C) sequence analysis were performed as described previously (Watson et al. 1990).

Amino Acid Composition Analysis

The reduced and carboxymethylated CU samples were hydrolysed and analyzed with a Durrum D-500 amino acid analyzer as described previously (Watson et al. 1990).

Infrared Spectroscopy

For infrared spectroscopy, samples were prepared by dissolving freeze-dried protein either in H_2O or ^2H_2O buffer (50 mM Hepes, pH or p^2H 7.0) at a concentration of 30-50 mg/ml. Samples were assembled between CaF_2 windows separated by a 50 μm (^2H_2O buffer) or a 6 μm (H_20 buffer) Teflon spacer. Fourier-transform infrared spectra at 2 cm^{-1} resolution were recorded at room temperature with a Digilab FTS-60 spectrometer. In order to eliminate spectral contributions of atmospheric water vapour, the instrument was continuously purged with dry nitrogen. Difference spectra were obtained by digitally subtracting buffer spectra from the corresponding sample spectra. For subtraction of the strong H_2O band around 1645 cm^{-1}, the criteria of no negative sidelobes and flat baseline between 1720 and 2000 cm^{-1} were used. The technique of Fourier self-deconvolution (Kauppinen et al. 1981) was used to resolve the overlapping infrared bands.

Raman Spectroscopy

Solutions in H_2O and ^2H_2O were made at concentrations of approximately 100 mg/ml in the absence of buffer. The H_2O solutions were filtered prior to Raman analysis and were examined in a quartz capillary Raman cell. For the sample in ^2H_2O the solution was placed in a 2 mm ID quartz capillary which was subsequently sealed to avoid exchange with atmospheric water vapour. The tube was inverted, centrifuged at 10,000 rpm for 10 min, reinverted and the clear supernatent subjected to the focused laser beam. Spectra of dry CU toxin were obtained by tamping freeze-dried protein into the hollowed end of a 3 mm diameter steel rod and grazing the laser beam over the compressed flat protein surface.

All samples were examined in the usual 90^0 excitation-detection geometry, 100 mw and 70 mw of 514.5 nm excitation being used for the solution and solid samples, respectively. A stream of 5 C N_2 was directed over the samples to minimize the effect of laser heating. Spectra were recorded using a Jarrell Ash 0.75 m double monochrometer equipped with photon-counting as described previously (Kumar and Carey 1975). The spectral slit width was set at 8 cm^{-1} and peak positions (accurate to ± 2 cm^{-1}) were calibrated using rare gas discharge lines.

Results and Discussion

N-terminal Sequence

The N-terminal sequence of the first 30 residues of the ^{14}C-labeled carboxymethylated CU have been determined, and the results are shown in Fig. 1. The results confirm the identification of 26 residues out of the first 30 residues previously reported by Stevenson et al. (1979) and Stevenson and Takai (1982). The first residue was found to be serine, and not alanine. Feng et al. (1991) found also the first residue to be serine, and pointed out the presence of minor N-terminally truncated forms of CU. Because of heterogeneity due to the N-terminal proteolytic cleavages, it was not possible to obtain the sequence beyond Cys30 by N-terminal analysis.

Lys/C Peptides

N-terminal amino acid sequence analysis has shown that the single lysine residue is at position 13 (Fig. 1). Therefore it was expected that Lys/C enzyme would cleave CU into two fragments: K1 (1-13) and K2 (14-75). The two fragments were clearly separated on the TSK column (Fig. 2). The large fragment and the small fragment were identified as peptides K2 (14-75) and K1 (1-13), respectively, by N-terminal amino acid sequence and the amino acid composition analysis (Table 2).

Tryptic Peptides and Asp/C Peptides

The tryptic peptides (T1 to T4b) and Asp/C peptides (D1 to D7) were isolated by reverse-phase HPLC and sequenced. (Fig. 1). The tryptic peptides, Asp/C peptides, and Lys/C peptide K2 were aligned to give the amino acid sequence from Cys30 to the C-terminal residue Ile75. The amino acid composition of tryptic peptides and Asp/C peptides are shown in Tables 2 and 3, respectively.

Internal Sequence Homology

The sequence from Serl4 to Cys30 is very similar to the sequence from Ser54 to Cys69 as shown below:

```
14          17  18            22  23  24  25  26    28     30
Ser-X-X-Cys-Cys-X-X-X-Ile-Leu-Gly-Val-Ala-X-Leu-X-Cys-
54          57  58            62  63  64  65  66  67     69
Ser-X-X-Cys-Cys-X-X-X-Leu-Leu-Gly-Leu-Ala-Leu-X-Cys
```

This suggests that the second sequence may have been derived from the first sequence by gene duplication.

Sequence Comparison

The Genetics Computer Group Sequence Analysis Software Package, Version 7.1, March 1992 (see Devereux et al. 1984)) was used to compare CU sequence with nucleic acids and protein data banks (GenBank, EMBL, PIR-Protein, PIR-Nucleic, SwissProt). CU has no apparent homology with published protein sequences (see Barker et al. 1990). However, when compared with nucleic acid sequences, *Saccharomyces cerevisiae* chromosome III (Y3) (see Oliver et al. 1992) has one section of the DNA (locus:YSCCHRIII) which is very similar to CU (56-68) as shown below.

```
     56  57  58  59  60      62          65      67  68
CU   ARG-CYS-CYS-THR-LEU-ser-LEU-leu-gly-LEU-ala-LEU-VAL -
     82785                                          82797
Y3   ARG-CYS-CYS-THR-LEU-val-LEU-ser-cys-LEU-cys-LEU-VAL -
```

Amino Acid Composition

The amino acid composition of CU is shown in Table 1. Assuming 9 residues of leucine, the total number of amino acid residues is 75 which is in complete agreement with the number found from the amino acid sequence analysis shown in Figure 1. The residue number of individual amino acids found by the composition analysis agrees very well with that found by sequence analysis. Methionine and tryptophan were not detected as reported previously by Stevenson et al. (1979). Seven glutamic acid residues were found to be glutamine from the sequence analysis (Fig. 1) and no glutamate was found. One residue each of tyrosine, phenylalanine, histidine, and lysine are present in CU. These results agree reasonably well with the previously reported composition analysis (Stevenson et al. 1979), if 9 residues of leucine, instead of 16, are used to estimate the residue number of other amino acids. CU contains eight residues of cysteine which can form four disulphide bonds.

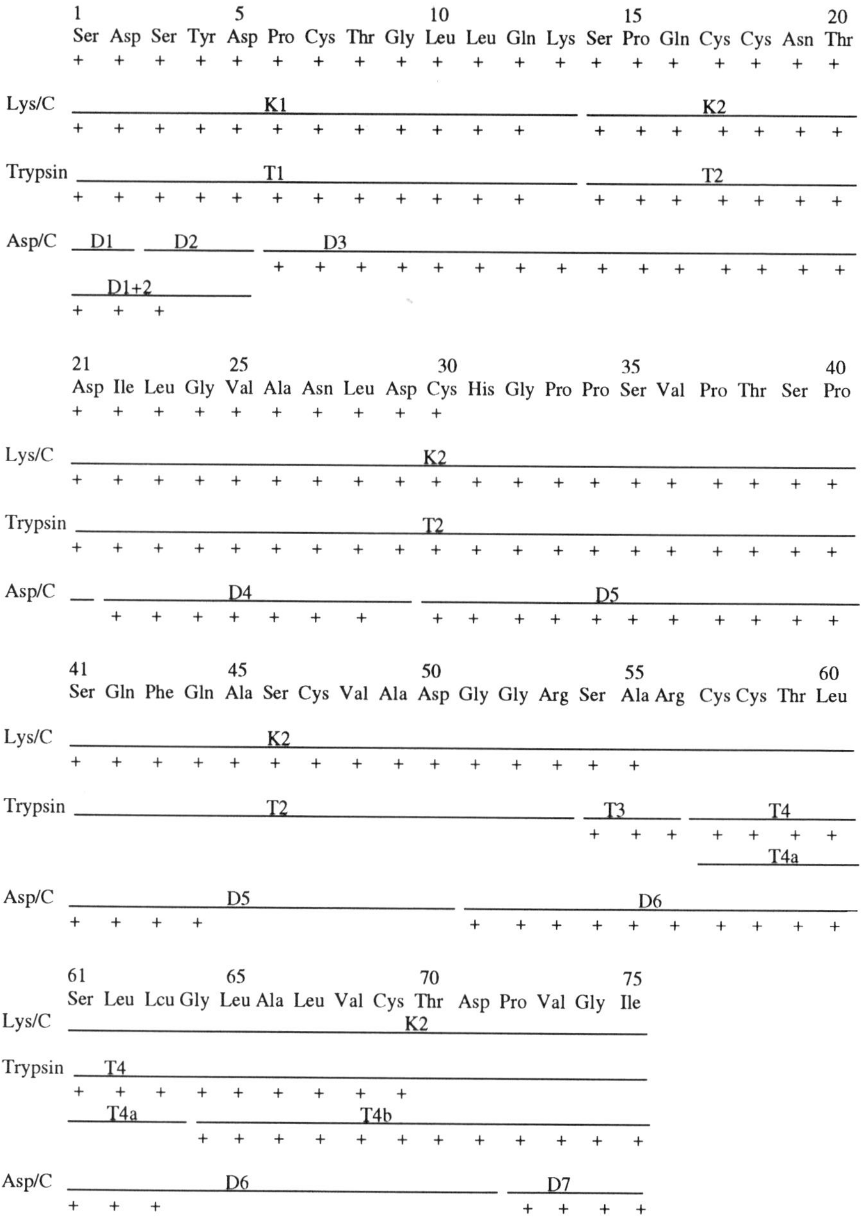

Figure 1. Amino Acid Sequence and Peptides of Cerato-ulmin. K denotes peptides obtained by Lys/C enzyme, T those obtained by trypsin, D those obtained by mild acid hydrolysis. + indicates residues positively identified directly by phenylthiohydantoin amino acid derivatives obtained by automated Edman degradation.

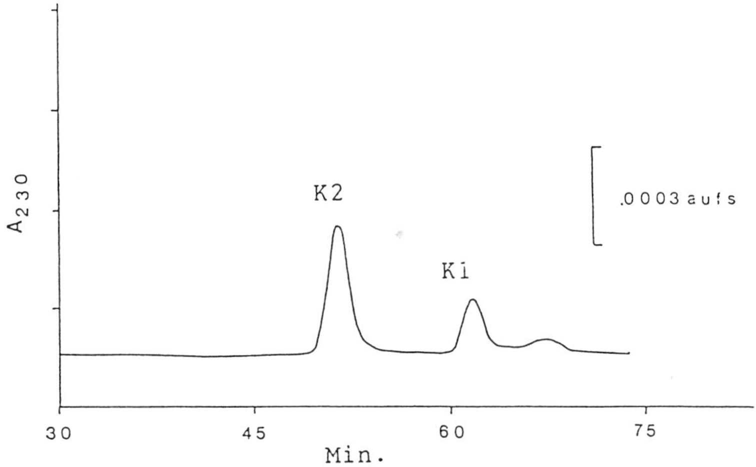

Figure 2. Isolation of Lys/C Peptides from Reduced and Carboxymethylated Cerato-ulmin with TSK SW2000 colums.

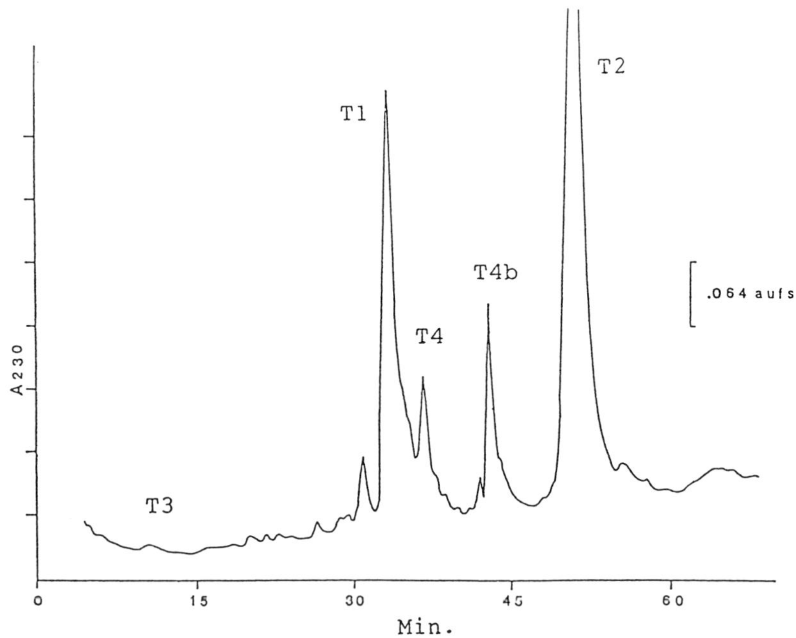

Figure 3. Isolation of Tryptic Peptides from Reduced and Carboxymethylated Cerato-ulmin by Reverse-Phase HPLC.

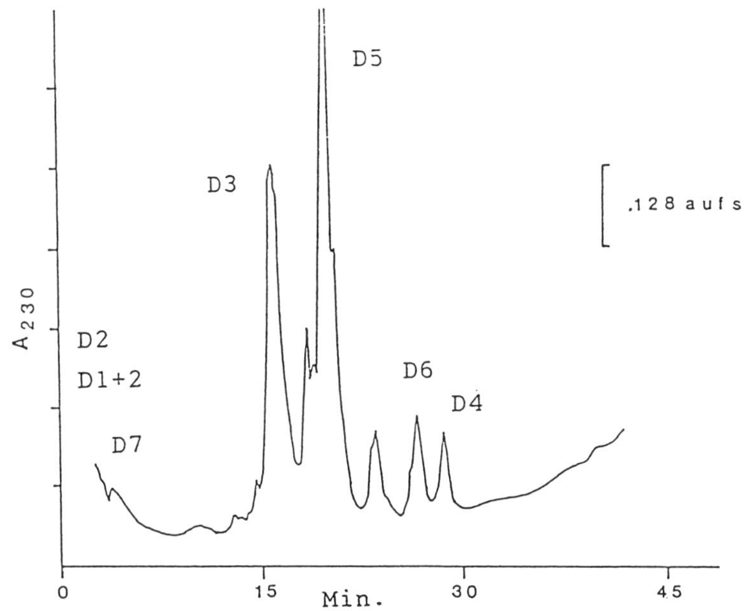

Figure 4. Isolation of Asp/C Peptides from Reduced and Carboxymethylated Cerato-ulmin by Reverse-Phase HPLC.

Table 1. Amino Acid Composition of Cerato-ulmin

amino acid				amino acid		
Asx	7.83	(8)		Met	0.00	(0)
Asp			(6)	Ile	1.83	(2)
Asn			(2)	Leu	9.00	(9)
Glx	4.02	(4)		Tyr	0.87	(1)
Glu			(0)	Phe	1.09	(1)
Gln			(4)	His	1.05	(1)
Thr	4.40	(5)		Lys	1.18	(1)
Ser	8.08	(9)		Arg	1.94	(2)
Pro	7.00	(7)		Trp	0.00	(0)
Gly	7.08	(7)		Cys/2	8.64	(8)
Ala	5.31	(5)				
Val	4.64	(5)		TOTAL............................(75)		

Nine residues of leucine are assumed. Cys/2 was determined as cysteic acid after performic acid oxidation. Trp was analyzed following hydrolysis with methanesulphonic acid. The values shown are the average of two determinations following hydrolysis with 6N HCl for 24 hours at 110°C. Losses due to the hydrolysis were not corrected for. The values in parenthesis are the number of residues determined by sequence analysis.

Table 2. Amino Acid Composition of Tryptic and Lys/C Peptides from Reduced and Carboxymethylated Cerato-ulmin (mol of residues/mol)

	Tl (=K2)	T2	T3	T4	T4b	K2 (T2+TT3+T4)
Residues	1-13	14-53	54-56	57-75	64 -75	14-75
Asx	1.83 (2)	4.63 (5)		1.07 (1)	1.01 (1)	6.20 (6)
Thr	0.90 (1)	1.96 (2)		1.92 (2)	0.93 (1)	3.85 (4)
Ser	1.51 (2)	4.14 (5)	1.26 (1)			7.12 (7)
Glx	0.82 (1)	2.93 (3)				3.24 (3)
Pro	1.09 (1)	5.48 (5)		0.88 (1)	0.96 (1)	5.75 (6)
Gly	1.25 (1)	4.43 (4)		2.93 (2)	1.90 (2)	5.95 (6)
Ala		2.60 (3)	1.00 (1)	0.84 (1)	0.98 (1)	4.56 (5)
Val		2.79 (3)		0.49 (2)	1.76 (2)	4.49 (5)
Ile		0.98 (1)		0.56 (1)	0.87 (1)	1.85 (2)
Leu	2.00 (2)	2.41 (2)		5.00 (5)	2.00 (2)	7.00 (7)
Tyr	0.97 (1)					
Phe		1.00 (1)				1.15 (1)
His		0.96 (1)				0.90 (1)
Lys	1.04 (1)					
Arg		0.74 (1)	1.00 (1)			2.54 (2)
Cys/2	(1)	(4)		(3)	0.53 (1)	4.49 (7)
Total	(13)	(40)	(3)	(19)	(12)	(62)

Values in parentheses indicate the number of residues determined from sequence analysis.

Disulfide Bridges

Native CU was treated by mild acid hydrolysis, trypsin, and chymotrypsin to isolate fragments which contain intact disulfide bonds. The peptides were fractionated on a reverse-phase C3 HPLC column. The amino acid composition of the fragments, after performic acid oxidation, indicates that there are two kinds of cystine containing peptides. The first group of peptides was eluted early from the reverse-phase column and one of the peptides had the following composition $Cys_{(4.0)}$, $Asx_{(1.0)}$, $Thr_{(2.0)}$ and $Ser_{(1.1)}$. This corresponds to a peptide (Ser46-Cys47)-S-S-(Cys57-Cys58-Thr59)-S-S-(Cys69-Thr70-Asp71) (Fig. 5). Another cystine containing peptide eluted later and had the following composition $Cys_{(4.0)}$, $Asx_{(1.4)}$, $Thr_{(2.3)}$, $Ser_{(3.3)}$, $Glx_{(1.7)}$, $Pro_{(4.2)}$, $Gly_{(3.7)}$, $Val_{(1.7)}$, $Leu_{(1.0)}$ and $His_{(1.3)}$. This peptide contains the single histidine residue at position 31, and is rich in proline, glycine, and serine. This composition appears to fit (Pro7-Cys7-Thr8-Gly9-Leul0)-S-S-(Cysl7Cysl8-Asnl9)-S-S-(Cys30-His31-Gly-Pro-Pro-Ser35-Val36-Pro-Pro-Thr-Ser-Pro-Ser-Gln42). Four possible disulfide bridges are shown in Figure 5. The peptide bonds between Cysl7-Cysl8 and Cys57-Cys58 have not been specifically cleaved, and it is still not clear whether Cys7 is bonded to Cys17 or Cys18. It is also uncertain as to whether Cys47 is linked to Cys57 or Cys58.

Table 3. Amino Acid Composition of Asp/C Peptides from Reduced and Carboxymethylated Cerato-ulmin (mol of residues/mol)

	D1+2	D2	D3	D4	D5	D6	D7
Residues	1-5	3-5	6-21	22-29	30-50	51-71	72-75
Asx	1.57 (2)	0.28 (1)	0.87 (2)	1.19 (2)	0.30 (1)	0.32 (1)	
Thr			1.81 (2)		0.95 (1)	1.79 (2)	
Ser	1.87 (2)	1.00 (1)	0.94 (1)		3.03 (4)	2.04 (2)	
Glx			2.04 (2)		2.28 (2)		
Pro			2.11 (2)		4.12 (2)		0.94 (1)
Gly			1.13 (1)	1.25 (1)	1.38 (1)	3.41 (3)	1.10 (1)
Ala				1.06 (1)	1.93 (2)	2.24 (2)	
Val				0 67 (1)	1.84 (2)	0.89 (1)	0.99 (1)
Ile				0.77 (1)			1.00 (1)
Leu			2.00 (2)	2.00 (2)		5.00 (5)	
Tyr	1.00(1)	1.00(1)					
Phe					1.00 (1)		
His					1.00 (1)		
Lys			0.98 (1)				
Arg						1.73 (2)	
Cys/2			1.72 (3)		0.68 (2)	1 46 (3)	
Total	(5)	(3)	(16)	(8)	(21)	(21)	(4)

Values in parentheses indicate the number of residues determined from sequence analysis. The low recovery of Asx is due to the loss of C-terminal Asp by the mild acid hydrolysis.

```
 1               5       7               11
Ser-Asp-Ser-Tyr-Asp-Pro-Cys-Thr-Gly-Leu-Leu
                         S               |
                         |              Gln
 23                  18  S               |
Leu-Ile-Asp-Thr-Asn-Cys-Cys-Gln-Pro-Ser-Lys
 |                   S   17              13
Gly                  |
 |                   S                               38
Val-Ala-Asn-Leu-Asp-Cys-His-Gly-Pro-Pro-Ser-Val-Pro-Thr
 25                  30                  35           |
                                                      |
     51                  47              43           |
    Gly-Asp-Ala-Val-Cys-Ser-Ala-Gln-Phe-Gln-Ser-Pro-Ser
     |               S                               39
    Gly              |
     |               S   58                  63
    Arg-Ser-Ala-Arg-Cys-Cys-Thr-Leu-Ser-Leu-Leu
     53              57  S                   |
                         |                   |
                         S                   |
Ile-Gly-Val-Pro-Asp-Thr-Cys-Val-Leu-Ala-Leu-Gly
 75              71      69                  64
```

Figure 5. Proposed Location of Disulfide Bonds in Cerato-ulmin.

The proposed model in Figure 5 shows that the N-terminal half (1-30) and C-terminal half (47-75) have a similar structure and they are connected by a proline/serine-rich "linker" peptide (31 to 46). The linker peptide contains no disulfide bond.

Molecular Weight

The molecular weight of CU calculated from the amino acid sequence of 75 residues (Fig. 1, Table 1) with 4 disulfide bonds (Fig. 5) is 7618.7. The molecular weight as determined by Ion-Spray Mass Spectrometry using a Sciex API III spectrometer is reported to be 7618.4 (Feng et al. 1991). It is clear, therefore, that CU is not a glycoprotein, and previously reported 4.4% carbohydrate found in CU preparations (Stevenson et al. 1979) must be bound non-convalently.

The molecular weight determined by SDS/polyacrylamide gel electrophoresis (SDS/PAGE) was reported to be about 13,000 (Stevenson et al. 1979). However, when diluted samples of CU were applied to SDS/PAGE in Pharmacia's PhastSystem, we found the MW to be about 8000. Since CU is known to aggregate in aqueous solution, the high molecular weight was probably due to the presence of CU dimers. High speed sedimentation equilibrium centrifugation indicated a MW of 8480 (Stevenson et al. 1979).

Secondary Structure Analysis from IR Data

The infrared spectra of the toxin were investigated both in H_2O (Fig. 6) and in ^2H_2O (Fig. 7) solution. The spectral region shown in Figures 6 and 7 covers the so-called amide I and amide II bands; however, our analysis of protein secondary structure is based mainly on the amide I band (between approximately 1600 and 1700 cm^{-1}), the conformational sensitivity of which is well established (Susi 1969; Byler & Susi 1986; Surewicz & Mantsch 1988; Krimm & Bandekar 1986).

The original broad amide I band contour of the toxin in H_2O buffer (top trace in Fig. 6) exhibits a maximum at 1644 cm^{-1} and two well defined shoulders above 1680 and below 1625 cm^{-1}. The maximum of the amide II band is at 1550 cm^{-1}. In the ^2H_2O buffer, the intensity of the amide II band at 1550 cm^{-1} decreases markedly and the maximum of the amide I band shifts to 1639 cm^{-1}. These differences reflect the H-^{2}H replacement of exchangeable amide groups in the protein backbone.

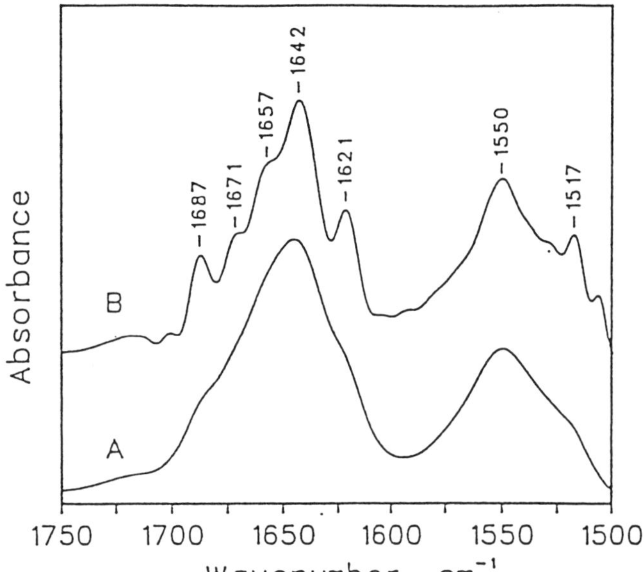

Figure 6. Infrared Spectrum in the amide I and amide II region of Cerato-ulmin in H_20 buffer (A) and same spectrum after band narrowing using the method of Fourier self-deconvolution (B).

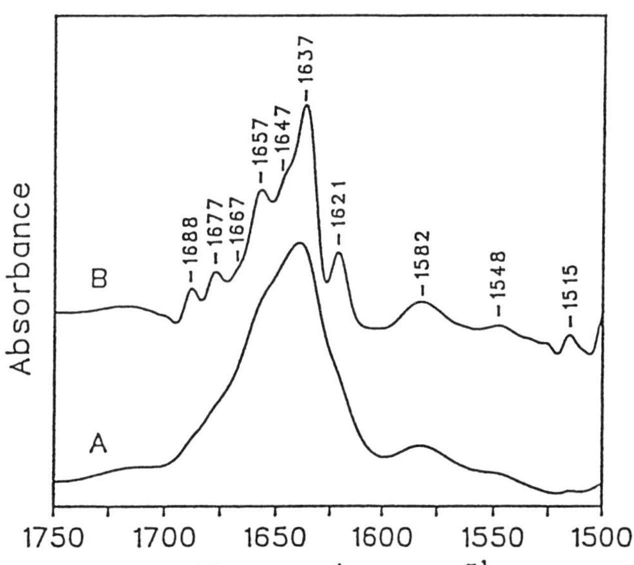

Figure 7. Infrared Spectrum in the amide I and amide II region of Cerato-ulmin in $^2H\ _2O$ buffer (A) and same spectrum after band narrowing using the method of Fourier self-deconvolution (B).

Whereas the position of the original amide I band contours clearly indicates the dominance of ß-sheet structure in the toxin (Susi 1969; Byler & Susi 1986; Krimm & Bandekar 1986), a more detailed interpretation of the spectra is hampered by the well-known overlapping of the component bands that contribute to the amide I mode (Byler & Susi 1986; Surewicz & Mantsch 1988). Deeper insight into the conformation of the protein may be obtained from resolution-enhanced infrared spectra. Application of the computational procedure of band narrowing by Fourier self-deconvolution allows the identification of otherwise hidden amide I sub-bands that represent various components of the secondary structure (Byler & Susi 1986; Surewicz & Mantsch 1988). The deconvolved spectra of the toxin in H_2O and ^2H_2O buffers are shown in Fig. 6B and Fig. 7B, respectively. These spectra reveal the presence of five (spectrum in H_2O) or seven (spectrum in ^2H_2O) bands in the amide I region, the frequencies of which are given in Table 4. The assignment of these bands is based on previously established spectra-structure correlations (Byler & Susi 1986; Surewicz & Mantsch 1988; Krimm & Bandekar 1986; Haris et al. 1986; Alvarez et al. 1987; Surewicz et al. 1987) and can be summarized as follows. The most prominent component band in the ^2H_2O spectrum of the toxin at 1637 cm^{-1} or its counterpart in the H_2O spectrum at 1642 cm^{-1} is highly characteristic of a ß-structure. Beta-strands are also represented by a band at 1621 cm^{-1}. These two "beta-components" most likely reflect the presence of ß-type segments with a somewhat different pattern of hydrogen bonding and a different exposure to water. While the structures represented by a 1637/1642 cm^{-1} band are readily accessible to hydrogen-deuterium exchange, there is no indication of such an exchange within the amide groups represented by the band at 1621 cm^{-1}. The latter band shows remarkable insensibility to the substitution of H_2O by ^2H_2O; its position remains unaltered even after prolonged (at least up to 24 h) incubation of the protein in ^2H2O. Bands seen in the ^2H_2O spectrum of the toxin at 1647 and 1657 cm^{-1} can be assigned to unordered polypeptide fragments in which the amide groups are accessible to isotopic exchange and to α-helices, respectively. In the spectrum measured in H_2O , the above two structures are represented by a single band at 1657 cm^{-1}. The band at 1667 cm^{-1} (1671 cm^{-1} in the H_2O spectrum) most likely represents turns. The assignment of the two minor components in the ^2H_2O spectrum of the toxin at 1677 and 1688 cm^{-1} (or a single component at 1687 in the H_2O spectrum) is somewhat less certain as both bands due to turns and high frequency vibrations of ß-strands may contribute to the spectral region between 1670 and 1690 cm^{-1}.

Table 4. Wave Number Positions and Fractional Areas of the Amide I Component Bands of Cerato-ulmin in ^2H_2O.

Band Position cm^{-1}	Fractional area %	Band assignment
1688	1	
1677	5	Turn/ß-structure
1667	11	Turn
1657	18	α-helix
1647	22	Unordered
1637	33	ß-structure
1621	10	ß-structure

The quantitative contribution of each component band to the total amide I contour may be obtained by curve fitting procedure; for this the essential input parameters, i.e. the number of component bands and their positions are obtained from deconvolved spectra, and the fitting is performed by iterative adjustment of the heights and half-widths of these component bands. The fractional areas of the fitted components (in the spectrum recorded in ^2H_2O) have been proposed to provide an estimate of the relative populations of the conformational structures represented by these component bands (Surewicz & Mantsh 1990), although such quantitative estimation of protein secondary structure is based on a number of assumptions and approximations (Surewicz & Mantsch 1990) and should be used with caution. Nevertheless, it is notable (Table 4) that the components representing beta-structure account for as much as 44-48% of the total area of the amide I band contour, depending on which of the bands between 1670 and 1690 cm^{-1} is chosen as a high-wavenumber "beta-component". The content of α-helix is much lower, as indicated by the modest intensity of the "α-helical band" at 1657 cm^{-1} (Table 4).

Analysis of Raman Data

The Raman spectrum of the solid protein is shown in Fig. 8. The data are compared to those for the CU in H_2O and in ^2H_2O in Table 5. wherein band assignments are also made. Features due to amino acid side chains and to the peptide backbone can be identified in the Raman spectrum. The amide I and amide III features can be used to provide a semiquantitative estimate of secondary structure whilst the features due to the S-S linkages and the tyrosine side chain provide information on the conformation and environment of these linkages.

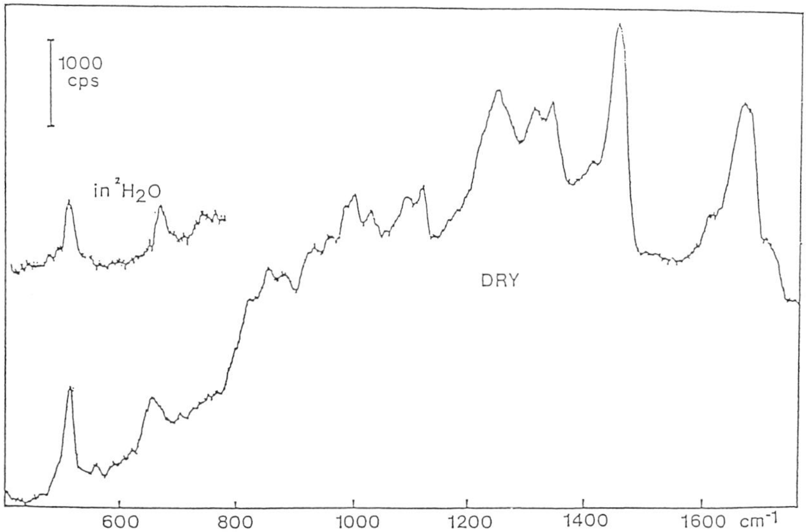

Figure 8. Raman Spectrum on the Solid Cerato-ulmin in H_20 and in $^2H_20$

Using the method of Lippert at al. (1976) the amide III profile of CU in H_20 and the amide I profile of CU dissolved in $^2H_20$ have been used to form an estimate of secondary structure. The results are 8% α-helix, 50% ß-sheet and 42% random structure. Given the approximations involved in both methods this is considered to be in good agreement with the values estimated from the FTIR data. The essential conclusion is that CU consists predominantly of ß-sheet structure, with a relatively small amount of α-helix.

A careful search revealed no evidence for a S-H stretching vibration near 2580 cm^{-1} but an intense S-S stretching mode is apparent at 514 cm^{-1} (Fig. 8). Thus, we conclude that all eight cystine residues are linked in cysteine bonds for CU in solution and in the solid state. Within the limitations of the signal-to-mode ratio achieved, $\upsilon_{s\text{-}s}$ appears as a single symmetric feature at 514 cm^{-1} in the solid and 512 cm^{-1} in solution (insert to Fig. 8 shows $\upsilon_{s\text{-}s}$ for CU in ^2H_2O). Since the exact position of $\upsilon_{s\text{-}s}$ is conformationally dependent (Harada & Takeuchi 1986), we are able to use $\upsilon_{s\text{-}s}$ to estimate the C-C-S-S-C-C- dihedral angles. The single symmetric feature at 514 cm^{-1} strongly suggests that the conformation about all four S-S linkages is GGG (Harada & Takeuchi 1986). The GGG conformation has a C-S-S-C dihedral angle in the 65-100 range with C-C-S-S values also lying within these limits.

Table 5. Raman Frequences, Relative Intensities (in parentheses) and Assignments for Cerato-ulmin.

SOLID	SOLUTION H_20	SOLUTION ^2H_2O	ASSIGNMENT
514 (3.2)	512 (6.5)	512 (4.3)	$\upsilon_{S\text{-}S}$
567 (0.5)			amide VI; skeletal [T,E] helix [G]
663 (1.9)	662 (2.4)	662 (2.5)	C-S (Cys)
761 (0.5)	764 (1.2)	739 (1.1)	C-S
	807 (2.0)sh	807 (1.3)sh	?
833 (2.7)	838 (3.7)	835 (3.9)	Y
859 (3.3)	864 (4.2)	859 (4.2)	Y
887 (2.6)	889sh	887 (3.5)	G
	924sh		
940 (2.6)	939 (2:3)	938 (5.1)	V,L
964 (3.3)	962 (2.1)	964 (5.5)	V,L (amide III; in ^2H_2O)
993 (4.2)		987 (5.8)	T
1005 (4.5)	1005 (4.5)	1005 (5.9)	F
1031 (3.6)	1031 (2.5)	1039 (2.6)	F,G
	1069 (1.0)		K,E,(S)
1104 (3.4)	1105 (2.4)		A
1127 (4.0)	1127 (4.0)	1130 (2.0)br	V,L,I,E,D,$\upsilon_{c\text{-}c}$
	1170br,sh		V,L
		1203 (6.9)	F,Y
1256 (6.6)	1255 (8.4)	1263 (4.2)	amide III tin H_20)
1315 (5.7)	1314 (5.2)	1315 (3.7)sh	CH_2 twist/wag
1339 (6.0)	1340 (6.0)	1347 (4.0)	CH_2 twist/wag
1414 (4.1)	1400 (5.2)	1416 (3.8)	$-COO^-$
1454 (10.0)	1454 (10.0)	1453 (10.0)	CH_2 scissor, CH_2, asym. def.
1624 (2.1)sh		1622 (2.9)sh	Y
1667 (7.5)	water	1656 (6.7)	amide I
1720sh		1720sh	-COOH?

The two peaks at 833 and 859 cm^{-1} (Fig. 8) are due to modes of the tyrosine side chain and have been designated the tyrosine doublet (Carey, 1982). The relative intensities of these features are sensitive to the H-bonding pattern of the sidechain's OH group. The relative intensitites recorded in Table 5 are typical for the -OH group acting as a medium- to weak H-bond donor or receptor to a protein or water residue. Thus the present Raman data for CU cannot distinguish between a buried, H-bonded to protein or exposed tyrosine OH bound to the solvent.

The final side chain features of note are those due to the acidic residues. The feature near 1410 cm^{-1} is assigned to $-COO^-$, whereas the broad shoulder near 1710 cm^{-1} (solid spectrum Fig. 8 and in ^2H_2O Table 5). could be due to side chains -COOH. If the latter assignment is correct this would suggest that some of the Glu and Asp residues have unusual pKa's since the Raman data were obtained near neutral pH.

Nuclear Magnetic Resonance

Two-dimentional NMR studies show that presence of two antiparallel ß-sheets. One sheet involves the strands made of residues 16-20 and 28-32. The other is made up of residues 56-61 and 65-71 with a bulge at residues 67-68. A helical structure is indicated in the region of residues 43-49. The disulfide bonds between Cys7 and Cysl7 and between Cysl8 and Cys30 (Fig. 5) were confirmed by the NMR data (R. Muhandiram, H. Hunter and E. O. Purisima (unpublished results) from Biotechnology Research Institute, Montreal) .

Conclusion

Cerato-ulmin is a single polypeptide chain. It has 75 amino acid residues with four disulfide bonds. It molecular weight is 7618 and it is not a glycoprotein. Cerato-ulmin consists predominantly of ß-sheet structure, with a relatively small amount of α-helix. Cerato-ulmin has two structually similar domains. The first (1-30) domain contains two disulfide bonds (Cys7-S-S-Cysl7-Cysl8-S-S-Cys30) and the second domain (47-75) has also two disulfide bonds (Cys46-S-S(Cys57-Cys58)-S-S-Cys69). A proline and hydroxyamino acid rich peptide (31-46) forms a linker between the two disulfide containing domains.

Acknowledgements

We thank Dr. E. O. Purisima for NMR data, Dr. Warren Wakarchuk and Mr. David Watson for helpful discussion. This publication is NRCC 34262.

References

Alvarez, J., P. I. Haris, D. C. Lee and D. Chapman. 1987. Conformational changes in concanavalin A associated with demetallization and α–methylmannose binding studied by Fourier-transform infrared spectroscopy. Biochim. Biophys. Acta 916:5-12.

Barker, W. C., D. G. George, and L. T. Hunt. 1990. Protein sequence database. Methods in Enzymology 183:31-49.

Brasier, C. M. and J. N. Gibbs. 1973. Origin of the Dutch elm disease epidemic in Britain. Nature(London) 242:607-609.

Byler, M., and H. Susi. 1986. Examination of the secondary structure of proteins by deconvoluted FTIR spectra. Biopolymers 25:469-487.

Carey, P. R. 1982. Protein conformation from Raman and resonance Raman spectra. In Biochemical Applications of Raman and Resonance Raman Spectroscopies. pp. 71-98, Academic Press, New York.

Devereux, J., P. Haeberli and O. Smithies. 1984 A comprehensive set of sequence analysis programs for the VAX. Nucleic Acids Research 12:387-395.

Feng, R., Y. Konishi and A. W. Bell. 1991. High accuracy molecular weight determination and variation characterization of proteins up to 80 ku ionspray mass spectrometry. J. Am Soc. Mass Spectrom. 2:387-401.

Harada, I. and H. Takeuchi. 1986. Raman and ultraviolet resonance Raman spectra of proteins and related compounds. In Spectroscopy of Biological Systems, 13:113-175 (R J. H Clark and R. E. Hester eds.) John Wiley & Sons.

Haris, P. I., D. C. Lee and D. Chapman. 1986. A Fourier-transform infrared investigation of the structural differences between ribonuclease A and ribonuclease S. Biochim. Biophys. Acta 874:255-265.

Kauppinen, J. K., D. J. Moffatt, H. H. Mantsch and D. G. Cameron. 1981. Fourier self-deconvolution: a method for resolving intrinsically overlapped bands. Appl. Spectrosc. 35 :271-277.

Krimm, S. and J. Bandekar. 1986. Vibrational spectroscopy and conformation of peptides, polypeptides, and proteins. Adv. Protein Chem. 38:181 -364.

Kumar, K. and P. R. Carey. 1975. Resonance Raman spectra and structure of some complex nitroaromatic molecules. J. Chem. Phys. 63:3697-3707.

Lippert, J. L., D. Tyminski and P. J. Desmeules. 1976. Determination of the secondary structure of proteins by laser Raman spectroscopy. J Am. Chem. Soc. 98:7075-7080.

Oliver, S. G., Q. J. M. van der Aart, M. L. Agostoni-Carbone, M. Aigle, L. Alberghina, D. Alexandraki, G. Antoine, R. Anwar, J. P. G. Ballesta, P. Benit, G. Berben, E. Berganito, N. Biteau, P. A. Bolle, M. Bolotin-Fukuhara, A. Brown, A. J. P. Brown, J. M. Buhler, C. Carcano, G. Carignani, H. Cederberg, R. Chanet, R. Contreras, M. Crouzet, B. Daignan-Fornier, et al. 1992. The complete DNA sequence of yeast chromosome III. Nature 357:38-46.

Richards, W. C. and S. Takai. 1984 Characterization of the toxicity of cerato-ulmin, the toxin of Dutch elm disease. Can. J. Plant Pathol. 6:291-298.

Richards, W. C. and S. Takai .1988. Production of cerato-ulmin in white elm following inoculation with *Ceratocystis ulmi.* Physiol. Mol. Plant Pathol. 38:279-285.

Russo, P. S., F. D. Blum, J. D. Ipsen, Y. J. Abul-Hajj and W. G. Miller. 1981 The solubility and surface activity of the *Ceratocystis ulmi* toxin cerato-ulmin. Physiol. Plant Pathol. 19:113-126.

Russo, P. S., F. D. Blum, J. D. Ipsen, Y. J. Abul-Hajj and W. G. Miller. 1982. The surface activity of the phytotoxin cerato-ulmin. Can. J. Bot. 60:1414-1422.

Stevenson, K. J. , A. Slater and S. Takai. 1979. Cerato-ulmin - a wilting toxin of Dutch elm disease fungus. Phytochemistry 18:235-238.

Stevenson, K. J., and S. Takai. 1982. Structural studies on cerato-ulmin - a wilting toxin of Dutch elm disease fungus, *Ceratocystis ulmi.* In: "Proceedings of the Dutch Elm Disease Symposium and Workshop, Winnipeg, Manitoba, October 5-9, 1981 (Kondo, E. S., Y. Hiratsuka and W. B. G. Denyer, eds.) pp. 298-321 Manitoba Dept. of Natural Resources, Manitoba, Canada.

Surewicz, W. K. and H. H. Mantsch. 1988. New insight into protein secondary structure from resolution-enhanced infrared spectra. Biochim. Biophys. Acta 952:115-130.

Surewicz, W. K. and H. H. Mantsch. 1990. The conformation of proteins and peptides in a membrane environment: An infrared spectroscopic approach. In: Protein Engineering - Approaches to the Manipulation of Protein Folding (Narang, S., ed.) pp. 131-157, Butterworth.

Surewicz, W. K., A. G. Szabo and H. H. Mantsch. 1988. Conformational properties of azurin in solution as determined from resolution-enhanced Fourier-transform infrared spectra. Eur. J. Biochem. 167:519-523.

Susi, H. 1969. In: Structure and Stability of Biological Macromolecules (Timasheff, S. N. and G. D. Fasman, eds.) pp. 575-663, Marcel Dekker, Inc., New York.

Takai, S. 1974. Pathogenicity and cerato-ulmin production in *Ceratocystis ulmi.* Nature 252:124-126.

Takai, S. 1978. Cerato-ulmin, a wilting toxin of *Ceratocystis ulmi:* Cultural factors affecting cerato-ulmin production by the fungus. Phytopathol. Z. 91:147-158.

Takai, S. 1980. Relationship of the production of the toxin cerato-ulmin to synnemata formation, pathogenicity, mycelial habit and growth of *Ceratocystis ulmi* isolates. Can. J. Bot. 58:658-662.

Takai, S., and Richards, W. C. 1978. Cerato-ulmin, a wilting toxin of *Ceratocystis ulmi:* Isolation and some properties of cerato-ulmin from the culture of *C. ulmi.* Phytopathol. Z. 91:129-146.

Takai, S., Richards, W. C., and Stevenson, K. J. 1983. Evidence for the involvement of cerato-ulmin, the *Ceratocystis ulmi* toxin, in the development of Dutch elm disease. Physiol. Plant Pathol. 23: 275-280.

Watson, D. C., M. Yaguchi, and K. R. Lynn. 1990. The amino acid sequence of chymopapain from *Carica papaya*. Biochem, J. 266:75-81.

13
Strategies for the Production of Disease-Resistant Elms

Mark G. Bolyard and Mariam B. Sticklen

Abstract

This work was supported by the National Park Service of the Department of the Interior (MBS; CA-3040-9-8003).

The possibility exists for the application of genetic engineering toward the production of Dutch elm disease resistant American elms. Plant genetic engineering experiments require that three processes be completed in the development of transgenic plants. Tissue culture systems must be developed for regeneration of the plant of interest. We have previously reported the regeneration of American elms from leaf explants, and these findings will also be reported by others. In addition, genes must be isolated or produced which elicit a desired effect when introduced in the proper molecular environment into the plant of interest. Cerato-ulmin (CU), a DED toxin produced by *O. ulmi,* has been produced in an expression vector in *E. coli* and has been shown to have biological activity. Work is underway to make altered forms of this toxin to be tested as non-toxic competitive inhibitors of CU. DNA encoding effective inhibitors can then be engineered into elms for expression of the inhibitors.

Finally, the ß-glucuronidase reporter gene is being used to establish a transformation system for the American elm. Although this phase of the project has not yet yielded transgenic elms in the greenhouse, promising preliminary results have been obtained.

Additional keywords. ß-glucuronidase, cerato-ulmin, electroporation, polymerase chain reaction, recombinant DNA, woody plant tissue culture.

Introduction

Dutch elm disease (DED) research is taking on a new role, reducing its focus on the solution of a great practical problem. American elms (*Ulmus americana* L.) are no longer requested by nurserymen and homeowners because of their reputation for demise without radical intervention. While this trend may reverse with the release of DED resistant elms, one current exception to this philosophy has come from the National Park Service, which sustains numerous American elms on the National Mall in Washington, D.C. This is one of the few locations from which there is a strong demand for DED-resistant American elms. DED research has now become a tremendous intellectual challenge to researchers in the field, and serves as a model for the application of various strategies against plant pathogens. This second symposium on DED research will serve to showcase the wide variety of approaches being undertaken by the top researchers in this area. The following report describes approaches that we have taken to produce DED-resistant American elms.

Plant biotechnology consists of three phases, with the achievement of each phase being required for the production of transgenic plants. The first two phases must be completed before moving on to the third phase. The first phase is the establishment of a system of plant regeneration. This has been achieved for American elms by several investigators using cell suspension cultures (Durzan and Lopushanski 1975), callus cultures (Karnosky et al. 1982) and from leaf tissue (Bolyard et al. 1991). Regeneration of other species of elm from tissue culture has also been achieved, providing opportunities to genetically alter these species as well. These include the hybrid Pioneer elm (*Ulmus* x 'Pioneer'; Sticklen et al. 1986) and Chinese elm (*Ulmus parvifolia* Jacq.; Bolyard et al. 1991). Discussion of elm tissue culture will be a major topic by other researchers during this symposium, although it is important to note that this first step toward the production of transgenic American elms has been achieved.

The second phase is the identification, isolation, and/or development of genes which, once integrated into the genome of the target plant, will elicit the desired effect, in this case resistance to *Ophiostoma ulmi,* the fungus which causes DED. In this area as well, several approaches are being undertaken which will be described by other investigators during this symposium. Our approach has been to focus on a toxin produced by *O. ulmi* known as cerato-ulmin (CU; Takai 1974).

This toxin has been implicated as a direct causative agent in the progression of DED, and much effort has been put into its analysis. For example, the amino acid sequence of CU has been solved, and is being reported for the first time at this symposium (Yaguchi et al., this volume). Our research has been based on this previously unpublished sequence, and we are indebted to Dr. Yaguchi for sharing this information with us through Dr. Shozo Takai. Research on the toxin will enable us not only to understand more about the mode of action of the protein, but may also provide a key into the development of DED-resistant American elms. Preliminary data (S. Takai and W. Richards; M. Bolyard, and M. Sticklen) have indicated that the toxicity of CU is related to its properties as a water-insoluble surfactant. When CU is reduced and denatured to become water-soluble, the toxin appears to lose its activity. However, reduced CU, allowed to refold without denaturation, and unreduced, denatured CU retain their surfactant properties. This indicates that some or all of the four pairs of disulfide bonds within the toxin are involved in maintaining structure and activity.

We have assembled synthetic oligonucleotides, based on the amino acid sequence of CU, in order to express the toxin in *Escherichia coli* as a fusion protein with glutathione S-transferase (GST) (Bolyard and Sticklen 1992). This fusion facilitates production and purification of the toxin, which can later be separated from GST following cleavage with the serine protease thrombin. During synthesis of the oligonucleotides, specific mutations were made to change the cysteine residues of CU to serines at random locations within each oligonucleotide. The result was a mixture of assembled oligonucleotides containing some or all of the cysteine residues. The goal of this project is to test altered CU molecules containing various combinations of cysteine and serine residues for activity as well as in competition with native CU to identify competitive inhibitors. A number of the sets of assembled oligonucleotides have been sequenced and will be inserted into the expression vector at a later date.

Figure 1. Shoot regenerated from Chinese elm culture.

Recombinant CU was produced, purified using a glutathione affinity column, and tested on American elm callus cultures following incorporation into the growth medium. The recombinant toxin lowered the growth rate of the callus, with greater decreases being noted as the concentration of recombinant CU was increased, although the recombinant toxin did not appear to be as toxic as native CU. However, the recombinant toxin preparation was not absolutely pure, which could have had an effect on the comparison. Details of these experiments are being published elsewhere (Bolyard and Sticklen 1992). It will be important to establish rapid and reproducible screening techniques to analyze the numerous altered CU molecules which could be produced using this system. Perhaps other methods, such as electrolyte leakage analysis (Mezzetti et al. 1988), can be standardized for use in testing these recombinant toxins.

The third phase of plant genetic engineering is the transformation of the plant of interest (American elm) for which the regeneration system has been developed (step one) with the gene which has been identified and put into an appropriate molecular environment, including promoter and terminator sequences (step two). Although the ideal competitive inhibitor of CU has not yet been identified, other reporter genes are available to optimize the transformation of American elm. We have used the ß-glucuronidase (GUS) system because of its commercial availability as well as its ease of screening. Plants which are transformed with the GUS gene are capable of producing a blue precipitate in the presence of the appropriate substrates. These experiments will be examined in detail in the remainder of this paper.

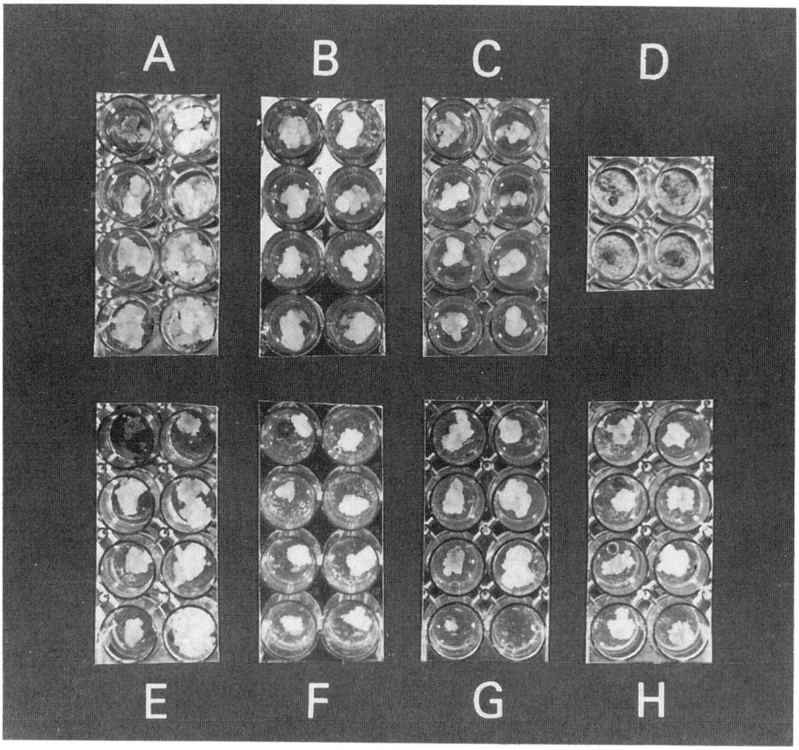

Figure 2. American elm callus cultures treated with: A) 0.1X recombinant CU; B) 1X recombinant CU; C) 10X recombinant CU; D) 100X recombinant CU; E) untreated; F) 3.5% ethanol; G) E. coli proteins; and H) 1X native CU.

Materials and Methods

American elm seeds were germinated and the second mature leaf was sterilized in 20% Chlorox as described (Bolyard et al. 1991). Leaves were cultured first on modified Murashige and Skoog (1962) media containing 0.1 μM thidiazuron (Nor-Am Chemicals) as well as several concentrations of kanamycin (0, 15, 30, 45, 60 mg/L) to establish an optimal concentration of antibiotic for selection.

A second parameter to be tested was the voltages which the elm leaf strips could withstand during electroporation. The basic protocol was a modification of the work of Dekeyser and his colleagues (1990) in which rice tissues were electroporated. Elm leaf strips were cultured in Magenta GA-7 vessels on regeneration medium for two days to test for contamination. Following this pre-culture, leaf strips from each leaf were gently inserted (stacked on top of each other in piles of 5-6 strips) into a 4 mm gap electroporation cuvette (Bio-Rad) which contained 1 ml of 1X EPR buffer (10% glucose, 4 mM $CaCl_2$, 10 mM HEPES, pH 7.2) with 0.2 mM spermidine (free base). Leaf strips were incubated in the buffer for one hour, then electroporated at several voltages (0 - 400 volts in 50 volt increments) at a constant capacitance of 910 μF using a BTX model 600 electroporator.

Once appropriate voltages were determined, leaf strips were electroporated with DNA. After the 1 hour pre-incubation as described above, the solution was removed and replaced with fresh buffer, to which was added 10 μg of pBl 121 (Jefferson 1987) plasmid DNA (Clontech) for another 1 hour incubation.

This plant expression vector contains a selectable kanamycin resistance gene (neomycin phosphotransferase II.), and the ß-glucuronidase (GUS) gene driven by the cauliflower mosaic virus 35S promoter. ß-glucuronidase produced in transgenic cells cleaves an externally added substrate to turn transformed tissues a blue color. Following the 1 hour incubation with the DNA solution, NaCl was added to a final concentration of 165 mM and the solution was gently mixed. Leaf strips were then electroporated at the appropriate voltage and returned to the regeneration medium for two days. After this time, leaf strips were transferred to regeneration medium containing the appropriate concentration of kanamycin.

After approximately one month on selection medium, small shoots (5 cm) were excised from the explant and divided into two sections (or in a subsequent experiment transferred to modified MS medium containing 3 μM gibberellic acid). One section of the immature shoots was treated for GUS enzymatic activity via the histochemical assay according to Wang et al. (1988) using the substrate X-gluc (Research Organics, Inc.). The other section was used for DNA extraction following the method of Rogers and Bendich (1985). Extracted DNA was amplified via the polymerase chain reaction (PCR; Mullis and Faloona 1987) using synthetic oligonucleotides complementary to two regions of the GUS gene (MB 21-2 and MB 22-2; Zhong et al. 1993).

Amplification products were separated by agarose gel electrophoresis and blotted onto a 0.45 μm Nytran membrane (Micron Separations, Inc.) which was subjected to Southern analysis (1975) according to a modification recommended by Schleicher and Schuell. The membrane was hybridized using a 1.87 kb Bam HI-Sst I fragment from pBl 121 labeled with 32p-dCTP (New England Nuclear) using random hexamers according to the method of Feinberg and Vogelstein (1983). Hybridized membranes were then exposed to X-omat AR5 film (Kodak) for autoradiography.

Results

Electroporation and selection parameters were tested initially to establish the optimal conditions for the production of transformed American elms. The lowest concentration of antibiotic from which there was no regeneration of shoots from control American elm leaf explants was 30 mg/L kanamycin in modified Murashige and Skoog medium containing 0.1 μM thidiazuron (Bolyard et al. 1991). This concentration was used for subsequent transformation experiments.

A range of voltages was also tested, using a constant capacitance setting of 910 μF, to determine the viability of the leaf explants in terms of ability to regenerate shoots after electroporation. The highest voltage treatment which permitted regeneration of shoots was 100 volts (250 v/cm). This voltage was therefore determined to be optimum for electroporation experiments.

Using these electroporation parameters, 14 American elm leaves were sterilized and cut into strips (six strips per leaf). Each set of leaves was precultured in a separate culture vessel then transferred to another culture vessel containing the selection medium after electroporation. Ten shoots were regenerated from the explants following electroporation, while no shoots were produced from control plants on selection medium. Each shoot was cut into two pieces, and total DNA was extracted from one of the halves. Figure 3 shows the results of the PCR analysis of the ten samples, as well as various controls. Figure 4 is a Southern blot of the gel illustrated in figure 3, demonstrating that each of the bands from the ten samples, as well as the positive control (using pBl 121 DNA as a template), hybridize to the GUS probe. These results indicate that the GUS gene was present in each of the ten shoots that were regenerated, making this a very efficient system. However, these results do not address the issue of the percent of cells transformed in each sample, because PCR is a very sensitive technique. For this reason, GUS enzymatic activity was also analyzed using histochemical staining techniques.

Results of the histochemical assay were not as conclusive as those of the PCR reactions. Two of the ten shoot sections showed small patches of blue color, suggesting that only a small percentage of the cells had been transformed. However, this is not an accurate analysis, as detailed sectioning of the explants was not performed. Figure 5, although not in color, shows a patch of cells (arrowhead) that demonstrated a strong blue signal. Negative controls (shoots regenerated without electroporation; not shown) showed no blue coloration at all.

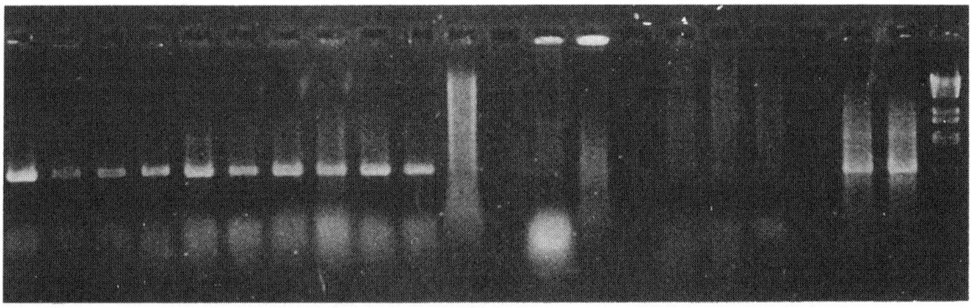

Figure 3. PCR reactions of DNA from the regenerated American elm samples. The first ten lanes (left to right) are the electroporated samples, while the second and third lanes from the right are positive controls. The other lanes contain various negative controls, such as a "no DNA" control and untransformed elm DNA controls. The lane farthest to the right is λ DNA digested with Bst Ell.

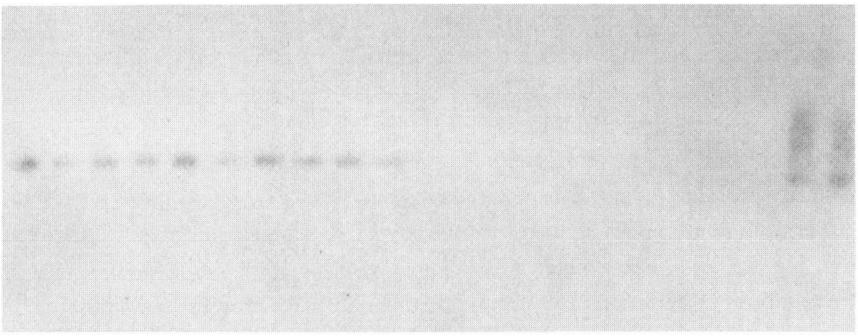

Figure 4. Southern blot of figure 3. The lanes are the same as above, except the λ DNA lane has been omitted.

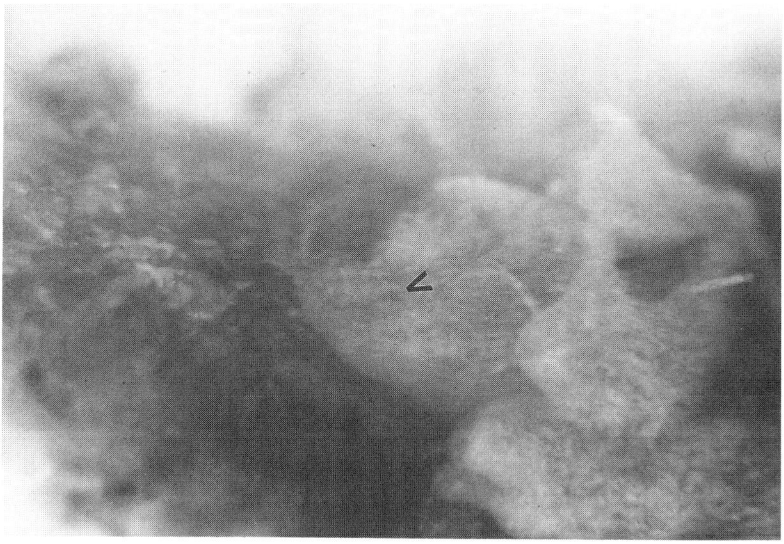

Figure 5. Small area of blue coloration from a PCR-positive shoot.

Because the first set of shoots regenerated on selection medium had been sacrificed for analysis, the experiment was repeated to determine whether selected shoots could be transferred to soil, acclimated to the greenhouse, and remain positive by PCR. Twenty-four leaves were cultured and electroporated as described above. From these cultures, nine shoots were produced after two months in culture, and were elongated on medium containing gibberellic acid until the nutrients were depleted and roots developed (three months). Shoots were then transferred to soil, and acclimated to a greenhouse. PCR analysis was performed on DNA extracted from leaves one year after electroporation. None of the nine plants assayed were positive by PCR analysis, although the reason for this is not clear. Perhaps only a few cells were transformed during electroporation, and these did not divide as vigorously as non-transformed cells after the selective pressure (transfer to soil) was removed. This would yield a chimeric plant containing only a few transformed cells. Another possibility is that the shoots were "escapes," although results of the previous experiment, in which all of the regenerated shoots were positive by PCR, makes this unlikely.

Discussion

We have described several experiments which we hope will lead to the production of DED resistant American elms using the technology of genetic engineering. The next step in our research involving recombinant CU is the development of non-toxic competitive inhibitors of the toxin. This is being addressed by producing mutants which have cysteine residues randomly converted to serines. The rationale behind this approach is that reduced and denatured CU loses toxicity, while reduced or denatured CU can be treated to regain its toxicity. Our goal is to screen altered recombinant CU molecules with the goal of finding a non-toxic version of CU which will continue to bind to cellular receptors within susceptible elms. The long-term objective would then be to insert the DNA encoding these competitors into the elms to confer a type of "acquired immunity". The major obstacle in this process is the development of a rapid, reproducible, and accurate screening method to evaluate the toxicity of the altered CU molecules. While the callus screening method is suitable for small scale experiments such as those described here, it is very time consuming to perform as a routine assay.

Transformation of plant protoplasts via electroporation has become a fairly routine practice (Fromm et al. 1985), while transformation of intact cells and tissues using this method has been reported infrequently (Lindsay and Jones 1990) and regeneration of transformed plants using this technique has not been reported to the best of our knowledge. It will be important to continue to refine the parameters for electroporation of American elm tissues, as preliminary evidence suggests that transformation with *Agrobacterium tumefaciens* may not be straightforward (unpublished observations). However, all methods should be investigated in order to establish an optimal protocol for the genetic engineering of American elm, as new genes are discovered and developed which may help protect these trees from DED.

References

Bolyard, M.G., Srinivasan, C., Cheng, J., and Sticklen, M.B. 1991. Shoot regeneration from leaf explants of American and Chinese elm. HortScience 26:1554-1555.

Bolyard, M.G., and Sticklen, M.B. 1992. Expression in *E. coli* of a modified Dutch elm disease toxin. Mol. Plant-Microbe Interact. 5:520-524.

Dekeyser, R.A., Claes, B., De Rycke, R.M.U., Habets, M.E., Van Montague, M.C., and Caplan, A.B. 1990. Transient gene expression in intact and organized rice tissues. Plant Cell. 2:591-602.

Durzan, D.J., and Lopushanski, S.M. 1975. Propagation of American elm via cell suspension culture. Can. J. For. Res. 5:273-277.

Feinberg, A.P., and Vogelstein, B. 1983. A technique for radiolabeling DNA restriction endonuclease fragments to high specific activity. Anal. Biochem. 132:6-13.

Fromm, M., Taylor, L.P., and Walbot, V. 1985. Expression of genes transferred into monocot and dicot plant cells by electroporation. Proc. Natl. Acad. Sci. USA 82:5824-5828.

Jefferson, R.A. 1987. Assaying chimeric genes in plants: The GUS gene fusion system. Plant Mol. Biol. Rep. 5:387-405.

Karnosky, D.F., R.A. Mickler and D.D. Lange. 1982. Hormonal control of shoot and root induction in hypocotyl callus cultures of American elm. In Vitro 18:275.

Lindsey, K., and Jones, M.G.K. 1990. Electroporation of cells. Physiol. Plant. 79:168-172.

Mezzetti, B., L. Mittempergher, and P. Rosati. 1988. *Ophiostoma ulmi* metabolites and elm cell membrane permeability. Possible use in early tests of resistance. Eur. J. For. Pathol. 18:77-84.

Mullis, K.B., and Faloona, F.A. 1987. Specific synthesis of DNA *in vivo* via a polymerase catalyzed chain reaction. Meth. Enzymol. 155:335-350.

Murashige, T., and Skoog, F. 1962. A revised medium for rapid growth and bio-assays with tobacco tissue culture. Physiol. Plant. 15:473-497.

Rogers, S.O., and Bendich, A.J. 1985. Extraction of DNA from milligram amounts of fresh, herbarium, and mummified plant tissues. Plant Mol. Biol. 5:69-76.

Sticklen, M.B., Domir, S.C., and Lineberger, R.D. 1986. Shoot regeneration from protoplasts of *Ulmus* x 'Pioneer'. Plant Sci. 47:29-34.

Southern, E.M. 1975. Detection of specific sequences among DNA fragments separated by gel electrophoresis. J. Mol. Biol. 98:503-517.

Takai, S. 1974. Pathogenicity and cerato-ulmin production in *Ceratocystis ulmi.* Nature 252:124-126.

Wang, Y.C., Klein, T.M., Fromm, M., Cao, J., Sanford, J.C., and Wu, R. 1988. Transient expression of foreign genes in rice, wheat, and soybean cells following particle bombardment. Plant Mol. Biol. 11:433-439.

Yaguchi, M., Pusztai, M., Roy, C., Surewicz, W.K., Carey, P.R., Stevenson, K.J., Richards, W.C., and Takai, S. 1992. Amino acid sequence, infrared and Raman spectroscopy of Dutch elm disease toxin cerato-ulmin. Proc.1992 International Dutch Elm Disease Workshop.

Zhong, H., Bolyard, M.G., Srinivasan, C., and Sticklen, M.B. 1993. Transformation of creeping bentgrass (*Agrostis palustris* Huds.) embryogenic callus via microprojectile bombardment. Manuscript in preparation.

14
Development of Model Systems to Screen Elms Resistant to Dutch Elm Disease

Subhash C. Domir, Lawrence R. Schreiber
and Paula M. Pijut

Introduction

Dutch elm disease (DED), a serious vascular wilt disease caused by *Ophiostoma ulmi* (Buisman) Nannf., has been responsible for destruction of elm trees in Europe and North America (Went 1954). The principal host in North America is the American elm, *Ulmus americana* L. The most practical means for suppressing this disease is the exploitation of polygenetically controlled host resistance (Heybroek 1970). Conventional breeding programs for improving disease resistance of elms require a long term commitment and resources to develop and evaluate germplasm. In various studies on host-parasite interactions to determine the nature of infection and mechanisms of resistance, tyloses formation, pit membrane alterations, cell wall breakdown, direct penetration of vessel cell walls and invasion of *U. americana* L. by *O. ulmi* have been reported to occur (Krause and Wilson 1972; MacDonald and McNabb 1974; Ouellette 1978a; Ouellette 1978b).

Current methods of screening for resistance involve inoculating elm seedlings or ramets with the pathogen and ranking disease severity. Tissue culture systems for studying host-pathogen interactions offer advantages over those utilizing intact plants (Brettell and Ingram 1979). Although concerns have been expressed regarding positive correlations between tissue culture and *in vivo* systems (Dixon 1980), such correlations have been demonstrated (Budde and Helgeson 1981; Deaton et al. 1982; Holliday and Klarman 1979). Ultrastructure studies have also confirmed such correlations (Miller and Maxwell 1983). Haberlach et al. (1978) noted that tissue morphology, incubation temperature, and phytohormone concentrations in the medium quantitatively affected the amount of tissue colonization. Thus, before a tissue culture system can be useful in studying host-pathogen interactions, one must identify the level of genetic control and the influence of environmental factors. Initial work carried out in our laboratory to study host-pathogen interaction involved use of callus cultures and culture filtrate from aggressive and nonaggressive strains of *O. ulmi*. Later on we used the fungus itself on callus cultures and *in vivo* to determine host-pathogen interactions. These studies were conducted to determine if a reliable tissue culture screening system to detect *O. ulmi* resistance could be developed.

Studies With Culture Filtrates

Production of a toxin, cerato-ulmin, and a wilt-inducing glycopeptide (peptidorhamnomannan) by the fungus, *O. ulmi*, is well known (Nordin and Strobel 1981; Takai 1974). These compounds were present in the culture filtrate used in these studies. The plant material used included susceptible American elm and two other American elm selections, Delaware #2 (DEL2) and #8630, classified as intermediate-resistant and resistant, respectively. The fungal isolates used were PMPl and H961, two aggressive isolates; and PG442, a nonaggressive isolate. Leaf tissue was used for development of callus cultures. Callus samples of each selection were tested on a medium containing 1, 5, 10, 20, and 50% culture filtrate with a final volume of half strength modified Murashige and Skoog (MS) medium (Murashige and Skoog 1962). Samples of callus for transmission electron microscopy (TEM) were prepared according to established protocol. The details for these procedures have been described elsewhere (Pijut et al. 1990a; Pijut et al. 1990b).

The results indicated that the culture filtrates of all isolates tested, suppressed callus growth of the susceptible elm, as concentration of culture filtrate was increased. Growth was reduced about one-half when the filtrates were added as 50% of the total medium volume for all isolates. Growth inhibition was less severe at 1 or 5% culture filtrate concentration; however, callus growth was also significantly reduced when 10 or 20% filtrate was added.

The culture filtrate of the aggressive isolate, H961, significantly reduced callus fresh weight at all concentrations. Both the nonaggressive PG442 and the PMPl isolates significantly reduced callus growth at 10, 20, and 50% dilutions. Callus fresh weight of DEL2 was significantly reduced at 50% filtrate concentration of isolate H961. However, isolates PMPl and PG442 had no discernible effect on callus growth. Callus fresh weight of American elm #8630 was not significantly decreased in the presence of culture filtrate from any isolate at any concentration.

Callus cultures of susceptible elm exposed to culture filtrate of PMPl, displayed a decrease in cell viability. No such loss in viability of cells from DEL2 and #8630 was observed. Stem cuttings of the susceptible elm displayed DED wilt-like symptoms 48 h after cut ends were placed in culture filtrate. Leaves wilted, curled inward, became necrotic and desiccated. Cuttings from DEL2 showed some symptoms, whereas #8630 did not exhibit any DED foliar symptoms.

TEM studies were carried out with callus tissue of susceptible elm and resistant elm #8630. Callus tissues were exposed to culture filtrate at 50% concentration. Cultured cells from control susceptible and resistant American elms were similar in composition to meristematic parenchyma cells, with a complement of organelles, defined cell walls, membranes, middle lamella, large central vacuole, etc. Analysis of cells from susceptible elm exposed to culture filtrate revealed that the cytoplasm was disrupted and there was an increase in vacuolation. Although the nucleus seemed intact, the mitochondria were swollen, the outer membrane was disrupted, and cells were devoid of organelles. Plasmolysis was evident, and the middle lamella pulled away from the cell wall, which was sometimes incomplete. As with the susceptible elm, treated cells of resistant elm #8630 were highly vacuolated and lacked organization. Approximately 40% of the cells contained an electron-dense precipitate in the vacuoles. Starch grains were abundant and there was evidence of damage to the cell wall. An unusually heavy precipitate was present along the plasmalemma and frequently found along the exterior of the cell wall.

The damage to mitochondria and loss of internal structure may be attributed to the fact that they were the probable sites of toxin action (Aldrich et al. 1977; Miller and Koeppe 1971). Ouellette (1978a) showed that infection by *O. ulmi* consistently results in cell wall disruption and breakdown. Presence of phenolic-like deposits, which appeared as electron dense particles, has been suggested as a resistance response (Krause and Wilson 1972; Ouellette 1978a; Ouellette 1978b) and as a defensive mechanism.

Studies with the fungus *O. ulmi*

As a continuation of the above work with culture filtrates, we studied the direct interaction between *O. ulmi* and callus tissue. Preliminary studies used callus from American elm genotypes with varying degrees of susceptibility to DED, resistant Siberian elm (*Ulmus pumila* L.) and a non-host petunia (*Petunia* x hybrida) to determine *in vitro* criteria correlation with known levels of resistance and susceptibility of intact plants. These experiments established reproducible techniques and a rating system to determine the effect of variables on interaction between the fungus and host callus.

The fungal isolates of *O. ulmi* included aggressive PMPl and nonaggressive TN. The rate of fungal growth on callus cultures was studied to determine the role of variables such as, inoculum level, temperature, explant source for callus initiation, presence (MMSC) or absence of coconut milk (MMS) in the medium, and frequency of subculturing of callus tissue. The experimental procedures used in these studies have been described elsewhere (Domir et al. 1991; Domir et al. 1992).

Results showed that isolate PMPl grew most rapidly on resistant American elm #8630 followed by the susceptible American elm and Siberian. Increase in inoculum level resulted in an enhanced rate of fungal growth on callus tissue and on medium without callus tissue. Of the three temperatures evaluated, 22 C was the optimum temperature for fungal growth followed by 16 and 28 C. Significant interaction was evident between callus line and temperature. At 16 and 28 C rates of fungal growth remained constant; however, at 22 C PMPl grew more rapidly on susceptible American elm than on #8630 or Siberian.

In the absence of callus, PMPl grew most rapidly at 22 C followed by 16 and 28 C. Fungal growth rate was equal at 16 and 28 C on both susceptible elm and resistant #8630; but with Siberian elm, the callus tissue was colonized more rapidly at 16 than at 28 C. Thus, temperature differentially affects growth depending on the callus line. Even though isolate PMPl grew more rapidly on resistant #8630 than on susceptible American elm, fungal density was greater on the latter. Callus from #8630 turned lighter in color after inoculation. Such color changes have been reported by others in association with a hypersensitive, resistance reaction (Haberlach et al. 1978; Miller et al. 1984).

Two explant sources, leaf and stem tissue, were used to develop callus cultures. Stem tissue was used as an explant source for callus since DED is a vascular and not a foliar disease. Explant source did not affect the rate of fungal growth on callus tissue. But significant interaction was observed between explant source and the elm selection. Fungal growth was more rapid on callus derived from stem tissue than from leaf tissue of Siberian elm, but no such differences were observed with susceptible American elm and resistant #8630. This suggests that significant genetic differences may exist among various selections overshadowing the response to explant source. In experiments using scanning electron microscopy (SEM) no difference in fungal density was observed on callus from leaf or stem tissue.

The fungal growth rate on callus tissue from susceptible American elm or Siberian elm cultured either on MS medium with coconut milk (MMSC) or without coconut milk (MMS) was not significantly different. However, the growth rate on #8630 cultured on MMSC was significantly less than on MMS. SEM studies indicated that maximum fungal sporulation occurred on susceptible American elm followed by Siberian elm, and was absent on #8630. In addition mycelial growth was less dense on Siberian elm than on susceptible American elm or #8630. Isolate PMPl colonized #8630 more densely when cultured on MMS than on MMSC.

TEM analysis revealed that uninoculated callus cells, from susceptible elm had a normal complement of organelles. Cell walls appeared normal with intact plasmalemma, nucleus, and mitochondria. Inoculated susceptible American elm callus had cells with extensive vacuolation, phenolic-like deposits, and plasmolysis. Conidia were present in abundance on the callus surface. SEM observations revealed that the fungus penetrated to a depth of more than 75 cell layers within 60 hr after inoculation. Hyphae penetrated directly into callus cells. Inoculated resistant #8630 showed some cell wall damage and phenolic-like deposits. However, cell membranes, mitochondria, and nucleus appeared intact. SEM showed that cells had not collapsed and remained unaffected even at the inoculation site. The fungus did not penetrate the cells and sparse hyphae were noticed in the intercellular spaces. The fungus penetrated to less than 20 cell layers. Conidial formation was minimized in close proximity to the phenolic-like deposits.

Interactions between frequency of subculturing and growth of *O. ulmi* on callus tissue.

Researchers have reported mansonones, a type of phytoalexin, are produced in elms in response to fungal infection; and that American elms can be induced to develop resistance to infection by *O. ulmi* as a result of production of these compounds (Elgersma and Overeem 1971; Hubbes and Jeng 1981). Mansonones are present in higher amounts in resistant Siberian elm than in susceptible American elm (Duchesne et al. 1985). Hubbes and Jeng (personal communication) indicated that frequent subculturing of elm callus cultures results in loss of mansonones. Therefore a study was conducted to determine if frequent subculturing of callus would affect growth of *O. ulmi* isolates PMPl and TN.

One set of callus tissue was maintained at 4 C and not subcultured; the other set was maintained at 22 C and subcultured at six-week intervals. Six months after initiation, callus lines were inoculated. The results indicated that no definite trends existed regarding growth of PMP1 and TN isolates between regularly subcultured and nonsubcultured callus tissue generated from susceptible American elm, DEL2, #8630, and Siberian elms. Fungal growth on callus tissue is independent of the frequency of subculturing. Thus, we cannot substantiate that biochemical changes occurring in frequently subcultured tissue reduce production of antifungal products. The data indicated that callus culture is not homogeneous and may undergo somaclonal variation whether callus tissue is subcultured or not.

High performance liquid chromatography (HPLC)

Analysis of soluble aromatic metabolites by HPLC showed that an extract from leaf-derived callus tissue of susceptible American elm had some, but not all, characteristics found with leaf mesophyll, leaf mid-rib, and stem tissue. A major metabolite, eluting at 11 min, was detected in all four samples. Chromatograms from explants had features distinct from each other as well as from the leaf-derived callus. However, chromatograms from leaf- and stem-derived callus showed that they produced similar secondary metabolites. Analysis of extracts from PMPl-infected callus tissue of the resistant #8630 indicated presence of a chemical eluting at 26.5 min. This chemical was absent from susceptible American elm. It is hypothesized that this compound may be related to resistance in #8630. However, it is not related to mansonones which were eluted between 11-19 min (unpublished data). In summary, it appears that in order to use a tissue culture system to correlate with an *ex vitro* system, environmental and biological variables must be standardized. Additional studies are in progress to determine the feasibility of using this technology to predict the resistance in mature trees.

References

Aldrich, H.C., V.E. Gracen, D. York, E.D. Earle and O.C. Yoder. 1977. Ultrastructural effects of *Helminthosporium maydis* race T-toxin on mitochondria of corn roots and protoplasts. Tissue Cell. 9:167-177.

Brettell, R.I.S. and D.S. Ingram. 1979. Tissue culture in the production of novel disease-resistant crop plants. Biol. Rev. 54:329-345.

Budde, A.D. and J.P. Helgeson. 1981. Phytoalexins in tobacco callus tissues challenged by zoospores of *PhytoPhthora Parasitica* var. *nicotianae*. Phytopathology 71:206.

Deaton, W.R., G.J. Keyes, and G.B. Collins. 1982. Expressed resistance to black shank among tobacco callus cultures. Theor. Appl. Genet. 63:65-70.

Dixon, R.A. 1980. Plant tissue culture methods in the study of phytoalexin induction. In Tissue Culture Methods for Plant Pathologists. Ingram, D.S. and Helgeson, J.P. (Eds.). pp. 185-196. Blackwell Scientific. Oxford.

Domir, S.C., L.R. Schreiber, and J.M. Ichida. 1991. Factors affecting growth of *Ophiostoma ulmi* on elm callus tissue. J. Environ. Hort. 9:211-215.

Domir, S.C., L. R. Schreiber, J. M. Ichida, and S.M. Eshita. 1992. Effect of elm selection, explant source, and medium composition on growth of *Ophiostoma ulmi* on callus cultures. J. Environ. Hort. 10:59-62.

Duchesne, L.C., R.S. Jeng, and M. Hubbes. 1985. Accumulation of phytoalexins in *Ulmus americana* in response to infection by a nonaggressive and an aggressive strain of *Ophiostoma ulmi*. Can. J. Bot. 63:678-680.

Elgersma, D.M. and J.C. Overeem. 1971. The relation of mansonones to resistance against Dutch elm disease and their accumulation, as induced by several agents. Neth. J. Plant Pathol. 77:168-174.

Haberlach, G.T., A.D. Budde, L. Sequeira, and J.P. Helgeson. 1978. Modification of disease resistance of tobacco callus tissue by cytokinins. Plant Physiol. 62:522-525.

Heybroek, H.M. 1970. Three aspects of breeding trees for disease resistance. In 2nd World Consultation on Forest Tree Breeding, Proceedings 1, FAO, Rome. pp. 519-535.

Holliday, M.J. and W.L. Klarman. 1979. Expression of disease reaction types in soybean callus from resistant and susceptible plants. Phytopathology 69:576-578.

Hubbes, M. and R.S. Jeng. 1981. Aggressiveness of *Ceratocystis ulmi* strains and induction of resistance in *Ulmus americana*. Eur. J. For. Pathol. 11:257-264.

Krause, C.R. and C.L. Wilson. 1972. Fine structure of *Ceratocystis ulmi* in elm wood. Phytopathology 62:1253-1256.

MacDonald, W.L., and H.S. McNabb. 1974. Electron microscope observations of *Ceratocystis ulmi* - induced tylosis development in *Ulmus*. Eur. J. For. Pathol. 4:2-10.

Miller, R.J. and D.E. Koeppe. 1971. Southern corn leaf blight: Susceptible and resistant mitochondria. Science 173:67-69.

Miller, S.A. and D. P. Maxwell. 1983. Evaluation of disease resistance. In Handbook of Plant Cell Culture. Vol. 1, Techniques for propagation and breeding. D.A. Evans, W.R. Sharp, P.V. Ammirato, and Y. Yamada (Eds). Macmillan, NY. pp. 853-879.

Miller, S.A., L.C. Davidse, and D.P. Maxwell, 1984. Expression of genetic susceptibility, host resistance, and nonhost resistance in alfalfa callus tissue inoculated with *Phytophthora meqasperma*. Phytopathology 74:345-348.

Murashige, T. and F. Skoog. 1962. A revised medium for rapid growth and bioassay with tobacco tissue cultures. Physiol. Plant. 15:473-497.

Nordin, J.H. and G.A. Strobel. 1981. Structural and immunochemical studies on the phytotoxic peptidorhamnomannan of *Ceratocystis ulmi*. Plant Physiol. 67:1208-1213.

Ouellette, G.B. 1978a. Fine structural observations on substances attributable to *Ceratocystis ulmi* in American elm and aspects of host cell disturbances. Can. J. Bot. 56:2550-2566.

Ouellette, G.B. 1978b. Ultrastructural observations on pit membrane alterations and associated effects in elm xylem tissues infected by *Ceratocystis ulmi*. Can. J. Bot. 56:2567-2588.

Pijut, P.M., S.C. Domir, R.D. Lineberger and L. R. Schreiber. 1990a. Use of culture filtrates of *Ceratocystis ulmi* as a bioassay to screen for disease tolerant *Ulmus americana*. Plant Sci. 70:191-196.

Pijut, P.M., R.D. Lineberger, S.C. Domir, J.M. Ichida, and C.R. Krause. 1990b. Ultrastructure of cells of *Ulmus americana* cultured *in vitro* and exposed to the culture filtrate of *Ceratocystis ulmi*. Phytopathology 80:764-767.

Takai, S. 1974. Pathogenecity and cerato-ulmin production in *Ceratocystis ulmi*. Nature 252:124-126.

Went, J.C. 1954. The Dutch elm disease summary of fifteen years hybridization and selection work (1937-1952). Tijdschr. Plziekt 60:109-127.

15
Biological Control of Dutch Elm Disease by Exploiting Resistance in the Host

D.M. Elgersma, T. Roosien and R.J. Scheffer

Introduction

Biological control of Dutch elm disease, caused by the ascomycete *Ophiostoma ulmi* (Buisman) Nannf., has been extensively studied in the Netherlands using *Pseudomonas spp.,* non-aggressive isolates of the Dutch elm disease fungus or isolates of selected fungi that are non-pathogenic for elms (Scheffer 1983; Scheffer 1984; Scheffer 1989; Scheffer 1990; Scheffer et al. 1989; Scheffer et al. 1980; Scheffer et al. 1989). This article reviews the results of the Dutch projects till the recent approval by Dutch authorities of a commercial product. Using *Pseudomonas* bacteria in attempts to control Dutch elm disease it was concluded that only preventive treatments had a prophylactic effect against Dutch elm disease, but that the methods employed, and the elm clone or species distinctly influenced the ultimate effect of the treatment. In Commelin elms which are resistant against the non-aggressive isolate of the pathogen, a good control was achieved, but in field elm (*U. carpinifolia)* no suppression of Dutch elm disease symptoms was obtained (Scheffer 1983; Scheffer 1984). The bacterial isolates used varied from a potent producer of antimycotic substances *in vitro* to a low producer, in this respect without differences in the capacity of suppressing the development of Dutch elm disease symptoms. This observation and the clone depending effect suggest that the major mechanism explaining this biological control is most probably related to induced resistance in the host tree rather than to a direct antibiosis between bacterium and pathogen. The effective suppression of Dutch elm disease symptoms after preventive treatment by a non-aggressive strain of *Ophiostoma ulmi* in elms which are resistant to this strain but not to the aggressive strain (*O. novo-ulmi*) and the equally effective suppression of symptoms by using a strain of *Verticillium dahliae* in the very susceptible field elm, considering the methods used, can only lead to the conclusion that induced resistance in the tree is the cause of this strong protective effect (Scheffer 1990). Stimuli inducing resistance and mechanisms causing this effect in elm are, however, still unknown. An important question yet to be answered was: is the treatment still effective during the following years. To obtain insight in a possible long lasting effect a field plot was designed using the very susceptible Belgica clone, diameter about 10 cm at breast height. The *V. dahliae* strain was used as a control agent. Methods and disease index recording were performed according to Scheffer (Scheffer 1990). In addition to a challenge inoculation one week after treatment with *V. dahliae,* a group of trees was inoculated with *O. novo-ulmi* one year after treatment with *V. dahliae* (table 1).

Table 1. Clone Belgica trees were challenged after one week or one year after treatment with *V. dahliae* . Disease index recorded after two months and one year (numbers between brackets).

Treatments	Number of trees	Disease index 0	1	2	3	4	5
water, *O. novo-ulmi* after 1 week	12	0 (3)	0 (0)	0 (0)	2 (0)	8 (1)	2 (8)
V. dahlia, O. novo-ulmi after 1 week	12	7(10)	3 (1)	2 (0)	0 (0)	0 (0)	0 (1)
water, *O. novo-ulmi* after 1 year	11	11 (0)	0 (0)	0 (0)	0 (0)	0 (0)	0 (11)
V. dahliae, water	12	10 (12)	2 (0)	0 (0)	0 (0)	0 (0)	0 (0)
water, water	8	8 (8)	0 (0)	0 (0)	0 (0)	0 (0)	0 (0)
V. dahliae, O. novo-ulmi after 1 year	12	9 (0)	2 (0)	1 (0)	0 (0)	0 (0)	0 (12)

It was shown clearly that one year after treatment with *V. dahliae* no protective effect existed anymore. Another important observation was that the protective agent (*V. dahliae*) could not be isolated from the tree after one year, even not at the size of application. Discoloration of the xylem caused by *V. dahliae* was limited to the annual ring of the year of treatment. The promising results of biological control of Dutch elm disease by preventive treatments with the *V. dahliae* isolate WCS850 and the low costs and limited (if any) environmental impact compared to classical (chemical) approaches, was reason for the Dutch Heidemij company to apply for registration of this isolate for control of Dutch elm disease in the Netherlands. In April 1992 this resulted in the approval of the authorities to market a product from the 1992 growing season on. In order to obtain this registration the Dutch Heidemij Company was demanded to perform experiments with trees of different clones and at various locations in the Netherlands. Results of three of these experiments are reported here. Table 2 shows the results of an experiment at Dronten, using large Vegeta trees, 45-50 cm diameter at breast height. Trees were treated with V. *dahliae* as described earlier (Scheffer 1990), but with the modification that the distances between the injections were either 7.5,15 or 30 cm. Challenge inoculations were performed one week later into two or four wrist thick shoots in the crown of the trees. Disease indices were recorded after two months and after one year. After one year, however, several trees were already removed by the local authorities because of severe disease development.

Table 2. Experiment performed with Vegeta trees at Dronten. Challenge inoculation with *O. novo-ulmi* (isolate H106) was performed one week after treatment with *V. dahliae* (isolate WCS 850). Disease recordings were made after two months and one year (numbers between brackets). Distances between *V. dahliae* injections were either 7.5, 15 or 30 cm.

		Disease index						
treatments	Number of trees	0	1	2	3	4	5	trees removed
water, *O. novo-ulmi*	20	0 (4)	2 (0)	7(3)	10 (5)	1 (2)	0 (0)	6
water, -	20	14 (15)	5 (1)	1(0)	0 (0)	0 (1)	0 (1)	2
V. dahliae (15 cm),	20	17 (18)	2 (0)	0 (0)	1 (1)	0 (0)	0 (0)	1
V. dahliae (7.5 cm),	20	20 (19)	0 (1)	0 (0)	0 (0)	0 (0)	0 (0)	0
V. dahliae (30 cm), *O. novo-ulmi*	15	4(8)	1 (2)	10 (1)	0 (4)	0 (0)	0 (0)	0
V. dahliae (15 cm), *O. novo-ulmi*	20	14(18)	4(0)	2 (1)	0(1)	0(0)	0(0)	0
V. dahliae (7.5 cm), *O. novo-ulmi*	15	15(14)	0 (1)	0 (0)	0 (0)	0 (0)	0 (0)	0
V. dahliae (15 cm), *O. novo-ulmi* *	14	3(12)	8(1)	3(0)	0(1)	0 (0)	0(0)	0

* Four instead of two shoots were inoculated.

This experiment at Dronten showed a strong protective effect of the *V. dahliae* treatment. Shorter distances (7.5 cm) between *V. dahliae* injection places had a better effect than further apart (30 cm). Inoculation of four shoots instead of two had no influence on the protective effect caused by *V. dahliae*. Disease development after *V. dahliae* treatment alone was caused by rootgraft infection for which *V. dahliae* gives no protection. Table 3 shows the results obtained with clone Vegeta at Delft. Diameter of these trees was about 30 cm at breast height. Methods used were the same as in Dronten. Injections of *V. dahliae* were performed only at distances of 15 cm at breast height. Challenge inoculation was after one week and disease indices were recorded after two months. Directly after this recording of disease symptoms the trees were removed.

This experiment too showed the protective effect of *V. dahliae* treatment. Only 2 out 20 trees became unacceptably diseased. Table 4 shows the results obtained in Amsterdam, again with Vegeta trees, diameter at breast height about 30 cm.

This experiment was performed as in Delft, with injections of *V. dahliae* at 15 cm intervals. Disease indices were recorded after two months.

Table 3. Experiment performed at Delft with Vegeta trees. Distances between *V. dahliae* injections were 15 cm at breast height. Challenge inoculation one week after. Disease indices were recorded after one month.

treatments	number of trees	Disease index 0	1	2	3	4	5
-, *O. novo-ulmi*	20	0	1	6	12	1	0
V. dahliae, O. novo-ulmi	20	8	2	8	1	1	0
-, *V. dahliae*	20	19	1	0	0	0	0
-, water		10	0	0	0	0	0

Table 4. Experiment performed in Amsterdam with clone Vegeta. Methods according to the experiment at Delft (Table 3). Disease indices were recorded after two months.

treatments	number of trees	Disease index 0	1	2	3	4	5
-, *O. novo-ulmi*	15	0	0	0	5	10	0
V. dahliae, O. novo-ulmi	20	18	2	0	0	0	0
-, *V. dahliae*	20	19	1	0	0	0	0
-,water	20	20	0	0	0	0	0

This experiment showed a perfect protection against Dutch elm disease.

Conclusion

All experiments confirmed our earlier results that the *V. dahliae* isolate WCS 850 had a protective effect against a later challenge inoculation. Two decisive prerequisites exist, however: the treatment must be performed before infection occurs and the infection must be in the crown of the tree. In other words, symptoms due to root contact infections are not reliably controlled. If these two prerequisites are met, a nearly perfect control was achieved in experiments carried out since 1985.

References

Scheffer, R.J. 1983. Biological control of Dutch elm disease by *Pseudomonas* species. Ann. Appl. Biol. 103:21-30.

Scheffer, R.J. 1984. Dutch elm disease. Aspects of pathogenesis and control. Thesis, University of Amsterdam, the Netherlands.

Scheffer, R.J. 1989. *Pseudomonas* for biological control of Dutch elm disease. III. Field trials at various locations in the Netherlands. Neth. J. Plant Pathol. 95:305-318.

Scheffer, R.J. 1990. Mechanisms involved in biological control of Dutch elm disease. J. Phytopathol. 130:265-276

Scheffer, R.J., Elgersma, D.M. and Strobel, G.A. 1989. *Pseudomonas* for biological control of Dutch elm disease. II. Further studies on the localization, persistence and ecology of *Pseudomonas* isolates injected into elms. Neth. J. Plant Pathol. 95:293-304.

Scheffer, R.J., Heybroek, H.M. and Elgersma, D.M. 1980. Symptom expression in elms after inoculation with combination of an aggressive and non-aggressive strain of *Ophiostoma ulmi* . Neth. J. Plant Pathol. 86:315-317.

Scheffer, R.J., Elgersma, D.M., De Weger, L.A., and Strobel, G.A. 1989. *Pseudomonas* for biological control of Dutch elm disease. I. Labeling, detection and identification of *Pseudomonas* isolates injected into elms; comparison of various methods. Neth. J. Plant Pathol. 95:281-292.

16
Cloning of Pathogenesis-Related Genes From *Ulmus americana*

Ravindra K. Hajela, Lawrence Graham and
Mariam B. Sticklen

Abstract

Challenge by pathogens causes the induction of a variety of genes in plants. These genes code for several pathogenesis related (PR) proteins that are believed to have roles in disease resistance. Very little is known about molecular events following pathogenic challenge in elm (*Ulmus* spp.) by *Ophiostoma ulmi,* the causal organism of the Dutch elm disease (DED). We have used heterologous probes from poplar and bean to look for similar genes and their expression in elms. Preliminary evidence suggests that the elm genome has sequences related to some plant chitinase-like genes and that these are expressed. We have isolated several cDNA clones from a cDNA library, made from *O. ulmi* induced *Ulmus americana* NPS3-487 callus tissue, using these heterologous chitinase gene-like sequences as probes. Efforts are currently directed at the basic characterization of these clones.

Introduction

Dutch Elm Disease (DED) is caused by the fungal pathogen *Ophiostoma ulmi*. Brought from Europe in the early twentieth century, DED has resulted in the virtual destruction of most stands of the American elm (*Ulmus americana*) along the Eastern seaboard and Midwestern United States (Karnosky and Mickler 1986). Once an important part of the American timber industry, the American elm has proven one of the most susceptible elms to the DED pathogen. The virulence of the pathogen is directly related to its production of a toxic peptide, cerato-ulmin (CU; Takai 1974). CU is able to diffuse away from the site of infection, causing wilting and tissue death systemically. Several other species of elm, notably the Chinese elm (*U. parvifolia*) and the Siberian elm (*U. pumila*) are relatively resistant to DED (Karnosky and Mickler 1986), but the cause of this resistance is not understood.

Many plants respond to the presence of a fungal infection with the production of several pathogenesis-related proteins, among them peroxidases (Wyatt et al. 1991), thaumatin-like proteins (Pierpoint 1990), ribosomal inhibitors (Leah et al. 1991), glucanases (Meins and Ahl 1989), and chitinases (Benhamou et al. 1990; Graham and Sticklen 1992; Mauch et al. 1988a; Roby et al. 1988). Chitinases have been shown to be antifungal proteins. In herbaceous plants, chitinase activity is induced strongly during infection. In cucumber, steady state levels of chitinase mRNA rise as high as 600-fold in response to infections (Métraux and Boller 1986). Ethylene, a plant hormone which is produced during fungal infections, also induces chitinase activity (Mauch and Staehelin 1989; Memelink et al. 1990; Samac et al. 1990). Chitinase, alone or in concert with ß-1,3-glucanase, lyses the hyphal tips of invading fungi which possess chitin as part of their cell walls (Mauch et al. 1988b). This lysis has been demonstrated in vitro (Boller et al. 1983; Dunsmuir and Suslow 1989; Mauch et al. 1988b) and *in vivo* (Toyoda et al. 1991). Chitinases are not present solely in herbaceous plants, however. Chitinases have also been found in cell cultures of the spruce *Picea abies* in response to cell wall preparations from *Amanita muscaria* (Sauter and Hager 1989). Additionally, several wound induced mRNAs with homology to Class I chitinases (nomenclature of Shinshi et al., 1990) accumulate in poplar trees in response to physical wounding (Parsons et al. 1989). It is very possible, then, that chitinases fulfill the same proposed antifungal role in woody plants that they do in herbaceous plants.

Little is known about the molecular processes that occur within the American elm during the course of infection. For instance, it is not known whether *O. ulmi* infections can induce the accumulation of pathogenesis-related proteins in American elm as they do in tobacco or cucumber. It is possible that the resistance of Chinese and Siberian elms is the result of the local or systemic production by these elms of antifungal proteins such as chitinase. To address this hypothesis, we report here the isolation and preliminary characterization of several *O. ulmi*-induced *U. americana* cDNAs which hybridize to cDNAs encoding Class I (basic) chitinases from bean and poplar.

Callus tissue was used as the source of RNA because of the ease of recovering RNA relative to mature leaves. Callus lines were established as follows. Young elm leaves (*Ulmus americana* NPS3-487) were first surface sterilized with 20% commercial bleach for 20 min. then rinsed several times with sterile distilled water. Sterilized leaves were cut into 5 mm wide strips and laid on culture media (1/2 strength MS with vitamins (Sigma # M5519), 40g/L maltose, 100 mg/L casein hydrolysate, 100 mg/L thiamine, 2.5 μM 2,4,D, 1 μM kinetin, pH 5.7, supplemented with 100 ml/L coconut milk (Sigma) and solidified with 6.5g/L phytagar (BRL). In three to four weeks when the microcalli were formed at the edges of leaves, they were subcultured on fresh media and the process continued until fast growing, friable callus lines were established. The established callus lines were used as the source of RNA.

For putative induction, vigorously growing callus (3-4 weeks old plates, with about 10 gm tissue/plate) was overlaid with three to four day old cultures of *O.*

ulmi (virulent strain CFE16K and three avirulent strains Q311, Q412 and SSMF) on potato dextrose agar (PDA)-impregnated filter papers (PDA - Sigma, 39g/L) for 20-24 hrs. Treated calli were flash frozen in liquid nitrogen and pulverized to fine powder before being stored at -80 C.

Using this callus derived RNA, a unidirectional, cDNA library was custom constructed by Clontech in the following manner. Total RNA was isolated from this putatively 'induced' callus and poly A^+ mRNA fractionated using affinity matrix. Oligo dT (17 mer) attached to an Xho I linker was used to prime first strand cDNA synthesis. Double stranded cDNA was linkered with Eco RI linkers and the completed, end modified cDNA was cloned into Eco RI/Xho I opened bacteriophage λ Zap II (Stratagene). The recombinant phage was plated on *E. coli* Sure (Stratagene) cells and the library amplified one cycle.

Two heterologous probes were used to screen the 'induced' elm cDNA library. One probe was a 764 bp Eco RI cDNA cloned insert from win 6, a wound inducible Class I chitinase-like cDNA sequence from *Populus* (Parsosn et al. 1989). The other was a 1146 bp Eco RI *Phaseolus* Class I chitinase cDNA cloned fragment from pCH18 (Broglie et al. 1986). The deduced amino acid sequences of both these probes have considerable homology. Radiolabelled ($\alpha^{32}P$) probes were generated by the random oligonucleotide primed method (Feinberg and Vogelstein 1983) using a 14 mer based kit from NEN. Plaque lifts on Magna NT (MSI) were hybridized overnight in 6X SSPE at 60 C and washed at moderate stringency (same temperature in 3X SSPE). Autoradiography of washed membranes revealed about 0.2% of plaques giving positive signal of varying intensities.

Positive plaques were purified two to three cycles by plating at low density and doing plaque lift hybridization screening as described above until 99% of plaques showed positive signal. Twenty random selected plaques from each of the final screens were used to subclone the cDNA insert in pBluescript SK II- via biological rescue with the helper bacteriophage R418 (Stratagene). Four differently sized subclones from each were finally selected for study and named pHS1, 2, 3, 5, 8, 17, 18 and 19 in the chronological sequence of their isolation. Large scale plasmid preparations were made and purified using CsCl isopycnic gradients for use in probes and sequencing experiments. The subcloned inserts were excised by cutting with Eco RI/Xho I, gel purified by electroelution on NA45 DEAE membrane (S&S) and used for making radiolabelled probes.

Plaque lift screens yielded varying degrees of signal intensity. This observation is consistent with heterogeneity of cDNA insert size as well as different degrees of homology to the heterologous probe used. As it turned out later, both these factors may have been responsible for the observation. The inserts range in size from about 700 bp to almost 1300 bp (Figure 1). When Eco RI/Xho I excised fragments of all the clones were probed with the largest insert pHS2, all but pHS5 showed homology to the probe (Figure 2) at high stringency (0.1X SSPE at 65 C). All the inserts except pHS5 have an internal Eco RV site. In addition, only pHS17 has several Sac I sites (data not shown). These observations suggest that we have at least six families of clones as follows: 1, 2, 3, 5, 8/17, 18/19 (Fig. 3).

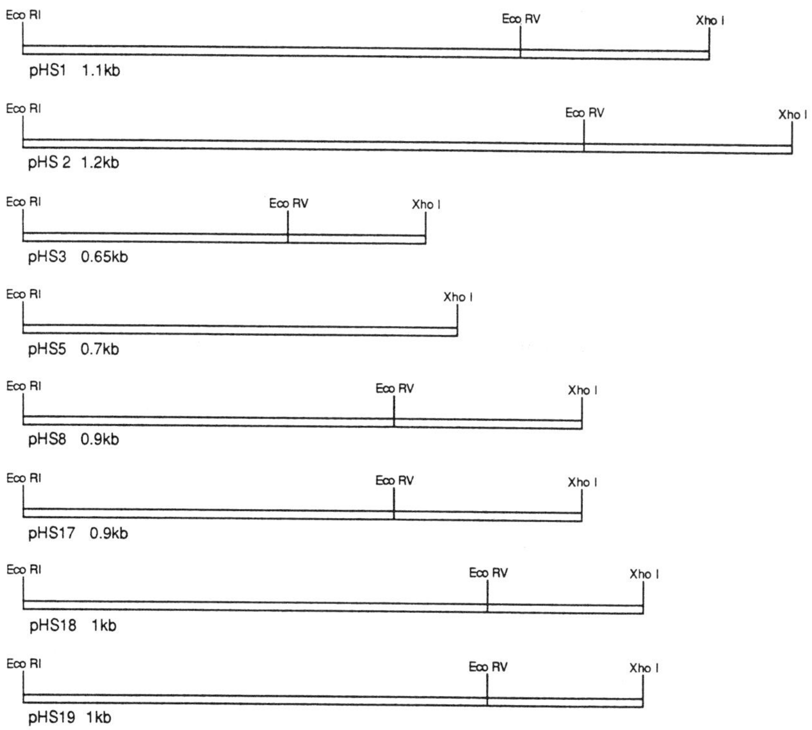

Figure 1. Relative sizes of the EcoRI/Xho I cDNA inserts subcloned in pBS SK II.

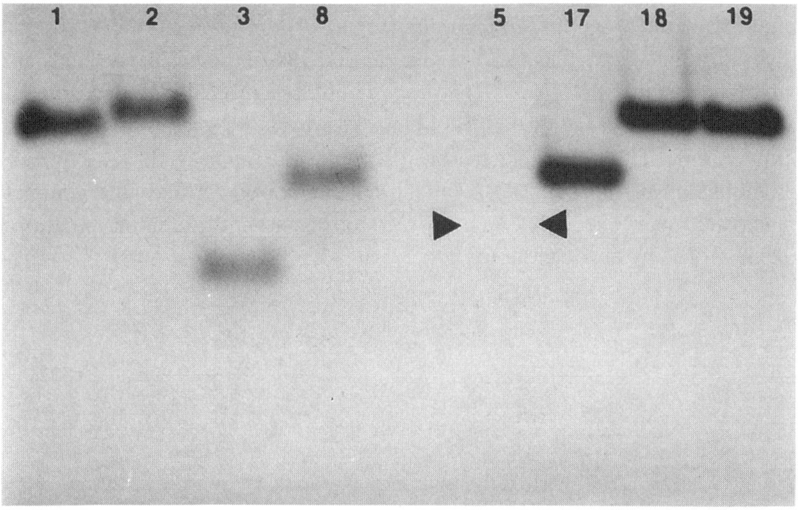

Figure 2. Autoradiogram of a Southern blot of EcoRI/Xho I excised inserts, probed with the insert from pHS 2.

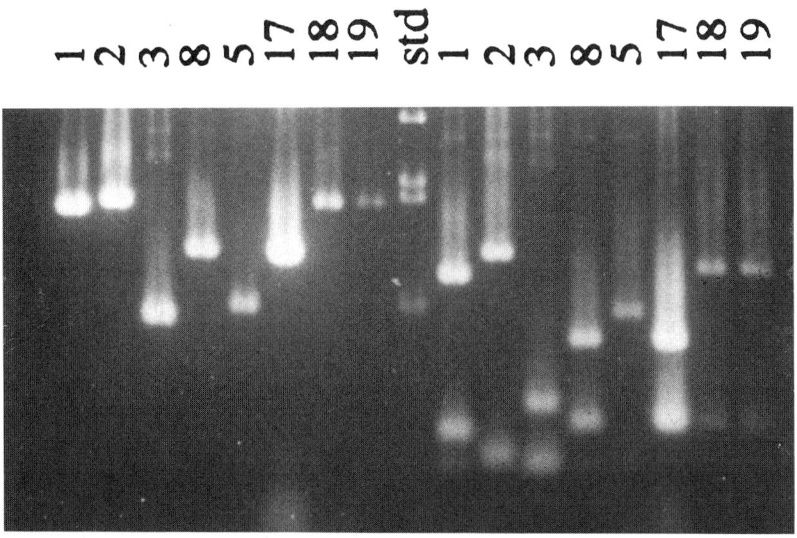

Figure 3. EcoRI/Xho I restriction pattern of clones pHS 1, 2, 3, 5, 8, 17, 18, 19 (left) and Eco R V digests of these EcoRI/Xho I excised inserts (right).

Total RNA was isolated from callus tissue of *Ulmus americana* NPS 3-487 and *U. pumila*, a DED resistant variety. Denaturing formaldehyde gel fractionated RNA was northern transferred to either Magna NT (MSI) or HybondMAP (Amersham) to maximize transfer of poly A+ RNA. The transfers were probed with insert from pHS2, which hybridized with an approximately 1200 nucleotide message in *U. americana* NPS3-487 and a larger, approximately 1600 nucleotide message in *U. Pumila* (Figure 4). However, there is some hybridization signal detected with the 28s and 18s ribosomal RNA bands also and the reason is not clear. The probe is not a ribosomal clone because the sequence of the 3' end of the pHS2 clone clearly shows a long (44 nucleotide) poly A tail. The only other reason for the hybridization could be sequences homologous to ribosomal RNA, a fact that can not be verified until complete nucleotide sequence is determined. Currently, northerns are in the process of being probed by the other inserts.

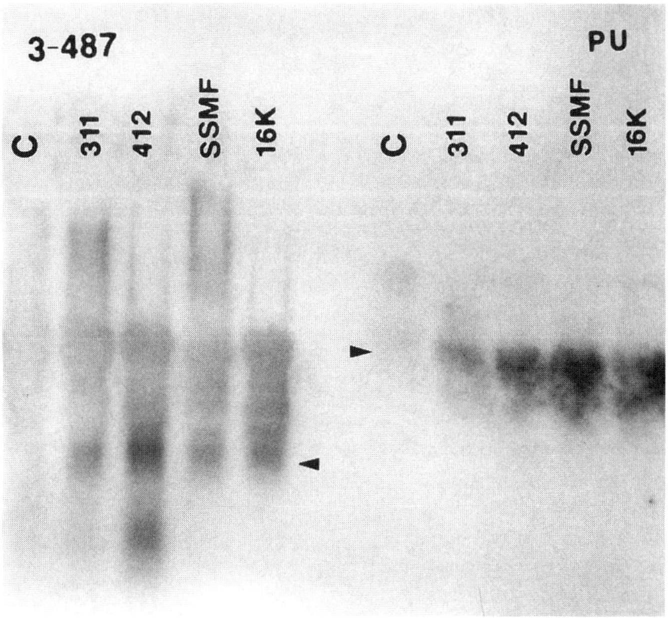

Figure 4. Autoradiogram of a northern blot of total RNA isolated from elm callus *U. americana* NPS 3-487 (left) or *U. pumila* (right): control (C) or 'induced' with *O. ulmi* isolates Q411, Q414. SSMF or CFE 16K. Gel was blotted on Hybond MAP poly T affinity membrane and probed with EcoRI/Xho I excised insert from pHS 2.

Elm DNA was isolated from callus tissue and sometimes from the tender, just opened first leaves of young twigs. These preparations were used for Southern analyses of various elms. When the insert from pHS2 was used as a probe (Figure 5) on Eco RI restricted DNA, American elm selections as well as *U. pumila* showed similarly sized band about 4 kb in size. However, Hind III digests of the same DNA samples showed heterogeneity (Figure 6). In this case, while all the hybridization signal was seen in very high molecular weight (larger than 9 kb) bands, the signal in *U. pumila* was with a slightly faster migrating band compared to American elms except NPS3-487 where an additional band slightly smaller than that in *U. pumila* was also seen. The significance of this observation is not clear and additional hybridizations with other probes are needed to determine if this heterogeneity of *U. pumila* and NPS3-487 compared to other elms is a generalized phenomenon.

DNA sequence determinations were made using purified double stranded plasmid DNA and Sequenase 2 (USB) sequencing kit using dideoxy chain termination method (Sanger et al. 1977) to resolve the differences and similarities between the clones. Both the 3' and 5' ends of inserts pHS 1, pHS2, pHS5 and pHS17 were sequenced (Figure 7). On the basis of the 3' sequence information we can confirm that the clones pHS1 and 2 are different and not varying lengths of same clone. Only 116 bp of pHS17 5' and 100 bp of the pHS2 5' have been sequenced to date. As the predicted length of mature chitinases is generally 30 - 35 kDa, it is highly probable that obtained sequences lie within the 5' untranslated region of the message.

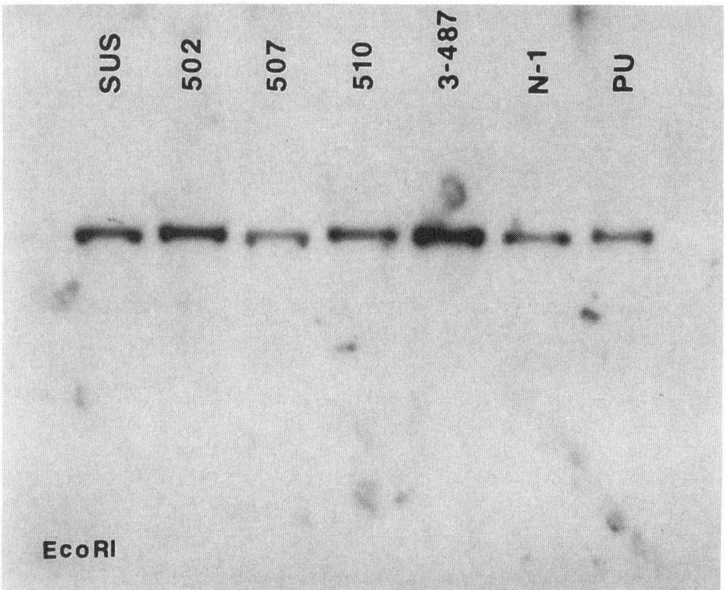

Figure 5. Autoradiogram of a Southern blot of total elm DNA restricted with EcoRI/Xho I excised insert from pHS 2.

The sequence at 3' end of the pHS1 clone shows a putative polyadenylation signal, AATAGG, is present 64 nucleotides from the start of the poly A tail (Figure 8). This sequence is preceded by a YAYTG-like motif (CACTT) that may orient the U4 snRNP for cleavage site selection (Joshi 1987). A YGTGTTYY-like motif (TGTGTGTA), which appears necessary for efficient polyadenylation (Joshi 1987), is present 37 nucleotides downstream of the 5' A of the AATAGG. The sequence AATAGG has rather low homology to the standard AATAAA or the AATAAG when one considers that these are highly conserved sequences, and this sequence is not seen in the 46 analyzed by Joshi (1987); thus, it is possible that this is not the true polyadenylation signal.

However, the existence of the other two motifs and their spatial arrangement argue that this is the correct signal.

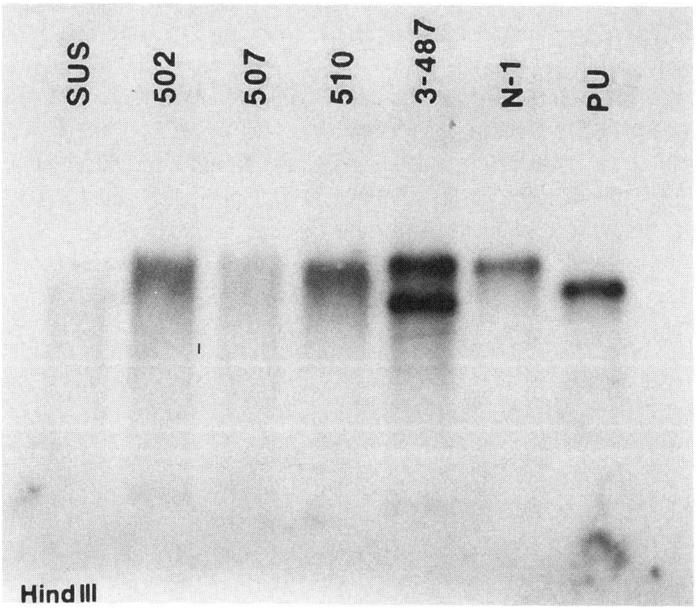

Figure 6. Autoradiogram of a Southern blot of total elm DNA restricted with Hind III and probed with EcoRI/Xho I excised insert from pHS 2.

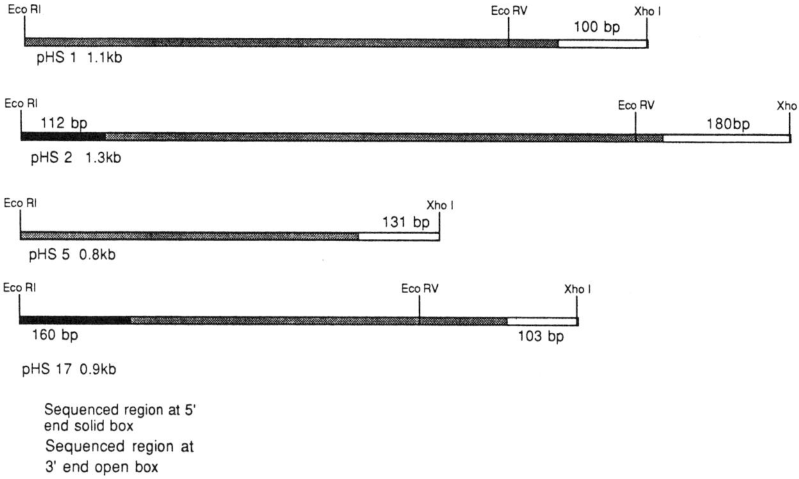

Figure 7. Diagrams of the four clones that have been sequencedfrom the 5' and 3' ends.

pHS1 3'

```
1  CCATCAACACTTGAATAGGTGGCGGTTAGCT

32 CATAGCTCATCAGAGAAAGATGTGTGTAGTC

63 ACTAATACTTTAATGAAAAAAAAAAAAAAAA
                     ***** *
          ****       TGTGTGTA
          CACTT      YGTGTTYY
          YAYTG      TGTGTAGT
                     *****  *
```

Figure 8. 3' terminal sequence of pHS 1 showing the putative polyadenylation signals.

Sequencing of the pHS2 3' end revealed a similar arrangement of polyadenylation motifs (Figure 9). The polyadenylation signal AATAAG is located 25 nucleotides upstream from the poly A tail. The polyadenylation signal AATAAG is not unusual in that it is also found in the maize zein gene (Joshi 1987). This signal is immediately preceded by CAATG, which has 80% homology to a YAYTG sequence. Additionally, a TGTGTGCAC sequence which is 75% homologous to the YGTGTTYY motif begins 15 nucleotides upstream of the poly A tail. This sequence. By visual inspection, the spatial arrangement of these three motifs in pHS2 appear more compact than those from many other plant genes.

pHS2 3'

```
1   AGATGTGTAGTTGTTATAATATCTGTGTTTTGAGGGTTTGGCTGG

46  TCGTTGTCGTCACTTCTTCGGTGACAATTTCGTTGTGTTATAGTT

91  ACTATATATGTCAATGGAATAAGAACAAAGGATGTGTGCACTACA

136 ACAAAAAAAAAAAAAAAAAAAAAAAAAAAAAAAAAAAAAAAAAAA

          ** **            ***** *
          CAATG            TGTGTGCA
          CAYTG            YGTGTTYY
```

Figure 9. 3' terminal sequence of pHS 2 showing the putative polyadenylation signals.

Analysis of the 3' end of clone pHS17 is somewhat more problematic (Figure 10). An AATAAG sequence is found beginning thirteen nucleotides from the start of the poly A tail, and three CAYTG-homologous sequences are present at nine, twenty, and fifty-two nucleotides from the 5' A of the AATAAG. The AATAAG, however, is much closer to the poly A tail than is found in most other plant genes (Joshi 1987) and there is a YGTGTTYY-homologous sequence upstream rather than downstream of the AATAAG; thus, it is probable that this AATAAG is not the true polyadenylation signal and that more extensive sequencing, already underway, is necessary to resolve the ambiguities. The most distal two YAYTG-homologous motifs are part of a perfect repeat of TAGTCACTA; the significance of this repeat is not known.

The evidence at this point that the clones are chitinase or chitinase-like genes, then, is solely that these clones hybridized to Class I chitinase clones from bean and poplar. If it is true that these are chitinase genes, then these enzymes possibly play as important a role in the antifungal defense of the elm tree as they do in the well-studied herbaceous plants. Further experiments are required to determine the degree, if any, to which chitinase mRNA accumulates locally and systemically to *O. ulmi* infections, and to what degree chitinase activities concomitantly increase as part of a pathogenesis response.

Chitinase has been an attractive enzyme to researchers wishing to modify plants genetically for increased disease tolerance. The protein is relatively small and compact, and acts directly on the invading organism, rather than producing a secondary metabolite with an antifungal activity (as do the enzymes responsible for lignin or phytoalexin production). Elevated levels of chitinase in transgenic tobacco do not appear to affect the plants' vigor, growth, or viability under noninfecting conditions. Furthermore, the production of disease-resistant plant varieties is becoming easier as several different chitinase genes have been cloned within the last three years (Table 1).

pHS17 3'

```
 1 CCGTTAGTCACTAGCCACTACGAGAAAGATGTGT

35 ATTAGTCACTAATACTTTAATGAGAAAATAAGTT

69 TTTCGAAAAAAAAAAAAAAAAAAAAAAAAAAAAA


        ****
        CACTA            ***** *
        YAYTG            TGTGTGCA
        TAATG            YGTGTTYY
        ** **
```

Figure 10. 3' terminal sequence of pHS 17 showing the putative polyadenylation signals.

Table 1. Isolated chitinase genes

Source of gene	Reference
Arabidopsis thaliana	Samac *et al.*, 1990
Azuki bean	Ishige *et al.*, 1990
Barley	Leah *et al.*,1991; Swegle *et al.*, 1989
Bean	Margis-Piniero *et al.*, 1991;Broglie *et al.*, 1989; Broglie *et al.*, 1986
Cucumber	Métraux *et al.*, 1989
Peanut	Herget *et al.*, 1990
Petunia	Linthorst *et al.*, 1990
Potato	Gaynor and Unkenholz,1989; Laflamme and Roxby, 1989; Stanford et al., 1989; Gaynor, 1988
Rice	Huang *et al.*, 1991
Tobacco	Payne *et al.*, 1990; Neale *et al.*, 1990; Linthorst *et al.*,1990; Shinshi *et al.*, 1990; Shinshi *et al.*, 1987

Several laboratories have produced transgenic plants bearing new chitinase genes driven by constitutive or tissue-specific promoters, and many of these plants show marked increases in resistance to fungal pathogens. For instance, transgenic tobaccos were produced by Broglie et al. (1989) using a CaMV 35S promoter fused to the bean CH5B gene. The transformed plants displayed a wide range of chitinase activities, with expression in some as much as 50-fold higher than others. The highest activities were found in the roots and leaves. In tests of homozygous transformants vs. the soil-borne pathogen *Rhizoctonia solani* , transgenic tobacco expressing high levels of chitinase grew faster, lost at least three-fold less root weight (15% vs. 46%), and had a lower seedling mortality rate (37% vs. 53%) than untransformed seedlings. Bacterial chitinase can also be expressed in flowering plants (Lund et al. 1989; Suslow et al. 1988). A *Serratia marcesens* exochitinase gene (ChiA; predicted MW = 58,000) has been expressed in tobacco under control of promoters from the CaMV 35S RNA gene, *Agrobacterium tumefaciens* nopaline synthase (nos) gene, petunia ribulose biphosphate carboxylase gene (Jones et al. 1988; Lund et al. 1989) or petunia chlorophyll a/b binding protein gene (Jones et al. 1988). Under the control of promoters that increased total chitinase activity by 30 - 40% over normal levels, ChiA conferred to transgenic tobaccos increased resistance to *Alternaria longipes* as seen by reduced necrotic lesions and chlorosis during infection (Suslow et al. 1988).

Despite the antifungal effect of many purified chitinases, there are difficulties to be overcome in designing overexpressing plant varieties. For example, the antifungal effect is dependent upon the species of fungus tested. Mauch et al. (1988b) found that hyphae of *Trichoderma viridae* alone are susceptible to lysis by a pea basic chitinase (as tested by the size of a clear zone around a filter disc containing 1.3 μg chitinase on a fungus-inoculated agar dish). Pathogens with commercial importance, such as *Fusarium solani*, *Alternaria solani*, and

Fusarium oxysporum are affected only by combinations of chitinase and ß-1,3-glucanase. A Class I chitinase from *Arabidopsis* is effective against *Trichoderma reesei*; it is not effective alone, however, against *Fusarium oxysporum, Alternaria solani, Sclerotina sclerotorium, Gaeummanomyces graminis* or *Phytophthora megasperma* (Verburg and Huynh 1991).

Also problematic is the discovery that not all plants with genetically engineered chitinase genes exhibit the desired increase in resistance to fungal pathogens. To test the hypothesis that overexpression of chitinase would enhance a plant's resistance to fungal pathogens, Neuhaus et al. (1991) placed a Class I chitinase gene from tobacco under control of the CaMV 35S promoter and transformed the construct into *Nicotiana sylvestris* in hopes of conferring resistance to *Cercospora nicotianae*, the Frog Eye disease pathogen. Transformants have chitinase activities ranging from 1-fold to more than 25-fold that of control plants while the activity of a control enzyme, ß-1,3-glucanase, is similar in controls and transformants. Transformants also appear to grow normally, though slightly more slowly than controls. The protection conferred by the transformed gene, however, are inconsistent, with some overexpressing plants demonstrating no increased resistance. Neuhaus et al. (1991) advanced several explanations for this lack of chitinase-mediated protection. The inoculum used was well above concentrations which would be encountered in the field; this alone could easily have overwhelmed even enhanced chitinase expression. Possibly extracellular rather than intracellular chitinases are more important in limiting this infection. Additionally, there is ample evidence that the effectiveness of chitinase lies in its synergistic action with ß-1,3-glucanase, which could have been a limiting factor.

Another possibility for the inconsistent results may lie with the choice of a tobacco chitinase for overexpression in tobacco. It is possible that, during the course of evolution, tobacco-specific fungal pathogens developed defenses specific to tobacco chitinases. The choice of a gene for a "foreign" chitinase, such as the *S. marcescens* chiA gene used in the study by Broglie et al. (1989), may prove to be a better choice.

It will become increasingly important to examine the response of American and resistant elms to challenges by *O. ulmi* to determine the difference, if any, in the production of pathogenesis-related proteins. In this way we can more fully understand the nature of the resistance that Siberian and Chinese elms, among others, have towards the DED pathogen. It may be prudent also at this time to perform initial studies geared towards the production of genetically modified American elms bearing 'foreign' chitinase genes. This approach is feasible now that callus lines of various elm species have been established and transformation techniques such as the biolistic system are becoming routine. There is still much to be discovered, however, about the complete physiological role chitinase plays in plant defense, a role that is still not fully elucidated in the best-studied systems. More work will be necessary, for example, to determine the most active chitinases, the ones that are located within the tissue to the best effect, and the ones to which *O. ulmi* is most susceptible. Despite these caveats, the successful increase in resistance of chitinase-enhanced transgenic plants in the hands of others points to a new and potentially powerful direction in elm resistance research.

References

Benhamou, Nicole, Matthieu H. A. J. Joosten, Pierre J. G. M. De Wit. 1990. Subcellular localization of chitinase and of its potential substrate in tomato root tissues infected by *Fusarium oxysporum* f.sp. radicis-lycopersici. Plant Physiol. 92:1108-1120.

Boller, T., A. Gheri, F. Mauch, and U. Vögeli. 1983. Chitinase in bean leaves: induction by ethylene, purification, properties, and possible function. Planta 157:22-31.

Broglie, K., Gaynor, J. J. and Broglie, R. M. 1986. Ethylene-regulated gene expression: Molecular Cloning of the genes encoding an endochitinase from *Phaseolus vulgaris*, Proc. Natl. Acad. Sci. USA 83:6820-6824.

Broglie, Karen E., Phyllis Biddle, Robert Cressman, and Richard Broglie. 1989. Functional analysis of DNA sequences responsible for ethylene regulation of a bean chitinase gene in transgenic tobacco. Plant Cell 1:599-607.

Dunsmuir, Pamela and Trevor Suslow. 1989. Structure and regulation of organ- and tissue-specific genes: Chitinase genes in plants. In Cell Culture and Somatic Genetics of Plants, Vol. 6, Academic Press, Inc. pp. 215-227.

Feinberg, A. P. and Vogelstein, B. 1983. A technique for radiolabelling DNA restriction endonuclease fragments to high specific activity. Anal. Biochem. 132:13.

Gaynor, John J. 1988. Primary structure of an endochitinase mRNA from *Solanum tuberosum*. Nucleic Acids Research 16(11):5210.

Gaynor, John J. and Karl M. Unkenholz. 1989. Sequence analysis of a genomic clone encoding an endochitinase from *Solanum tuberosum*. Nuc. Acids Res. 17(14):5855-5856.

Graham, Lawrence S. and Mariam B. Sticklen. 1992. Plant chitinases. In preparation.

Herget, Thomas, Jeff Schell and Peter H. Schrier. 1990. Elicitor-specific induction of one member of the chitinase gene family in *Arachis hypogaea*. Mol Gen Genet 224:469-476.

Huang, Jenq-Kuen, Lisa Wen, Mark Swegle, Hun-Cuong Tran, Tin H. Thin, Helen M. Naylor, S. Muthukrishnan and Gerald R. Reeck. 1991. Nucleotide sequence of a rice genomic clone that encodes a class I endochitinase. Plant Mol. Biol. 16:479-480.

Ishige, Fumiharu, Ken'ichi Yamazaki, Hitoshi Mori and Hidemasa Imaseki. 1991. The effects of ethylene on the coordinated synthesis of multiple proteins: accumulation of an acidic chitinase and a basic glycoprotein induced by ethylene in leaves of Azuki bean, *Vigna angularis*. Plant Cell Physiol. 32(5):681-690.

Jones, Jonathan D. G. Caroline Dean, David Gidoni, David Gilbert, Diane Bond-Nutter, Rino Lee, John Bedbrook and Pamela Dunsmuir. 1988. Expression of bacterial chitinase protein in tobacco leaves using two photosynthetic gene promoters. Mol. Gen. Genet. 212:536-542.

Joshi, C. P. 1987. Putative polyadenylation signals in nuclear genes of higher plants: a compilation and analysis. Nuc. Acids Res. 15(23):9627-9640

Karnosky, D. F. and A. Mickler. 1986. Elms. In: Biotechnology in Agriculture and Forestry Vol. 1: Trees, Bajaj, Y. P. S., ed. Springer-Varlag Berlin Heidelberg, pp. 326-340.

Laflamme, Daniel and Robert Roxby. 1989. Isolation and nucleotide sequence of cDNA clones ncoding potato chitinase genes. Plant Molecular Biology 13:249-250

Leah, Robert, Henrik Tommerup, Ib Svendsen and John Mundy. 1991. Biochemical and molecular characterization of three barley seed proteins with antifungal properties. J. Biol. Chem. 266(3):1564-1573.

Linthorst, Huub J. M., L. C. van Loon, Clemens M. A. van Rossum, Andreas Mayer, John F. Bol, Jeroen S. C. van Roekel, E. Josien S. Meulenoff, Ben J. C. Cornelissen. 1990. Analysis of acidic and basic chitinases from tobacco and petunia and their constitutive expression in transgenic tobacco. Molecular Plant-Microbe Interactions 3(4):252-258.

Lund, Peter, Rino Y. Lee and Pamela Dunsmuir. 1989. Bacterial Chitinase is modified and secreted in transgenic tobacco. Plant Physiol. 91:130-135.

Margis-Pinheiro, Marcia, Marie Hélène Metz-Boutigue, Abalo Awade, Marc de Tapia, Monique le Ret and Gérard Burkard. 1991. Isolation of a complementary DNA encoding the bean PR4 chitinase: an acidic enzyme with an amino-terminus cyteine-rich domain. Plant Molecular Biology 17:243-253.

Mauch, Felix, Lee A. Hadwiger and Thomas Boller. 1988a. Antifungal hydrolases in pea tissue: I. Purification and characterization of two chitinases and two ß-1,3-glucanases differentially regulated during development and in response to fungal infection. Plant Physiol. 87:325-333.

Mauch, Felix, Brigitte Mauch-Mani and Thomas Boller. 1988b. Antifungal hydrolases in pea tissue: II. Inhibition of fungal growth by combinations of chitinase and ß-1,3-glucanase. Plant Physiol. 88:936-942.

Mauch, Felix, L. Andrew Staehelin. 1989. Functional implications of the subcellular localization of ethylene-induced chitinase and ß-1,3-glucanase in bean leaves. Plant Cell 1:447-457.

Meins, Frederick, Jr. and Patricia Ahl. 1989. Induction of chitinase and ß-1,3-glucanase in tobacco plants infected with *Pseudomonas tabaci* and *Phytophthora parasitica* var. nicotianae. Plant Sci. 61:155-161.

Memelink, Johan, Huub J. M. Linthorst, Rob A. Schilperoort, J. Harry C. Hoge. 1990. Tobacco genes encoding acidic and basic isoforms of pathogenesis-related proteins display different expression patterns. Plant Molecular Biology 14:119-126.

Métraux , J. P. and Thomas Boller. 1986. Local and systemic induction of chitinase in cucumber plants in response to viral, bacterial, and fungal infections. Physiol. Mol. Plant Path. 28:161-169.

Métraux, J. P., William Burkhart, Mary Boyer, Sandra Dincher, Wayne Middlesteadt, Shericca Williams, George Payne, Michael Carnes and John Ryals. 1989. Isolation of a complementary DNA encoding a chitinase with structural homology to a bifunctional lysozyme/chitinase. Proc. Natl. Acad. Sci. USA 86:896-900.

Neale, Alan D., Jill A. Wahleithner, Marianne Lund, Howard T. Bonnett, Alan Kelly, D. Ry Meeks-Wagner, W. James Peacock and Elizabeth S. Dennis. 1990. Chitinase, ß-1,3-glucanase, osmotin, and extensin are expressed in tobacco explants during flower formation. Plant Cell 2:673-684.

Neuhaus, Jean-Marc, Patricia ahl-Goy, Ursula Hinz, Susan Flores and Frederick Meins, Jr. 1991. High-level expression of a tobacco chitinase gene in *Nicotiana sylvestris*. Susceptibility of transgenic plants to *Cercospora nicotianae* infection. Plant Molecular Biology 16:141-151.

Parsons, Thomas J., Harvey D. Bradshaw, Jr. and Milton P. Gordon. 1989. Systemic accumulation of specific mRNAs in response to wounding in poplar trees. Proc. Natl. Acad. Sci. USA 86:7895-7899.

Payne, G., P. Ahl, M. Moyer, A. Harper, J. Beck, F. Meins and J. Ryals. 1990. Isolation of complementary DNA clones encoding pathogenesis-related proteins P and Q, two acidic chitinases from tobacco. Proc. Natl. Acad. Sci. USA 87:98-102.

Pierpoint, W. S., P. J. Jackson, Rachel M. Evans. 1990. The presence of a thaumatin-like protein, a chitinase, and a glucanase among the pathogenesis-related proteins of potato *(Solanum tuberosum)*. Physiol. Mol. Plant Path. 26:325-338.

Roby, Dominique, Alain Toppan and Marie-Thérèse Esquerré-Tugayé. 1988. Systemic induction of chitinase activity and resistance in melon plants upon fungal infection or elicitor treatment. Physiol. Mol. Plant Path. 33:409-417.

Samac, Deborah, Cathy M. Hironaka, Pater E. Yallaly and Dilip M. Shah. 1990. Isolation and characterization of the genes encoding basic and acidic chitinase in *Arabidopsis thaliana*. Plant Physiol. 93:907-914.

Sanger, F., Nicklen, S. and Coulson, A. R. 1977. DNA sequencing with chain-terminating inhibitors, Proc. Natl. Acad. Sci. 74:5463.

Sauter, Margret and Achim Hager. 1989. The mycorrhizal fungus *Amanita muscaria* induces chitinase activity in roots and in suspension cultured cells of its host *Picea abies*. Planta 179:61-66.

Shinshi, Hideaki, Debra Mohnen and Frederick Meins, Jr. 1987. Regulation of a plant pathogenesis-related enzyme: Inhibition of chitinase and chitinase mRNA acumulation in cultured tobacco tissues by auxin and cytokinin. Proc. Natl. Acad. Sci. USA 84:89-93.

Shinshi, H., J.M. Neuhaus, J. Ryals, and F. Meins, Jr. 1990. Structure of a tobacco endochitinase gene: evidence that different chitinase genes can arise by transposition of sequences encoding a cysteine-rich domain. Plant Molecular Biology 14:357-368.

Stanford, Anne, Michael Bevan, Don Northcote. 1989. Differential expression within an family of novel wound-induced genes in potato. Mol. Gen. Genet. 215:200-208.

Suslow, T. V., D. Matsubara, J. Jones, R. Lee and P. Dunsmuir. 1988. Effect of expression of bacterial chitinase on tobacco susceptibility to leaf brown spot. Phytopathology 78(12 pt. 1):1556.

Swegle, Mark, Jenq-Kuen Huang, Grace Lee and Subbaratnam Muthukrishnan. 1989. Identification of an endochitinase cDNA clone from barley aleurone cells. Plant Mol. Biol. 12:403-412 .

Takai, Shozo. 1974. Pathogenicity and cerato-ulmin production in *Ceratocystis ulmi*. Nature 252:124-126.

Toyoda, Hideoshi, Yoshinori Matsuda, Takako Yamaga, Seiji Ikeda, Masahiro Morita, Takayuki Tamai and Seiji Ouchi. 1991. Suppression of the powdery mildew pathogen by chitinase microinjected into barley coleoptile cells. Plant Cell Reports 10:217-220.

Verburg, John G. and Q. Khai Huynh. 1991. Purification and characterization of an antifungal chitinase from *Arabidopsis thaliana*. Plant Physiol. 95:450155.

Wyatt, S. E., S. Q. Pan and J. Kúc. 1991. ß-1,3-glucanase, chitinase, and peroxidase activities in tobacco naturally resistant and susceptible to blue mold. Plant Physiol. 96(1 supp.):6.

17
Mansonones, Elicitors and Virulence

Martin Hubbes

Abstract

Mansonones may constitute an effective systemic defense system of elms against *Ophiostoma ulmi,* the causal agent of Dutch elm disease. They are produced in response to fungal invasion, cause at 0.1 mM cellular ion leakage and reduce the rate of respiration of the pathogen. Changes in the ultrastructure of the fungus by these compounds include membrane disruption, ribosomal aggregation and aggregation of electron dense material in mitochondria. There seems to exist a correlation between pathogen virulence and mansonone production. The more virulent the strains are, the less mansonone they induce in elm seedlings. Fungal culture filtrates, cytoplasm and cell walls of *O. ulmi* contain molecules that strongly stimulate elm callus cells to produce mansonones. The production of mansonones is influenced by elicitor concentration, the callus culture conditions and callus subculturing frequency. A 21,000 dalton glycoprotein, identified as mansonone-inducing elicitor in elm callus, was purified from culture filtrate using both isoelectrofocusing and HPLC ion exchange chromatography. Western blotting with the polyclonal antiserum against the elicitor reacted only with the glycoprotein. Work on the sequence of the elicitor and its isolation from the genomic library of the fungus is in progress. Experiments are also carried out to determine whether cerato-ulmin, a toxin produced by *O. ulmi* plays a key role in pathogen virulence.

Introduction

During the past years, my students, research associates, cooperators and myself have attempted to identify factors that are directly or indirectly correlated with host resistance or pathogen virulence to understand and control crucial mechanisms in the pathogenesis of Dutch elm disease. I believe that effective control of this most devastating disease of elms can only be achieved by isolation and subsequent directed rearrangement of genes controlling the virulence of *O. ulmi* and those genes governing the elm's defense system. This believe is derived from our hypothesis which states that induction, identification and genetic manipulation of the host's defense system on the one hand, and control of pathogen virulence on the other hand, will lead to a long-term, well-balanced co-existence between the host and its pathogen (Hubbes 1981; 1986; 1988). This co-existence is possible through the phenomenon of induced resistance, which is the ability of pathogen strains with low virulence to induce and mobilize the elms defense system (Hubbes 1981). This latter system may protect the trees against more virulent strains of *O. ulmi*. If in addition non-virulent strains could be produced that can cross with those of high virulence, then virulence may be further lowered since it has been shown that the virulence of the F_1 progeny is lower than that of their parents (Brasier and Gibbs 1976).

Mansonones

Cross-protection against aggressive strains of *O. ulmi* was reported in *Ulmus hollandica* (Scheffer et al. 1980) and *Ulmus americana* (Hubbes 1981). This type of defense termed also as induced resistance is a well known phenomenon in the plant kingdom (Chessin and Zipf 1990) and trees are no exception (Hubbes 1986). The mechanisms of induced resistance are not fully understood. Phytoalexins as oxidative stress metabolites seem to play an active role in this process (Keen et al. 1972; Kuc 1976; Kuc and Caruso 1977; Nebert et al. 1990) and lignification is considered to be one of the final steps in this mechanism (Hammerschmidt 1982; Moerschbacher 1986).

Experiments in our laboratory indicated that presence and accumulation of fungitoxic compounds, against *O. ulmi* in the American elm may be related to induced resistance (Jeng et al. 1983). This assumption has been further substantiated by Duchesne et al. (1985; 1986). The active compounds, mansonones A, C, D and G have been isolated by our group from inhibitory sapwood extracts of elm seedlings inoculated with *O. ulmi* (Dumas et al. 1983; Dumas et al. 1986; Jeng et al. 1983). Mansonone E and F, originally isolated from the West African tree *Mansonia altissima* Chev., were first observed to accumulate in elms infected with *O. ulmi* by Elgersma and Overeem (1971). These authors were unable to correlate mansonone accumulation and resistance to Dutch elm disease. Wu et al. (1989) showed that these compounds cause at 0.1 mM cellular ionone leakage and reduce the rate of respiration of *O. ulmi*. Mansonones also change the ultrastructure of the fungus by disrupting its cell membrane. Ribosomal aggregation and aggregation of electron dense material in the mitochondria of the pathogen is another effect of these chemicals.

Duchesne et al. (1985; 1986) showed a possible correlation of mansonone accumulation and strainal virulence. The less virulent isolates induced more mansonones than the more virulent ones. The more virulent the isolates were the less mansonones they induced in *U. americana* seedlings (Duchesne 1990). Procter and Smalley (1988) also observed increased mansonone accumulation in elm seedlings following inoculation with *O. ulmi*. Experiments by Szczegola-Derkacz (1988) showed that tissue cultures, grown from the cotyledons of elm seed, responded to fungal inoculations with production of mansonones. Autoclaved spores of the yeast phase of the fungus produced the same effect.

Szczegola-Derkacz (1988) also showed that callus grown from different seeds produced different amounts of mansonones. There were cell lines that produced high amounts of mansonones and there were cell lines that produced low amounts, in particular mansonone F. Yang et al. (1989) demonstrated that fungal culture filtrates, cytoplasm and cell walls of *O. ulmi* contain molecules that elicit mansonone accumulation in elm calli. The elicitors from the non-aggressive isolate Q311 cause a more rapid reaction than those from the aggressive isolate MH75. On callus of *U. americana,* which is susceptible to the pathogen, the elicitors from both isolates caused higher amounts of mansonone production than on callus grown from *U. pumila* which is considered resistant to the fungus. These results appear inconsistent with the results obtained from the seedling inoculations. There are many facts that may be responsible for this discrepancy. One important factor is the length of subculture intervals of the cell culture lines. Yang et al. (1989) and Yang (1991) noticed that with the length of subculture intervals the elicitors induced different levels of mansonones. Furthermore, when callus cell lines are treated with elicitors not all cells in the cell culture receive an equal amount of elicitor and therefore each cell responds different. This is the reason why today suspension cultures of plant cells are widely used to study many aspects of plant biochemistry (Gotthardt and Grambow 1992). However, screening of cell suspension as well as tissue cultures for a particular biological marker may not always produce the expected results. The composition of growth media in particular growth hormones greatly influence expression of particular traits (Duchesne et al. 1985; Hammerschlag 1989).

Another possible explanation for the discrepancy is that the fungal strains used in our experiments underwent during the long period that they have been kept in culture slight changes that are expressed in their physiological behavior. Recent inoculation experiments showed that the aggressive isolate MH75 does not produces any wilting symptoms when inoculated into elm seedlings (Dumas, personal communication). Additional inoculation experiments are carried out to test whether its virulence can be restored. Unfortunately several years ago, strainal characterization by methods of molecular genetics was not so well advanced in our laboratory as today. Therefore it is difficult to trace whether any molecular changes that occurred in this and the other isolates are due to cultural conditions. Results of such investigations may also lead to the identification of factors of pathogen virulence.

Elicitors

Efforts were made to isolate, purify and characterize the elicitors of *O. ulmi*. Although the elicitor isolated from the cell wall induced higher amounts of mansonones than those of the cytoplasm and culture filtrate, we decided to further purify and analyze the culture filtrate (CF) elicitor (Yang et al. 1989).

First, the relation between the age of the culture filtrate and the mansonone F-inducing activity of the CF elicitor of the aggressive MH75 and non-aggressive Q311 isolate was tested. The results showed two peaks of mansonone F induction for both types of CF elicitors. A small peak was induced by the CF elicitors after 2 days incubation reaching its maximum in 4 days old culture filtrate. The second, large peak was measured with CF elicitors isolated from 12 to 14 days old culture filtrate. It is interesting to note that this second peak probably coincides with the period of mycelial lysis of the fungus when grown in liquid culture (Gagnon 1961) and as seen later slight changes of the elicitor structure influence greatly the activity of the elicitor. Furthermore, the CF elicitor isolated from 4 days old culture filtrate of the non-aggressive isolate Q311 induced more mansonone accumulation in both the resistant *(U. pumila)* and susceptible *(U. americana)* elm calli than those of the aggressive isolate MH75. From older fungal cultures (after 10 days), however, the CF elicitor of MH75 possesses much stronger mansonone-inducing activity than that of Q311. These results indicate that the CF elicitor isolated from the filtrate of younger cultures probably relates somewhat better to the experimental results of the seedling inoculations. The experiments also show that the age of the culture filtrate has an influence on the elicitor activity which may reflect structural changes of the elicitor.

Further experiments on the nature of the elicitor showed that it is heat stable and partially affected by periodate treatment, proteinase and sugar hydrolytic enzyme digestion (Yang et. al 1989).

The treatment with ß-glucosidase showed that this enzyme has a profound effect on the activity of the elicitor. As seen from Table 1, the enzyme treatment decreases for example the elicitor activity by 25% of the non-aggressive isolate Q311 on *U. americana* calli, in contrast, on callus lines of *U. pumila* its activity is increased by 80%. A similar fate happens to the activity of the elicitor of the aggressive isolate MH75. The enzyme treatment increases the mansonone inducing activity of this elicitor on calli of *U. americana* by 900% and by 200% on *U. pumila.* From these experiments we assume that each callus line of the two elm species has either a different receptor system or their systems differ in their sensitivity to the respective elicitors. We assume further that the elicitors are strainal specific.

Efforts were made to purify the CF elicitor. First it was shown that the CF elicitor binds to a concanavalin A (Con A) - sepharose 4B column (Yang et al. 1989). However, the Con A bound fraction was still a mixture of several glycoproteins. Therefore further purification was carried out with an ion exchange column (protein pack DEAE-SPW 7.5 mm x 7.5 cm, Waters) on an HPLC system. In this way the purified elicitors of isolates Q311 and MH75 were used for the production of polyclonal antisera in New Zealand rabbits.

Further purification and characterization by analytical IEF polyacrylamide gel, two dimensional polyacrylamide gel electrophoresis, and western blot analysis revealed that a 21,000 dalton protein band is a main mansonone-inducing elicitor of the culture filtrate of the fungus.

This glycoprotein has been subjected to amino acid sequence analysis. Once these results are available the corresponding primers will be constructed to detect and isolate the elicitor gene(s) from the genomic library of the fungus.

Table 1. Sensitivity of the culture filtrate (CF) elicitor of *Ophiostoma ulmi* to different enzyme treatments bioassayed on elm calli

	Mansonone F production as % of CF elicitor control							
Callus line	*U. americana*				*U. pumila*			
Treatment	Q311	Q412	16K	MH75	Q311	Q412	16K	MH75
Control	100	100	100	100	100	100	100	100
Pronase E	58	80	28	11	36	35	28	23
Proteinase	42	8	53	60	7	16	15	10
α-Glucosidase	51	100	83	71	18	37	26	58
ß-Glucosidase	**75**	nd	nd	**932**	**823**	nd	nd	**200**
ß-Galactosidase	54	38	71	4	59	44	26	32
α-Mannosidase	44	nd	nd	115	38	nd	nd	64

Virulence

In the early 1970s, the observation for variation in aggressiveness among strains of *O. ulmi* gave rise to numerous assumptions to explain pathogenicity and virulence (Bernier 1983; Gibbs et al. 1972). The non-aggressive isolates induce slower development of foliar symptoms during the first year of infection, but this difference tends to disappear during the second year (Schreiber and Townsend 1976).

Pathogenicity is thought to be under the control of nuclear genes (Bernier 1983; Brasier and Gibbs 1976). Takai (1980) showed that there exists a correlation of *in vitro* synthesis of a toxin, cerato-ulmin (CU), by *O. ulmi* and aggressiveness of wild type isolates. Ipsen and Abul-Hajj (1982) demonstrated that *O. ulmi* produces CU also *in vivo*. When Bernier (1988) crossed a number of wild type isolates in the laboratory and analyzed the F_1 progeny for its toxin production, isozyme and protein patterns, he could not find a correlation between CU production and virulence. The analysis showed further that the observed correlation between isozyme pattern and aggressiveness (Bernier et al. 1983), and protein pattern and aggressiveness (Jeng and Hubbes 1983) of wild type isolates could not be confirmed by these experiments. These results, therefore, left the question, whether CU is the main virulence factor wide open. The only way to prove the role of CU as a key factor of virulence is to produce a number of CU-negative mutants. Although Bernier (1988) obtained some low CU producing mutants, he did not detect any CU-negative ones. Therefore efforts were made to identify and isolate the gene responsible for CU production (Bawden et al. 1992, unpublished data). Work on this aspect is in progress.

Other metabolites of *O. ulmi*, such as phenolic compounds, glycopeptides, polysaccharides and a polydisperse peptidorhamnomannan of 105- 120 Kd were investigated as possible pathogenicity and virulence determinants (Claydon et al. 1980; Nordin and Strobel 1981; Scheffer 1983; Scheffer et al. 1987). The affirmation of these compounds as virulence factors awaits further confirmation. A wealth of studies reported *in vitro* synthesis of cell wall degrading enzymes and speculated about their possible role in pathogenicity (Beckman 1956; Husain and Dimond 1958; Miller and Elgersma 1976; Ouellette 1962; Pomerleau 1970; Svaldi and Elgersma 1982). None of these assumptions have been affirmed conclusively. With the advent of new methods of molecular biology it seems now more possible than ever to resolve and clarify aspects of pathogen virulence and host resistance to finally control this most devastating disease of elms.

References

Beckman, C.H. 1956. Production of pectinase, cellulase and growth-promoting substance by *Ceratocystis ulmi*. Phytopathology 46:605-609.

Bernier, L. 1983. Relationships between isoenzymes, cultural characters, and pathogenicity in *Ceratocystis ulmi* (Buism.) C. Moreau. M.Sc.F. thesis, Department of Forestry, University of Toronto.

Bernier, L. 1988. Induction, characterization and mapping of mutations in *Ophiostoma ulmi,* the causal agent of Dutch elm disease. Ph.D. thesis, Department of Forestry, University of Toronto.

Bernier L., Jeng, R.S., and Hubbes, M. 1983. Differentiation of aggressive and non-aggressive strains of *Ceratocystis ulmi* by polyacrylamide gel electrophoresis of intramycelial enzymes. Mycotaxon 17:456-472.

Brasier, C.M., and Gibbs, J.N. 1976. Inheritance of pathogenicity and cultural characters in *Ceratocystis ulmi*: Hybridization of aggressive and non-aggressive strains. Ann. Appl. Biol. 83:31-37.

Chessin, M., and Zipf, A.E. 1990. Alarm systems in higher plants. The Botanical Review 56: 193-235.

Claydon, N., Elgersma, D.M., and Grove, J.F. 1980. The phytotoxicity of some phenolic metabolic products of *Ophiostoma ulmi* to *Ulmus* species. Nether. J. of Plant Pathol. 86:229-237.

Daub, M.E. 1986. Tissue culture and the selection of resistance to pathogens. Ann. Rev. Phytopathol. 24:159-186.

Duchesne, L.C., Jeng, R.S., and Hubbes, M. 1985. Accumulation of phytoalexins in *Ulmus americana* in response to injection by a non-aggressive and an aggressive strain of *Ophiostoma ulmi.* Can. J. Bot. 63:678-680.

Duchesne, L.C., Hubbes, M., and Jeng, R.S. 1986. Mansonone E and F accumulation in *Ulmus pumila* resistant to Dutch elm disease. Can. J. For. Res. 16:410-412.

Duchesne, L.C., Jeng, R.S., Hubbes, M., and Sticklen, M.B. 1990. Accumulation of mansonones E and F in seedlings of *Ulmus americana* in response to inoculation with *Ophiostoma ulmi.* Trees 4:187-190.

Dumas, M.T., Strunz, G.M., Hubbes, M., and Jeng, R.S. 1983. Isolation and identification of six mansonones form *Ulmus americana* infected with *Ceratocystis ulmi.* Experientia 39:1089-1090.

Dumas, M.T., Strunz, G.M., Hubbes, M., and Jeng, R.S. 1986. Inhibition of *Ceratocystis ulmi* by mansonones A, C, D, E, F, and G isolated from *Ulmus americana.* Europ. J. For. Path. 16:217-222.

Elgersma, D.M., and Overeem, J.C. 1971. The relation of mansonones to resistance against Dutch elm disease and their accumulation as induced by several agents. Neth. J. Plant Pathol. 79:218-220.

Gagnon, C. 1961. Growth of *Ceratocystis ulmi* (Buisman) C. Moreau in liquid culture media. Can. J. Bot. 39:1087-1093.

Gibbs, J.N., Heybroek, H.M., and Holmes, F.W. 1972. Aggressive strain of *Ceratocystis ulmi* in Britain. Nature 236:121-122.

Gotthardt, U., and Grambow, J. 1992. Near-isogenic wheat suspension cultures: establishment, elicitor induced peroxidase activity and potential use in the study of host/pathogen-interactions. J. Plant Physiol. 139:659-665.

Hammerschlag, F.A. 1989. Screening peaches *in vitro* for resistance to *Xanthomonas campestris pv. pruni*. J. Amer. Soc. Hort. Sci. 113(1):164-166.

Hammerschmidt, R., and Kuc, J. 1982. Lignification as a mechanism for induced systemic resistance in cucumber. Physiol. Plant Path. 20:61-71.

Hubbes, M. 1981. New directions in Dutch elm disease control. In: Joint meeting of the Canadian Phytopathological Society and the Canadian Society of Microbiology, Ottawa, Ontario, June 21-25, 1981. Canadian Phytopathology Colloquium. Can. J. Phytopath. 3:258-259.

Hubbes, M. 1986. Mechanisms of induced resistance in trees. XVIIIth IUFRO World Congress, Division 2, vol. II., Forest Plants and Forest Protection. p. 786-798. Yugoslavia, September 7-21, 1986.

Hubbes, M. 1988. Pathogen virulence and host reaction in Dutch elm disease. Naturaliste can. (Rev. Ecol. Syst.), 115:157-161.

Hubbes, M., and Jeng, R.S. 1981. Aggressiveness of *Ceratocystis ulmi* strains and induction of resistance in *Ulmus americana*. Europ. J. Forest Path. 11:257-264.

Husain, A., and Dimond, A.E. 1958. The function of extracellular enzymes of Dutch elm disease pathogen. Natl. Acad. Sci. Proc. 44:594-601.

Ipsen, J.D., and Abul-Hajj, Y.J. 1982. Fluorescent antibody technique as a means of localizing *Ceratocystis ulmi* toxins in elm. Can. J. Bot. 60:724-729.

Jeng R.S., and Hubbes, M. 1983. Identification of aggressive and non-aggressive of *Ceratocystis ulmi* by polyacrylamide gradient gel electrophoresis of intramycelial proteins. Mycotaxon 17:445-455.

Jeng, R.S., Alfenas, A.C., Hubbes, M., and Dumas, M. 1983. Presence and accumulation of fungitoxic substances against *Ceratocystis ulmi* in *Ulmus americana*: possible relation to induced resistance. Europ. J. For. Path. 13:261-178.

Keen, N.T., Partridge, J.E., and Zaki, A.I. 1972. Pathogen produced elicitor of a chemical defence mechanism in soybeans monogenically resistant to *Phytophthora megasperma* var. *sojae*. Phytopathology 62:768.

Kuc, J. 1976. Phytoalexins. In: Encyclopedia of Plant Physiology. Vol. 4, pp. 632-652. (R.H. Heitefuss and P.H. Williams, eds.), Springer-Verlag, Berlin, Heidelberg and New York.

Kuc, J., and Caruso, F.K. 1977. Activated coordinated chemical defense against disease in plants. pp. 78-89. In: ACS Symposium Series No. 62, Host Plant Resistance to Pests (P.A. Hedin, ed.), Amer. Chem. Soc.

Miller, H.J., and Elgersma, D.M. 1976. The growth of aggressive and non-aggressive strains of *Ophiostoma ulmi* in susceptible and resistant elms, a scanning electron microscopical study. Neth. J. Pl. Path. 82:51-65.

Moerschbacher, B.K., Kogel, H., Noll, U., and Reisener, H.J. 1986. An elicitor of the hypersensitive lignification response in wheat leaves isolated from the rust fungus *Puccinia graminis f. sp. tritici*. I. Partial purification and characterization. Z. Naturforsch. 41c:830-838.

Nebert, D.W., Petersen, D.D. and Fornace, A.J., Jr. 1990. Cellular responses to oxidative stress: The [Ah] gene battery as a paradigm. Environmental Health Perspectives 88: 13-25.

Nordin, J.H., and Strobel, G.A. 1981. Structural and immunochemical studies on the phytotoxic peptidorhamnomomannan of *Ceratocystis ulmi.* Plant Physiol. 1208-1213.

Ouellette, G.B. 1962. Morphological characteristics of *Ceratocystis ulmi* (Buism.) C. Moreau in American elm trees. Can. J. Bot. 40:1463-1466.

Pomerleau, R. 1970. Pathological anatomy of the Dutch elm disease. Distribution and development of *Ceratocystis ulmi* in elm tissues. Can. J. Bot. 48:2043-2057.

Procter, R.H., and Smalley, E.B. 1988. Localized accumulation of mansonones E and F in elms following inoculation with *Ophiostoma ulmi.* Can. J. Plant Pathol. 10:371.

Scheffer, R.J. 1983. Toxins in Dutch elm disease. In: Research on Dutch elm disease in Europe (D.A. Burekin, ed.). Forestry Commission Bull. 60, London. pp. 82-85.

Scheffer, R.J., Heybroek, H.M., and Elgersma, D.M. 1980. Symptom expression in elms after inoculation with combinations of an aggressive and non-aggressive strain of *Ophiostoma ulmi.* Neth. J. Pl. Pathol. 86:315-317.

Scheffer, R.J., Liem, J.I., and Elgersma, D.M. 1987. Production *in vitro* of phytotoxic compounds by non-aggressive and aggressive isolates of *Ophiostoma ulmi,* the Dutch elm disease pathogen. Physiol. and Mol. Plant Path. 30:321-335.

Schreiber, L.R., and Townsend, A.M. 1976. Variability in aggressiveness, recovery and cultural characteristics of isolates of *Ceratocystis ulmi.* Phytopath. 66: 239-244.

Svaldi, R., and Elgersma, D.M. 1982. Further studies on the activity of cell wall degrading enzymes of aggressive and non-aggressive isolates of *Ophiostoma ulmi.* Eur. J. For. Pathol. 12:29-36.

Szczegola-Derkacz, M. 1988. Induction of mansonones by *Ophiostoma ulmi* in tissue cultures of *Ulmus americana* and *Ulmus pumila.* M.Sc.F. thesis, Department of Forestry, University of Toronto.

Takai, S. 1980. Relationship of the production of the toxin, cerato-ulmin, to synnemata formation, pathogenicity, mycelial habit, and growth of *Ceratocystis ulmi* isolates. Can. J. Bot. 58: 658-662.

Wu, W.D., Jeng, R.S., and Hubbes, M. 1989. Toxic effect of elm phytoalexin mansonones on *Ophiostoma ulmi,* the causal agent of Dutch elm disease. Europ. J. For. Path. 19:343-357.

Yang, D. 1991. Isolation, identification and characterization of phytoalexin elicitors from *Ophiostoma ulmi.* Ph.D. thesis, Department of Forestry, University of Toronto

Yang, D., Jeng, R.S., and Hubbes, M. 1989. Mansonone accumulation in elm callus induced by elicitors of *Ophiostoma ulmi,* and general properties of elicitors. Can. J. Bot. 67:3490-3497.

18 Characterization of Isolates of *Ophiostoma ulmi* by Molecular Techniques

Robert S. Jeng

Abstract

Ophiostoma ulmi (Buism.) Nannf. is the casual agent of Dutch elm disease. In nature, the pathogen exists as three distinct subpopulations: the weakly pathogenic nonaggressive subgroup and the highly pathogenic aggressive subgroup, which itself divides into two races, the Eurasian (EAN) and North American (NAN) races. These populations are originally separated based on the pathogenicity tests, cultural characteristics, physiological responses and sexual compatibilities. Results of isozyme patterns as well as two dimensional polypeptide mapping further support the view that these two groups are quite widely separated entities. Direct comparison of genome provides a means of identifying and comparing individuals, populations, and higher taxa. Within *O. ulmi* the mitochondrial genomes of the aggressive and nonaggressive groups have been shown to differ in both size and mtDNA restriction patterns. The nonaggressive isolates have a mt genome approximately 40% larger than their aggressive counterparts. At least six mtDNA restriction patterns have been detected for this pathogen. There is a close correlation between these mt types and nuclear DNA fingerprints as defined by the hybridization of a human minisatellite DNA sequence. The 1.7 kbp mtDNA fragment unique to the aggressive isolates has been sequenced and primers constructed for PCR amplification. The 8.8 kbp ribosomal DNA (rDNA) repeat is selected from a genomic library and restriction site and genic maps are constructed and compared for the nonaggressive and aggressive isolates. There are only three restriction site differences which distinguish two groups. Immuno-protein A-gold labeling technique can be used to localize and identify toxin cerato-ulmin (CU) and a polypeptide characteristic for nonaggresssive isolates. A DNA fragment coding for CU has been isolated from genomic library of aggressive and nonaggressive isolates, using PCR amplified DNA fragment as a probe.

Introduction

Ophiostoma ulmi (Buism.) Nannf. is the causal agent of Dutch elm disease. In nature, the pathogen exists as three distinct subpopulations: the weakly pathogenic nonaggressive subgroup, Eurasian (EAN) and North American (NAN) races, which are both considered as the highly pathogenic aggressive subgroup. Aggressive isolates can be differentiated from nonaggressive ones by a fast growth rate and fibrous striate colony on agar media (Brasier 1982; 1986). Isolates of the aggressive subgroup have a different temperature optimum for growth (Brasier et al. 1981), produce more synnemata (Hindal et al. 1981), have different mating-type frequencies (Brasier 1984), produce higher amounts of certain cell wall degrading enzymes (Svaldi and Elgersma 1982), and secrete more of the wilt toxin cerato-ulmin in culture (Takai 1980). These two subgroups also differ in electrophoretic patterns of soluble proteins and specific isozymes (Bernier et al. 1983; Jeng 1986; Jeng and Hubbes 1983; Jeng et al. 1988). Their mitochondrial and nuclear DNA polymorphisms (Bates et al. 1990; Hintz et al. 1989; Jeng et al. 1991), and mitochondrial DNA size (Jeng et al. 1991) are also different. Because of these differences, together with the considerable reproductive isolation between the aggressive and nonaggressive isolates, it has recently been reclassified that the aggressive subgroup be renamed as a new species, *O. novo-ulmi*, to distinguish it from the nonaggressive subgroup which is retained the older name *O. ulmi* (Brasier et al. 1991).

Recent development of molecular techniques in the field of polypeptides as well as nucleic acids have significantly advanced our understanding on genetic compositions of biological materials. The objectives of this presentation are not aiming to address the taxonomic problems of this pathogen, but to characterize isolates of each subpopulation by various molecular techniques.

Isolates

The data shown in this presentation is largely based on the results of 18 isolates of *0. ulmi* from different geographical origins. Twelve isolates were classified as an aggressive subgroup including EAN and NAN races whereas the remaining 6 isolates were nonaggressive (Jeng et al. 1991).

Isozyme and protein patterns

Gel electrophoretic patterns of soluble proteins and specific isozymes have received considerable attention and have shown some success in differentiating species, formae species, or physiological races of various fungi.

Isozyme patterns of alcohol dehydrogenase (ADH), peroxidase (PO), polyphenoxidase (PPO), esterase (EST) and hexokinase (HK) can be used to distinguish aggressive subgroup isolates from nonaggressive isolates of *O. ulmi* (Bernier et al. 1983). Zymograms of peroxidase and tetrazolium oxidase provide evidence for the occurrence of distinct differences between the EAN and NAN races (Fig. 1). This enzyme system is hence a valuable analytical tool for the rapid biochemical identification of *O. ulmi* isolates from the field.

Isoelectrofocusing (IEF), Sodium dodecyl sulfate - polyacrylamide gel electrophoresis (SDS-PAGE) and polypeptide mapping (2-D gel) are other powerful separation systems for total protein analysis. In *O. ulmi,* five distinct protein bands involved in subgroup difference have been detected using Coomassie blue stain after SDS-PAGE (Jeng and Hubbes 1983). However, this technique has the disadvantage of zone spreading as separation continues. A highly efficient method to overcome this problem is the method of electrofocusing in which the separated proteins are actively concentrated, or focused, at their isoelectric points, resulting in high-resolution profile. This is often followed by molecular size separation, which makes it possible to separate and compare several hundred proteins on a single gel. Results of IEF showed that the aggressive and nonaggressive isolates of *O. ulmi* each had about 60 Coomassie blue stained bands having isoelectric points from 3 to 7 with the majority of proteins focused between pH of 4.6 to 5.4. Both subgroups of this fungus had their own characteristic electrofocusing patterns. Nonaggressive isolates have five unique protein bands distributed from pH 4 to 6 whereas aggressive isolates possess two bands, one near the anode and one near the cathode. Two-dimensional electrophoresis of total soluble proteins depicted that there were at least 5 proteins found to be specific for nonaggressive isolates and 4 proteins for aggressive isolates. Moreover, the EAN race had a unique protein which is not present neither in NAN nor nonaggressive isolates. This analysis also shows that isolate #83 (NAN race) possesses some characteristic proteins specific for nonaggressive isolates. This isolate exhibited a very similar polypeptide mapping pattern to the nonaggressive isolates MAB and M273. Specific proteins through this analysis can then be western blotted and amino acid sequenced to obtain probes that can be used to screen a library for a clone of the gene that encodes the protein. Moreover, these proteins can also be used as the antigens in the immunocytological studies.

Immuno-gold-labeling

Immunocytochemistry has been readily used to identify and localize specific proteins (antigens) in biological materials. This is based on the detection of certain markers such as enzyme, fluorescent dye, or gold particle conjugated with protein-A and antigen-antibody complex. The high sensitivity and specificity of the antibody-antigen reaction allows the detection of low concentration of antigen. This technique provides a new way of studying the host and parasite interactions.

For *O. ulmi*, a high molecular weight glycoprotein, called peptido-rhamnomannan (PRM), has been isolated and located in the cell wall of spores (Benhamou et al. 1985). Further analysis on infected elm, using immuno-gold labeling technique revealed that this molecule is mainly deposited in the pit membrane and the area between the cell wall and plasma membrane of parenchymatous cells (Benhamou et al. 1985). The results of these analysis suggest that peptidorahmnomannan probably secretes from fungal spores into xylem fluid, penetrates through pit membrane and deposited outside the plasma membrane where molecular interaction takes place.

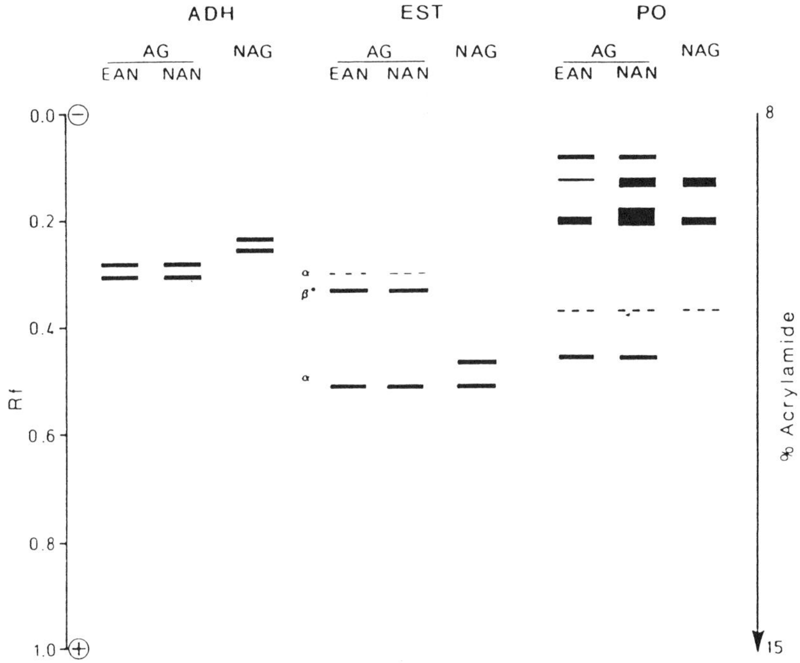

Figure 1. Electrophoretic characterization of subgroups and races in natural populations of *Ophiostoma ulmi.* The aggressive (AG) and nonaggressive (NAG) subgroups can be separated by their zymograms of alcohol dehydrogenase (ADH), esterase (EST), and peroxidase (PO). The Eurasian (EAN) and North American (NAN) aggressive races can be separated by their PO isozyme patterns. Broken rules indicate bands with faint enzymatic activity.

Using a similar method, toxin cerato-ulmin was detected on the surface of fungal spores by scanning electron microscope (Svircev et al. 1988) and found in vascular vessel of the infected elm by fluorescent antibody (Ipsen and Abul-Hajj 1982).

A soluble protein unique to the nonaggressive isolate Q412 has been isolated and localized in the cell wall of both conidia and mycelia. This protein has also been detected on the mycelial surface of other 4 nonaggressive isolates by scanning electron microscopy. All aggressive isolates tested are non-reactive. However, the nature and biological significance of this protein remain to be determined.

Ribosomal DNA polymorphisms

Restriction fragment length polymorphisms have been extremely useful for the determination of phylogenetic relationships within many fungi. RFLP analysis of the ribosomal DNA (rDNA) repeat has been particularly useful for the differentiation of organisms at the species and higher level taxa (Anderson et al. 1987; Hintz et al. 1989; Smith et al. 1990; Walsh et al. 1990).

Since the rDNA of *O. ulmi* was uncharacterized the 8.8 kpb rDNA repeat of *O. ulmi* was selected from a lambda DNA genomic library by screening with a heterologous probes for rDNA sequences. Using the cloned *O. ulmi* rDNA as a hybridization probe, detailed restriction site maps were constructed for both the aggressive and non-aggressive subgroups of this pathogen. The positions of the coding regions for the 18s, 5.8s and 26s rDNAs were determined by Southern hybridization to endonuclease digested nuclear DNAs with the rDNA-specific probes obtained from PCR amplification. The amplified DNAs corresponded to a portion of the 18s rDNA gene, the internal transcribed region including the 5.8s rDNA gene, and a region spanning the 5.8s and most of the 26s rDNA genes are shown in Fig. 2.

The rDNA restriction patterns were highly conserved for both the aggressive and nonaggressive subgroups. There were, however, restriction pattern differences which distinguished the aggressive from nonaggressive groups. These included an *Apa* I, *Sac* I and *Hinc* II site found only in the rDNA repeat of the nonaggressive isolates and not in the aggressive isolates (Fig. 2). The rDNA restriction pattern differences thus provided an easily assayed character to differentiate between the aggressive and nonaggressive subgroups.

MtDNA restriction fragment length polymorphisms

Unlike their mammalian counterparts, the mitochondrial genomes of fungi are highly variable in size, ranging from 18.9 kbp to 173.6 kbp (Hintz et al. 1988; Scazzocchia 1986). The majority of fungal species has circular mtDNAs. Another characteristic feature of fungal mtDNAs is the high AT content, thus mtDNAs can be purified by bisbenzimide-CsCl gradient centrifugation.

Agarose gel electrophoresis of native mtDNAs of *O. ulmi* show that some of the isolates have one or more plasmid-like DNAs. However, there is no relationship between pathogenicity and the presence of plasmid-like nucleic acids.

Restriction digests of the mtDNAs from the 18 isolates of *O. ulmi* using *Eco* RI, *Bgl* II and *Pvu* II yield a number of diagnostic bands for each group of isolates. The mtDNAs from the aggressive isolates restricted with *Eco* RI were characterized by three fragments of 10.0 kb, 5.9 kb and 1.7 kb whereas nonaggressive isolates were characterized by four fragments of 14.8 kb, 8.9 kb, 5.5 kb and 4.6 kb.

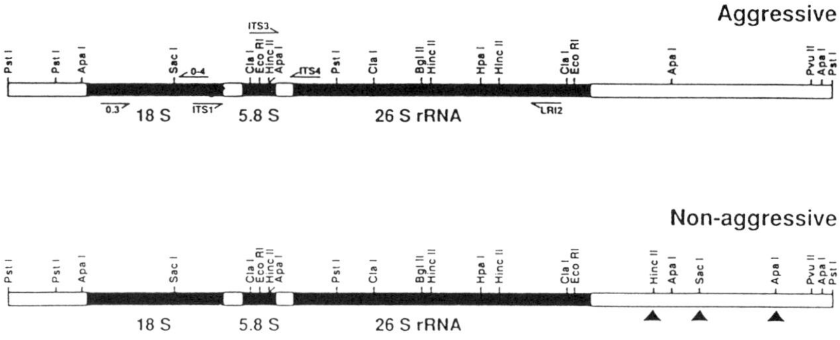

Figure 2. Ribosomal DNA repeat of *Ophiostoma ulmi*. The relative positions of restriction sites and the 18s, 5.8s, and 26s rDNA genes are indicated. There were three restriction-site polymorphism which distinguished the aggressive and non-aggressive isolates (solid arrows)

Mitochondrial DNAs from aggressive isolates restricted with *Bgl* II were characterized by two fragments of 6.6 kb and 4.8 kb whereas nonaggressive isolates were characterized by five fragments of 7.4 kb, 5.4 kb, 4.4 kb, 4.0 kb and 1.4 kb. When restricted with Pvu II, the nonaggressive isolates were characterized by four fragments of 23.0 kb, 6.7 kb, 2.7 kb and 2.3 kb. Comparison of the RFLP patterns within each group of isolates also indicated polymorphisms within the aggressive and nonaggressive subgroups. Six RFLP types were observed: 4 types (Type I, II III and IV) for aggressive subgroup and 2 types for nonaggressive subgroups (IV, V).

The mtDNA molecular size varied between the aggressive and nonaggressive subgroups of isolates. The nonaggressive isolates have an mt genome approximately 40% larger than their aggressive counterpart. There are no correlation between the mitochondrial genome types and the aggressiveness.

The restriction-site maps for four types of mtDNA genomes within aggressive subgroup are summarized in Fig. 3. The type IV mtDNA differed from the other mtDNAs by a 0.4 Kb length mutation and represents the largest mt genome with the aggressive subgroup. The Type I and Type II mt genomes differed from the Type III mt genome by discrete length mutation of 4.8 and 4.0 Kbp, respectively. It is unknown whether these length mutation resulted from insertions into or deletions from a progenitor mitochondrial genome.

The restriction-site maps for two types of mtDNA genomes within nonaggressive subgroup are under investigation.

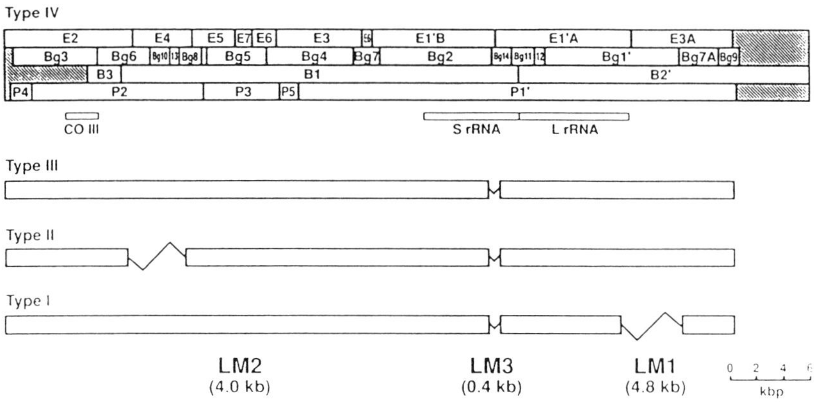

Figure 3. The mitochondrial DNA restriction-site map for the aggressive subgroup of *Ophiostoma ulmi*. The four mitochondrial types differ by the presence or absence of length mutations LM1, LM2, or LM3

Nuclear DNA fingerprinting

Nuclear DNA fingerprints were prepared using the SNAP DNA Fingerprint Kit from Molecular Biosystem Inc. (San Diego, CA). *O. ulmi* DNAs were digested with *Pst* I, *Hinf* I, or *Hae* III, electrophoresed, and capillary blotted onto nylon membrane.

Low-stringency hybridization of the human minisatellite probe DNA HVR 33.6 (Jeffreys et al. 1985) to restriction-digested *O. ulmi* DNA allowed for the detection of numerous cross-hybriding fragments. Fragments, ranging from approximately 1 to 20 kbp in size, were detected when the *O. ulmi* DNA was digested with the endonuclease *Pst* I which recognizes a 6-bp sequence. Certain fragments exhibited restriction polymorphisms. When the DNA was digested with either *Hinf I* or *Hae* III, both of which cleave at a 4-bp sequence, relatively small fragments were detected which should reflect the number of repeats per minisatellite (Jeffreys 1987). Between 9 to 15 intensely staining fragments were detected in these digest. The majority of the fragments were conserved between *O. ulmi* isolates but there was sufficient variation to sort the isolates according to differences in the hybridization patterns. The 18 *O. ulmi* isolates surveyed were divided into 6 groups on the basis of hybridization pattern with a human minisatellite DNA sequence: three types for aggressive subgroup (A, B, C) and three for nonaggressive (D, E, F). There was a close correlation between nuclear DNA fingerprints and mitochondrial DNA restriction patterns.

Polymerase chain reaction

The polymerase chain reaction (PCR) is an in vitro method for the enzymatic synthesis of specific DNA fragments, using two nucleotide primers that hybridize to opposite strands and flank the region of interest in the target DNA. A series of cycles involving template DNA denaturation, primer annealing, and the extension of the annealed primers by DNA polymerase results in the exponential accumulation of a specific fragment. This technique has shown an enormous impact in both basic and diagnostic aspects of molecular biology because it can produce fast and large amounts of a specific DNA sequences from small amounts of a crude DNA.

Attempts have also been made to amplify rDNA, mtDNA and genomic DNA using corresponding primers. Six isolates each representing different mtDNA type were used as the sources of template DNA. Each amplification reaction contains 50 ng of miniprep DNA. Agarose gel electrophoresis of amplified products shows that all six isolates have the same size of 26s and ITS rDNA fragments but with different NTS (IGS) rDNA. These results confirm the rDNA restriction-site map that there are few restriction-site differences between the isolates of the two subgroups. For mitochondrial DNA, the 1.7 kbp *Eco* RI-restricted fragment characteristic to aggressive isolates has been sequenced and primers constructed for PCR. Results of agarose gel analysis show that all 23 aggressive isolates tested possess this fragment. None of 11 nonaggressive isolates show the presence of this DNA fragment. This 1.7 kbp DNA has been routinely used in the laboratory as marker for the identification of aggressive isolates of *O. ulmi.*

RAPD (Random amplified polymorphic DNA) markers are capable of detecting polymorphisms in single copy and repetitive DNA, resulting from insertions, deletions, rearrangements, or single base changes (Williams et al. 1991). Amplification of total DNA with single RAPD primer has proven useful for genetic mapping (Williams et al. 1991), for breeding applications (Welsh et al. 1991) and for DNA fingerprinting (Welsh and McClelland 1991). This technique is particular useful when few morphological genetic markers were available.

Because of a need for large numbers of genetic markers, we are investigating the use of RAPDs for genetic analysis of Fl progenies of *O. ulmi.* Six RAPD markers were used singly in PCR amplification to screen for DNA polymorphosms from genomic DNA of 6 mtDNA types. Agarose gel electrophoresis of amplified products reveals that all 6 primers each amplified 3-10 prominent DNA fragments. All 6 primers generate 1 to 3 fragments which can be used to differentiate aggressive and nonaggressive subgroups. These RAPD markers will be used to study the genetic structure of *O. ulmi.*

Conclusions

Molecular markers such as certain proteins, isozyme polymorphisms and RFLPs of mtDNA, rDNA, and nDNA have proven useful not only for the identification or characterization but also for monitoring the population dynamic of a given species. They can also be used for obtaining genetic information and generating linkage maps.

Separation of *Ophiostoma ulmi* into the aggressive and non-aggressive subgroups were initially based on the pathogenicity tests, cultural characteristics, physiological responses and sexual compatibilities. In recent past the genetic structure of wild type as well as laboratory isolates of this pathogen have been determined through an analysis of RFLPs in the rDNA repeat, length mutations in the mtDNA, and DNA fingerprinting of the nuclear DNA using a minisatellite DNA probes. DNA polymorphisms amplified by arbitrary primers (RAPDs) are also been investigated.

The minor differences in the mitochondrial DNA restriction patterns suggest a relatively recent divergence of mt types within the aggressive subgroup. Preliminary analysis of restriction map for the nonaggressive subgroup also suggest a similar conclusion. For all isolates examined there was a close correlation between nuclear DNA fingerprints and mitochondrial DNA restriction patterns. There were 6 nuclear types over 100 isolates surveyed, with each type having a unique nuclear DNA fingerprint when hybridized with human minisatellite DNA. It is surprised to find a very close resemblance of nuclear DNA fingerprints from isolates from locations as diverse as in the North American, Europe, and East Asia. This could argue for the world-wide clonal propagation of a small number of extremely successful genotypes.

Restriction fragment length polymorphisms of rDNA have been extremely useful for the determination of phylogenetic relationship within many fungi. Analysis of the 8.8 kbp rDNA repeat of *O. ulmi* showed that the aggressive and non-aggressive subgroups have highly conserve rDNA restriction patterns. There were, however, restriction differences only in the NTS (non-transcribe-space) region of the rDNA repeat (Fig. 2). Because of the uniformity of restriction pattern of rDNA repeat within the subgroups, delineation of subpopulations in *O. ulmi* may be limited.

References

Anderson, J.B., Petsche, D.M., and Smith, M.L. 1987. Restriction fragment polymorphisms in biological species of *Armillaria mella*. Mycologia 79: 69-76.

Bates, M.R., Buck, K.W., and Brasier, C.M. 1990. Molecular variation in the Dutch elm disease fungus. In *Molecular Evolution* (ed. M.T. Clegg & S.J. O'Brien). UCLA Symposia on Molecular and Cellular Biology, New Series 122:171-178.

Benhamou, N., Lafontaine, J.G., Joly, J.R., and Ouellette, G.B. 1985. Ultrastructural localization in host tissues of a toxic glycopeptide produced by *Ophiostoma ulmi*, using monoclonal antibodies. Can. J. Bot. 63: 1185-1195.

Benhamou, N., Ouellette, G.B., Lafontaine, J.G., and Joly, J.R. 1985. Use of monoclonal antibodies to detect a phytotoxic glycopeptide produced by *Ophiostoma ulmi,* the Dutch elm disease pathogen. Can J. Bot. 63: 1177-1184.

Bernier, L., Jeng, R.S., and Hubbes, M. 1983. Differentiation of aggressive and nonaggressive strains of *Ceratocystis ulmi* by polyacrylamide gel electrophoresis of intramycelial enzymes. Mycotaxon 17:456-472.

Brasier, C.M. 1982. Occurrence of three sub-groups within *Ceratocystis ulmi. In* Proceedings of the Dutch elm disease symposium and workshop, Winnipeg, Oct. 5-9. 1981. Edited by E.S. Kondo, Y. Hiratsuka, and W.B.G. Denyer. Manitoba Department of Natural Resources, Man., Canada. pp. 298-321.

Brasier, C. M. 1984. Inter-mycelial recognition systems in *Ceratocystis ulmi:* their physiological properties and ecological importance. In The ecology and physiology of the fungal mycelium. Edited by D. Jennings and A.D.M. Rayner. Cambridge University Press, Cambridge. pp. 451-497.

Brasier, C.M. 1986. A comparison of pathogenicity and cultural characteristics in the EAN and NAN aggressive sub-groups of *Ophiostoma ulmi.* Trans. Br. Mycol. Soc. 87:1-13.

Brasier, C.M. 1991. *Ophiostoma novo-ulmi sp.* nov., causative agent of the current Dutch elm disease pandemics. Mycopathologia 115:151-161.

Brasier, C.M., Lea, J. and Rawlings, M.K. 1981. The aggressive and non-aggrssive strains of *Ceratocystis ulmi* have different temperature optima for growth. Trans. Br. Mycol. Soc. 76:213-218.

Hindal, D.F., Harner, E.J., and MacDonald, W.L. 1981. Cultural characteristics, pathogenicity predictions, and pathogenicity tests for some isolates of *Ceratocystis ulmi.* Can. J. Bot. 59:2543-2549.

Hintz, W.E., Anderson, J.B., and Horgen, P.A. 1988. Nuclear migration and mitochondrial inheritance in the mushroom *Agaricus bitorquis.* Genetics 119:35-41.

Hintz, W.E., Anderson, J.B., and Horgen, P.A. 1989. Relatedness of three species of *Agaricus* inferred from restriction fragment length polymorphism analysis of the ribosomal DNA repeat and mitochondrial DNA. Genome 32:173-178.

Hintz, W.E., Jeng, R.S., Hubbes, M., and Horgen, P.A. 1991. Identi~lcation of ·three populations of *Ophiostoma ulmi* (aggressive subgroup) by mitochondrial DNA restriction site mapping and nuclear DNA fingerprinting. Exp. Mycol 15:316-325.

Ipsen, J.D. and Abul-Hajj, Y.J. 1982. Fluorescent antibody technique as a means of localizing *Ceratocystis ulmi* toxins in elm. Can. J. Bot. 60:724-729.

Jeffreys, A.J. 1987. High variable minisatellites and DNA fingerprints. Biochem. Soc. Transact. 15:309-317.

Jeffreys, A. J., Wilson, V., and Thein, S.W. 1985. Hypervariable minisatellite regions in human DNA. Nature 314:67-73.

Jeng, R.S. 1986. Analytical electrofocusing and two-dimensional electrophoresis of proteins extracted from the mycelia of aggressive and nonaggressive strains of *Ophiostoma ulmi.* Can. J. Bot. 64:2073-2081.

Jeng, R.S., and Hubbes, M. 1983. Identification of aggressive and non-aggressive strains of *Ceratocystis ulmi* by polyacrylamide gradient gel electrophoresis of intramycelial proteins. Mycotaxon 17:445-455.

Jeng, R.S., Bernier, L., and Brasier, C.M. 1988. A comparative study of cultural and electrophoretic characteristics of Eurasian and North American races of *Ophiostoma ulmi.* Can. J. Bot. 66:1325-1333.

Jeng, R.S., Duchesne, L.C., Sabourin, M., and Hubbes, M. 1991. Mitochondrial DNAs restriction length polymorphisms of aggressive and non-aggressive isolates of *Ophiostoma ulmi.* Mycol. Res. 95:537-542.

Scazzocchia, C. 1986. The natural history of fungal mitochondrial genome. In Evolutionary biology of fungi (ed. Rayner, A.D.M., Brasier, C.M. and Moore, D.), pp. 52-73. Cambridge University Press.

Smith, M.L., Duchesne, L.C., Bruhn, J.N., and Anderson, J.B. 1990. Mitochondrial genetics in a natural population of the plant pathogen *Armillaria.* Genetics 126:575-582.

Svaldi, R., and Elgersma, D.M. 1982. Further studies on the activity of cell wall degrading enzymes of aggressive and non-aggressive isolates of *Ophiostoma ulmi.* Eur. J. For. Pathol. 12:29-36.

Svircev, A.M., Jeng, R.S., and Hubbes, M. 1988. Detection of cerato-ulmin on aggressive isolates of *Ophiostoma ulmi* by immunocytochemistry and scanning electron microscopy. Phytopathology 78:322-327.

Takai, S. 1980. Relationship of the production of the toxin cerato-ulmin to synnemata formation, pathogenicity, mycelial habitat and growth of *Ceratocystis ulmi.* Can. J. Bot. 58:658-662.

Walsh, S.R.A., Tyrrell, D., Humber, R.A., and Silver, J. C. 1990. DNA restriction fragment length polymorphisms in the rDNA repeat unit of Entomophaga. Exp. Mycol. 14:381-392.

Welsh, J., Honeycutt, R.J., McClelland, M. and Sobral, B.W.S. 1991. Parentage determination in maize hybrids using the arbitrarily primed polymerase chain reaction (AP-PCR). Theor. Appl. Genet. 82:473-476.

Welsh, J., Peterson, C., and McClelland, M. 1991. Polymorphisms generated by arbitrarily primed PCR in the mouse: application to strain identification and genetic mapping. Nucl. Acids Res. 19:306.

Williams, J.G.K., Kubelik, A.R., Livak, K., Rafalski, J.A. and Tingey, S.V. 1991. DNA polymorphisms amplified by arbitrary primers are useful as genetic markers. Nucl. acids Res. 18:6531-6535.

19 Genetic Diversity in Elms: What Molecular Data Tell Us

S. J. Wiegrefe, R. P. Guries, E. B. Smalley, and K. J. Sytsma

Abstract

Elms represent a valuable and diverse woody plant resource for traditional forest products as well as urban landscape use. Yet despite their obvious importance to urban and rural populations, we know relatively little about the extent and patterning of genetic variation in elms. Efficient utilization of genetic variation in future breeding programs, as well as integration of biotechnology with conventional plant improvement practices, requires a more complete and systematic assessment of genetic variation. Recent advances in molecular biology, especially the development of new sets of genetic markers (e.g. RFLP, RAPD, isozymes), hold great promise for enhancing our understanding of systematic affinities and the evolution of elms, as well as permitting the construction of detailed genetic maps for use in breeding and other forms of genome manipulation. Our evidence from chloroplast and nuclear ribosomal DNA restriction site analysis reveals that elms are rich in molecular variation, having a large number of mutations amenable to restriction site analysis. Examples of the utility of such data to clarify taxonomic problems are provided, as well as suggestions for future applications of molecular markers in breeding programs.

New Horizons in Elm Breeding

For generations, elms have been a valuable and durable part of the urban and rural landscape. Their graceful habit, adaptability to poor soils, and tolerance of air pollutants have made them a characteristic feature of the Eurasian and North American countryside. During the mid-part of the 20th century, a vascular wilt disease incited by *Ophiostoma ulmi* (Dutch elm disease) decimated elm populations leading to enormous economic losses in goods and services to millions of humans. The onset of Dutch elm disease (DED) also led to the initiation of research and breeding programs, first in Europe (Heybroek 1983), and later in the United States (Guries and Smalley 1986), to develop disease control strategies and new disease-resistant elms. Such programs, in place for several decades, have produced a wealth of information on both the disease and its hosts, but have not yet produced complete solutions to Dutch elm disease problems.

The domain of tree breeders (including elm breeders) traditionally has been the evaluation of germplasm and its subsequent use in developing improved varieties or cultivars. The large size, long generation intervals, and presence of large genotype x environment interactions pose formidable challenges to woody plant breeders, yet improved elm cultivars have been released (Guries and Smalley 1986; Heybroek 1983; Townsend et al. 1991, and others). While resistance to DED has been of special interest, selection for other characters including crown architecture, leaf size, adaptability, resistance to other pests, and numerous other traits have also been important. Progress in multi-trait breeding has been slow however, and conventional breeding practices hold little promise for efficient exploitation of most such variation in trees.

Identification of molecular polymorphisms circumvent the need for lengthy evaluation or screening trials if such markers are linked to traits of economic interest, thereby providing a means for indirect selection. Construction of genetic linkage maps has been accomplished for a number of plants, including maize (Dudley et al. 1992; Helentjaris et al. 1986), tomato (Paterson et al. 1988), *Brassica* (Song et al. 1991), loblolly pine (O'Malley 1992) and others (see Tanksley et al. 1989 for a review). Even complex genetic traits can be identified, mapped and selected via molecular markers which identify chromosome intervals containing the quantitative trait(s) (Lander and Botstein 1989). Applications of molecular markers to identify genes of economic interest, evaluate wild germplasm resources, improve the efficiency of selection programs, and explore the routes of crop plant evolution are becoming commonplace in plant science. While such work with woody plants lags behind crop plants for a variety of technical and economic reasons, opportunities do exist for employing molecular markers in tree breeding.

An understanding of genetic variation, its organization and value, is central to tree improvement practices. For programs which rely upon both native and exotic germplasm resources, obtaining and using such variation can be difficult, yet progress cannot be achieved without such knowledge. This paper will consider how our Wisconsin elm breeding program has taken advantage of molecular systematics to understand better the organization of genetic variation in *Ulmus* as an aid to our breeding work. In addition, we will suggest areas in which woody plant breeders could benefit from further molecular genetics research.

A Beginning: Characterizing Variation

In devising improvement programs, breeders have been able to take advantage of the considerable interfertility which exists among elm species. However, while such interfertility permits us to introduce diverse germplasm into locally adapted populations, it also poses the problem that one can never be certain that new germplasm collections are authentic. We have recognized that many *Ulmus* accessions received from Eurasia, particularly from botanical gardens or urban areas, actually represent hybrids or other introgressed forms. Initial efforts to exploit such germplasm frequently yielded unexpected results, especially with respect to adaptability. Only later did we recognize that expectations regarding performance were not realized because we were dealing with hybrid accessions segregating for a wide array of characteristics. A major reason for such confusion has been that basic systematics information available for elms is inadequate. Thus, we were anxious to improve our understanding of our own germplasm accessions via the application of molecular techniques to studies of the organization of genetic variation at a number of levels within *Ulmus* and the Ulmaceae. A better understanding of species relationships was a primary goal, but we also recognized that it might suggest new directions for breeding and the development of better elms

Chloroplast (cpDNA) and nuclear encoded ribosomal DNA (nrDNA) restriction site analysis were selected as initial tools for use in our elm systematics work because the expected level of variation was likely to be appropriate for studies at the interspecific and intergeneric level (Clegg and Zurawski 1992; Palmer et al. 1988; and others). In addition, the sampling required was compatible with the types of accessions available in our elm collections at Wisconsin. Most of our accessions, especially those received during the 1960's and 1970's include many Eurasian species, but only the most hardy sources have typically survived during our rigorous Wisconsin winters. During the past decade, we have received additional germplasm accessions from Asia, especially China; many species not hardy in Wisconsin are maintained as specimens in the greenhouse.

A complete chronology of the genetics studies underway in our labs is not possible here. However, two examples will demonstrate the utility of molecular techniques to address questions regarding genetic variation, as well as some of the limitations of such techniques to answer complex evolutionary and genetics questions.

Systematics of the Genus *Ulmus*

The genus *Ulmus L.* comprises approximately 45 woody species distributed throughout much of the north temperate regions of Eurasia and North America. While the genus is rather well defined, the delimitation of species and their taxonomic affinities has generated controversy (Fu 1980; Heybroek 1976; Richens 1983; and others). Most of the difficulties regarding taxonomic affinity within the genus can be attributed to a paucity of taxonomic characters in a group characterized by simple and highly reduced flowers and fruits, highly variable vegetative characters, few barriers to interspecific gene exchange, and limited chemical differentiation (Richens 1983). The historic movement of elms by migrating human populations and other anthropogenic factors further complicate the situation. In recent years, cladistic analysis of cpDNA restriction site data have proved useful in resolving similar systematics questions in plants (Clegg and Zurawski 1992; Palmer et al. 1988). We examined representatives of all proposed sections and subsections (See Wiegrefe 1992, for a complete discussion) using cladistic analysis to construct a phylogeny for comparison with various existing systems. Technical details regarding origin of accessions, DNA extraction and sample preparation, restriction enzymes employed, and data analysis are published elsewhere (Wiegrefe 1992).

Molecular diversity

Twenty-eight accessions representing 13 species revealed a total of 667 unique restriction sites following digestion with 25-30 restriction enzymes. One hundred and twenty-five mutations were detected within the genus *Ulmus* with another 23 unpolarized mutations detected between *Ulmus* and *Zelkova* (the outgroup). Of this total of 148 mutations, 83% were found to be phylogenetically informative. Despite the generally conservative nature of cpDNA relative to nuclear or cytoplasmic DNAs, and the relatively limited sampling of accessions available for this analysis, the large number of mutations detected represents a rich source of information for systematics studies and possible application to future use of markers in breeding. While the existence of such a wealth of markers in cpDNA does not prove that correspondingly high levels of variation will also be found in nuclear DNA (Soltis et al. 1992), there is good reason to be optimistic. Our past experience with elm isoenzymes, and previous work with flavonoids (Gianassi 1978; Sherman and Gianassi 1988), suggests that DNA restriction site data will be an increasingly fruitful source of markers.

Molecular phylogeny

Analysis of the *Ulmus* restriction site data using PAUP (Swofford 1991), a phylogenetic software routine employing the principle of Wagner parsimony, yielded 16 most parsimonious trees 168 steps in length with high consistency indices. When using character-state weighting to discriminate against the less likely parallel restriction site gains, a single shortest tree was created (Figure 1); this is also one of the 16 shortest trees identified in the unweighted analysis. As expected, all of the conspecific accessions were found to be more closely related to each other than to other accessions. To highlight the interspecific relationships, only one entry per species is shown (Figure 1) for the character-state weighted phylogeny.

Cladistic analysis revealed that cpDNA variation is both substantial and well distributed throughout the genus such that the phylogeny constructed here is well-resolved and has well supported branches. It provides clear evidence supporting Grudzinskaya's classification of the rock elms (Grudzinskaya 1974), as opposed to the more widely accepted scheme of Schneider (1916). Schneider's classification places *U. crassifolia* with *U. parvifolia,* and *U.*

thomasii with *U. alata*. Grudzinskaya removed *U. crassifolia* from section *Microptelea* and created a monotypic section *(Anisoptelea)* for it. She also recognized that *U. thomasii is* more closely related to *U. serotina* and placed them together in one section. Our data support the inclusion of her section *Anisoptelea* with *U. thomasii* and *U. serotina* in section *Trichoptelea.*

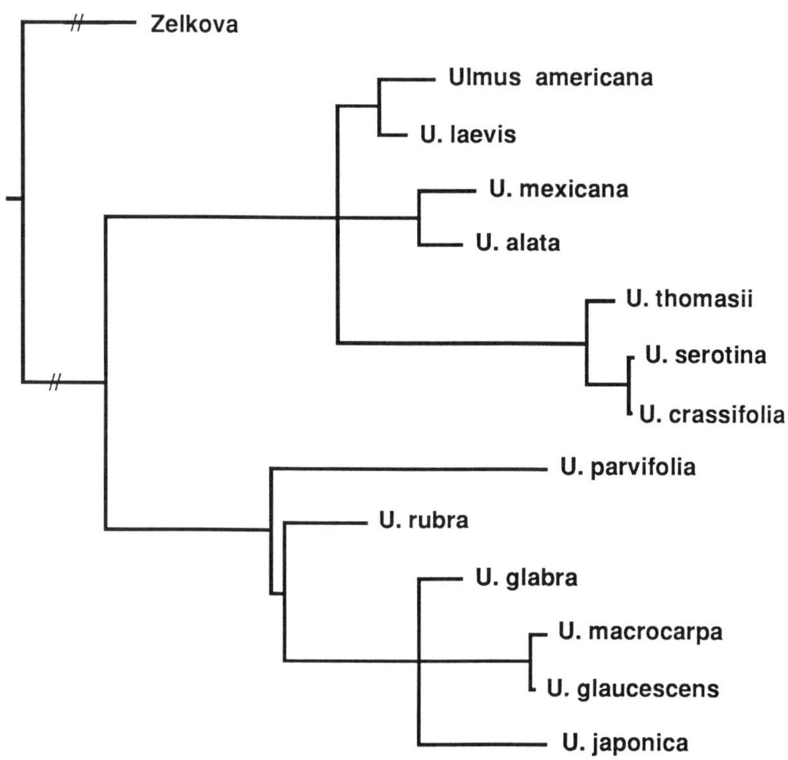

Figure 1. Phylogeny for the genus *Ulmus* reconstructed using cpDNA restriction site data with character-state weighting to discriminate against site gains. Branch lengths are proportional to the number of mutations uniting the taxa distal to that branch, with the exception of the branches uniting all *Ulmus* and all *Zelkova.* The allocation of mutations to these two branches could not be determined.

Hybridogenesis in the Elms

The possibility that hybridization has played a role in the evolution of the elms can be raised owing to the relative lack of barriers to interspecific gene exchange among most elms, as well as to the sympatric ranges of several species in the genus. Interspecific hybridization has been implicated in species formation in other wind-pollinated trees, including *Populus* (Smith and Sytsma 1990), *Quercus* (Hardin 1975), and *Pinus* (Wang et al. 1990).

Evidence of interspecific hybridization can be detected via comparisons of cpDNA data and those based on nuclear DNA mutations owing to the different mechanisms of inheritance (uni- vs. biparentally inherited genomes: see Rieseberg and Brunsfeld 1992 for a review). While the presence of species-specific markers in the putative parental species would provide stronger evidence of hybridogenesis, unlikely convergences identified during the combination of cpDNA and nrDNA data also may indicate potential hybridization in natural populations.

The case of the rock elms

The possibility of hybrid origin for two North American species, *U. thomasii* and *U. serotina,* was raised by Grudzinskaya (1974) based upon morphological evidence. We examined this question on a very limited scale via restriction site analysis of a single nuclear gene which encodes 3 of 4 nuclear ribosomal subunits. Data revealed two mutations, of apparent common origin, shared by accessions of *U. thomasii* and *U. alata* which were inconsistent with the cpDNA data. Three possible scenarios involving hybridogenesis could resolve the apparent discrepancy between cpDNA and nrDNA data: 1) ancient hybridization involving either the cross of progenitors of *U. mexicana* x *U. thomasii* to yield *U. alata,* or the cross *U. serotina* (or *U. crassifolia)* x *U. alata* to yield *U. thomasii;* 2) dispersed introgression due to the fact that the sampled individuals containing the mutations do not occur sympatrically (Figure 2); and 3) recent hybrid origin of the specific individuals examined. This last possibility is extremely remote given that the individuals of *U. thomasii* and *U. alata* used here were located more than 1,000 km apart. The other possibilities can not be so readily dismissed. In the scenario of ancient hybridization, fossils similar to *U. mexicana* have been found in Alaska, Washington and North Dakota (Burnham 1986), suggesting that sympatric populations of these species may have existed in the past. Unfortunately, we cannot make a definitive judgement regarding the likelihood that either scenario 1 or 2 is true owing to our limited sampling of the taxa involved. In addition, the morphological evidence for such an event is not compelling as the species in question appear to share few diagnostic characters. Still, our point is that molecular evidence can suggest past evolutionary origins that would not likely be considered based solely on current species ranges and limited morphological assessments. More detailed study of this and other possible examples of hybridogenesis in the elms appears warranted.

Molecular Markers in Elm Breeding

Applications for the use of molecular markers in plant breeding have expanded enormously during the past decade, but such growth has occurred primarily in a few crop plants for several reasons. The existence of a considerable amount of information on the identity, inheritance and linkage relationships of genes of economic importance undoubtedly has been a major factor here, but the potential for realizing financial gain in major crops has also been a motivating force. Aside from loblolly pine *(Pinus taeda)* and Douglas-fir *(Pseudotsuga menziesii)*, very few forest trees command a position of sufficient economic importance to generate adequate research funding. The fact remains that research in woody plant molecular biology is expensive, our knowledge of the underlying genetics of virtually all species is extremely limited, only a small number of organizations or industries are interested in supporting such research, and the perceived benefits are years, perhaps decades, removed. Faced with such a situation, why proceed at all?

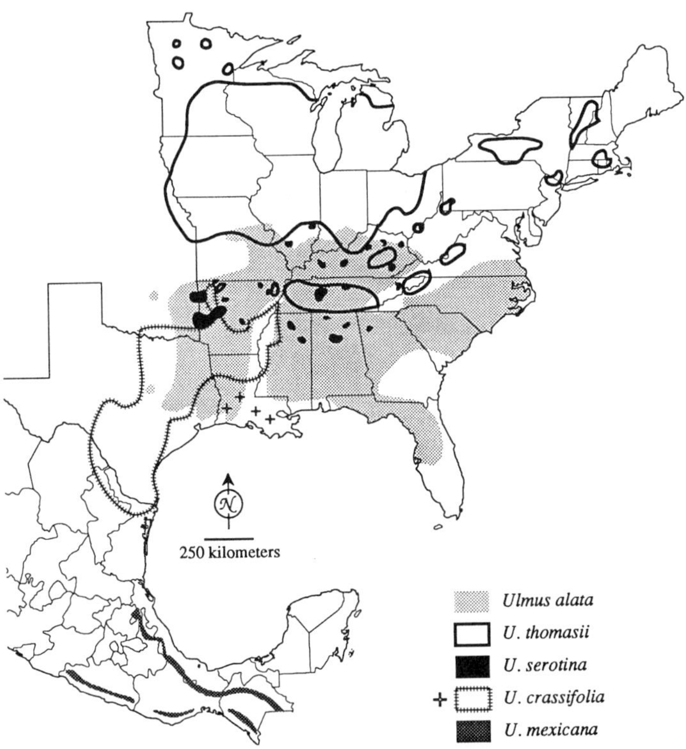

Figure 2. Distributions of five North American elm species which are possible participants in hybrid species-origin, based on cp- and nrDNA restriction site data.

Despite the concerns noted above, there are some good reasons to be optimistic about the prospect for future success in applying molecular technologies in elm breeding. Many of the features of woody plants that make traditional breeding problematic also make alternatives to such programs attractive. However, starting from a position of less knowledge and less funding, woody plant researchers will have to make careful choices regarding the kinds of technologies adopted, the specific problems attacked, and the species selected as useful models or targets for research. Our general inclination to simply grab the newest technology must be tempered, as well as our tendency to force woody plants to conform to a crop plant model. In general, such efforts will not (easily) succeed, and a few well-publicized successes are what we really need. With specific regard to elms and their breeding, let us suggest at least three directions in which research employing molecular markers could proceed.

Molecular systematics

The application of molecular markers to further our understanding of the evolution of elms, and to construct a "true" phylogeny, would be a very important direction for future research efforts. First, we know that such work can succeed; elms possess a rich store of molecular markers in cpDNA, nrDNA, and probably other nuclear encoded DNA as well. All evidence to date from our work indicates that elms are no more recalcitrant than other plant taxa in the application of such technologies. The playing field is even here, and the economic incentive for studying this large and diverse group of woody plants reinforces their intrinsic academic interest.

The existence of working collections of elm germplasm makes such work possible; continuous acquisition of new accessions from North America and Eurasia serves many purposes, including plant systematics. The potential for molecular systematics to resolve long-standing questions regarding the origin of certain species, or to suggest interspecific combinations of value for breeders, makes the work extremely useful. Certain species which serve as sources of genes for resistance to various pests (e.g. *U. pumila* or *U. parvifolia* as sources of resistance to DED) are essentially unstudied and deserve much more attention. For work at the level of sections, cpDNA may be usefully applied to even limited numbers of accessions. However, for work at the intra- or interpopulation level (provenances), or within sections, DNA which yields greater numbers of polymorphisms, such as the intergenic spacer region (IGS) of ribosomal DNA (Hamby and Zimmer 1992), RFLP's of single-to-low copy nuclear DNA, or randomly amplified polymorphic DNA (RAPD) will be more useful.

QTL mapping

The identification and mapping of genes of economic interest has occupied countless hours for plant breeders. Genetic maps for important agronomic crops were initially limited to small numbers of markers, but even this rather elementary tool was unavailable for any woody plant until recently. Ultimately, such maps were embellished with biochemical markers and disease-resistance traits enabling breeders to use indirect selection on markers (e.g. isozymes) to accelerate the screening and selection of desirable genotypes. The identification and mapping of quantitative trait loci (QTL) has been accomplished for several crop plants, including tomato (Paterson et al. 1988) and maize (Stuber et al. 1987). Molecular markers promise to facilitate selection and breeding for many organisms, but the rewards would be especially valuable in long-lived woody plants for which traditional approaches are time-consuming.

Not all trees will be amenable to such studies however, and the need for populations with at least moderate levels of linkage disequilibrium may permit application only to those species for which disequilibrium can be generated. For most short generation, inbred crops, F_1 backcross or F_2 populations are easily obtained, but this situation is rare in trees. However, we have generated numerous advanced generation elm progenies, backcrosses and F_2 populations, including interspecific combinations, in the course of evaluating exotic germplasm. It would be enormously time consuming and expensive to produce such progeny arrays for the sole purpose of QTL mapping, but their existence provides ready materials for such work. Whether all populations have the requisite levels of disequilibrium for efficient mapping remains to be seen, but the long-term value of maintaining such populations has become obvious.

The recent application of RAPD markers to the construction of a linkage map for loblolly pine placed over 190 markers in 12 groups (the base chromosome number for *Pinus)* and was completed in less than two months (O'Malley 1992) Such a highly saturated map for a woody plant was inconceivable even a decade ago. That it was accomplished for loblolly pine reinforces a point made earlier, that pioneering breakthroughs will be made largely with a few species of high economic interest. The next step with loblolly pine will be to map QTL's onto the RAPD map, beginning with a small number of apparently highly heritable traits in progenies specifically constructed for such screening. The measure of their success will determine whether work with other conifer species can succeed. Other woody plant breeders will need to identify which breakthroughs can profitably be adapted to their species and which cannot.

Bulked segregant analysis

The use of molecular markers could permit the construction of a saturated genetic map, but a current limitation in elms is the paucity of morphological or quantitative traits of known inheritance that can be combined with the molecular markers. While certain traits identified in our evaluations appear to have a simple genetic control (i.e. segregate in progenies of specific crosses), such as lobed leaf shape, or resistance to elm leaf spot, confirmation of inheritance is almost always lacking. For genes which govern growth, crown architecture, or other physiological or developmental traits, even less is known regarding inheritance. Despite the considerable cost involved in documenting mode of inheritance for certain traits of economic interest in elms, we may need to do this to facilitate marker-aided selection. Even for traits of obvious importance such as resistance to Dutch elm disease, no consensus exists regarding the number, nature or inheritance of the genes involved (Sticklen et al. 1992).

Plant breeders are taking advantage of a novel method of identifying molecular markers in specific genomic regions by evaluating pooled DNA samples from progenies which differ for a trait of interest (Michelmore et al. 1991). Rather than beginning with a molecular genetic map and attempting to map QTL's onto it, this procedure begins with segregating populations, partitions progenies into two pools (e.g. susceptible and resistant), and identifies molecular markers linked to the selected region. Markers which are polymorphic between the dissimilar pools will exhibit linkage to those genes used to create the pools; all other markers appear to be heterozygous. Such a procedure may work well in situations where genetic mapping is limited or incomplete, or for species for which near isogenic lines are not available. Bulked segregant analysis appears useful for a variety of woody plant problems, if progenies from a single cross can be accurately assigned to dissimilar classes for analysis. Certain interspecific elm combinations may lend themselves to such an approach, but as with other mapping approaches, linkage disequilibrium is requisite. The ability to make wide crosses in elms which yield large numbers of progenies in a short period of time, and which segregate for a variety of morphological and physiological traits, should permit us to ascertain the value of such an approach in woody plant breeding.

Conclusions

Applications of molecular markers in woody plants are expanding rapidly to address problems in systematics, evolutionary biology, and breeding. Many of the difficulties inherent in working with woody plants can be ameliorated by the judicious use of molecular techniques. Careful application of some technologies can form an important part of many research programs, but they are not a panacea. A marriage of appropriate techniques, both conventional and molecular, will likely succeed far better than the indiscriminate application of either alone.

References

Burnham, R. 1986. Foliar morphological analysis of the Ulmoideae (Ulmaceae) from the early Tertiary of western North America. Palaeontographica Abt. B. 201:135167.

Clegg, M., and G. Zurawski. 1992. Chloroplast DNA and the study of plant phylogeny: present status and future prospects. Pages 1-13 in: Molecular Systematics of Plants. P. Soltis, D. Soltis, and J. Doyle, eds. Chapman and Hall, New York.

Dudley, J., M. Saghai Maroof, and G. Rufener. 1992. Molecular marker information and selection of parents in corn breeding programs. Crop Sci. 32:301-304.

Fu, L. K. 1980. [A study of the genus *Ulmus* in China.] Journ. of North-eastern Forestry Inst. (Beijing). No. 3 (Sept.):26. [In Chinese].

Gianassi, D. 1978. Generic relationships in the *Ulmaceae* based on flavonoid chemistry. Taxon 27(4):331-342.

Grudzinskaya, I. 1974. [On taxonomic position and area of the section *Chaetoptelea,* genus *Ulmus.]* Bot. Zhurn. (Leningrad) 59:61-66. [In Russian].

Guries, R., and E. B. Smalley. 1986. Elms for today and tomorrow. Pages 214-218 in: Proc. 3rd Nat. Urban For. Conf. A. Philips and D. Gangloff, comps. Orlando, Florida.

Hamby, R., and E. Zimmer. 1992. Ribosomal RNA as a phylogenetic tool in plant systematics. Pages 50-91 in: Molecular Systematics of Plants. P. Soltis, D. Soltis, and J. Doyle, eds. Chapman and Hall, New York.

Hardin, J. 1975. Hybridization and introgression in *Quercus alba*.. J. Arnold Arbor. 56:336-363.

Helentjaris, T., M. Slocum, S. Wright, A. Schaefer, and J. Nienhuis. 1986. Construction of genetic linkage maps in maize and tomato using restriction fragment length polymorphisms. Theor. Appl. Genet. 72:761-769.

Heybroek, H. 1976. [Classification and nomenclature in the genus *Ulmus.*] Groen 32(8):237-240 (In Dutch with English summary.]

Heybroek, H. 1983. Resistant elms for Europe. Pages 108-113 in: Research on Dutch Elm disease in Europe. D. A. Burdekin, ed. Forestry Comm. Bull. #60, London.

Lander, E., and D. Botstein. 1989. Mapping Mendelian factors underlying quantitative traits using RFLP linkage maps. Genetics 121:185-199.

Michelmore, R., l. Paran, and R. Kesseli. 1991. Identification of markers linked to disease-resistance genes by bulked segregant analysis: A rapid method to detect markers in specific genomic regions by using segregating populations. Proc. Nat. Acad. Sci. (USA) 88:9828-9832.

O'Malley, D.1992. Personal communication.

Palmer, J., R. Jansen, H. Michaels, M. Chase, and J. Manhart. 1988. Phylogenetic analysis of chloroplast DNA variation. Ann. Missouri Bot. Gard. 75: 1180-1218.

Paterson, A., E. Lander, J. Hewitt, S. Peterson, S. Lincoln, and S. Tanksley. 1988. Resolution of quantitative traits into Mendelian factors by using a complete linkage map of restriction fragment length polymorphisms. Nature 335:721-726.

Richens, R. 1983. Elm. Cambridge Univ. Press, Cambridge.

Rieseberg, L., and S. Brunsfeld. 1992. Molecular evidence and plant introgression. Pages 151-176 in: Molecular Systematics of Plants. P. Soltis, D. Soltis, and J. Doyle. Chapman and Hall, New York.

Schneider, C. 1916. [Guide to knowledge of the genus *Ulmus. I.* Division of the genus and survey of the species] Oster. Bot. Zeitschr. 66:21-34. [In German].

Sherman, S., and D. Gianassi. 1988. Foliar flavonoids of *Ulmus* in eastern North Amcrica.13iochem. System. Ecol. 16:51-56.

Smith, R. and K. Sytsma. 1990. Evolution of *Populus nigra* (section *Aigeiros*): Interspecific hybridization and the chloroplast contribution of *Populus alba* (section *Populus).* Amer. J. Bot. 77:1176-1187.

Soltis, D., P. Soltis, and B. Milligan. 1992. Intraspecific chloroplast DNA variation: systematic and phylogenetic implications. Pages 117-150 in: Molecular Systematics of Plants. P. Soltis, D. Soltis, and J. Doyle, eds. Chapman and Hall, New York.

Song, K., J. Suzuki, M. Slocum, P. Williams, and T. Osborn. 1991. A linkage map of *Brassica rapa* (syn. *campestris)* based on restriction fragment length polymorphism loci. Theor. Appl. Genet. 82:296-304.

Sticklen, M., M. Bolyard, R. Hajela, and L. Duchesne. 1991. Molecular and cellular aspects of Dutch elm disease. Phytoprotect.72:1-13.

Stuber, C., M. Edwards, and J. Wendel. 1987. Molecular marker-facilitated investigations of quantitative trait loci in maize. II Factors influencing yield and its componcnt traits. Crop. Sci. 27:639-648.

Swofford, D. 1991. PAUP: Phylogenetic Analysis Using Parsimony. vers. 3.0r. Illinois Nat. Hist. Surv., Champaign, IL.

Tanksley, S., N. Young, A. Paterson, and M. Bonierbale. 1989. RFLP mapping in plant science; New tools for an old science. Bio/Technology 7:257-264.

Townsend, A. M., L. Schreiber, W. Masters, and S. Bentz. 1991. "Frontier" elm. Hortsci. 26:80-81.

Wiegrefe, S. J. 1992. Molecular Genetic Variation in the Ulmaceae: Phylogenetic Implications. Ph. D. Dissertation, Univ. Wisconsin-Madison., 166 pp.

Wang, X.-R., A. Szmidt, A. Lewandowski, and Z.-R. Wang. 1990. Evolutionary analysis of *Pinus densata* Masters, a putative Tertiary hybrid. I. Allozyme variation. Theor. Appl. Genet. 80:635-640.

20
Mechanisms Of Resistance: Can They Help Save Susceptible Elms?

Luc C. Duchesne

Abstract

This paper emphasizes the different mechanisms involved in host as well as nonhost resistance to Dutch elm disease. Resistance to Dutch elm disease is associated with the ability of elms to localize the pathogen. Several hypotheses have been advanced to explain this process: vascular occlusion, anatomical features of elms, compartmentalization associated with lignin and suberin, and synthesis of fungitoxic/fungistatic substances that may act as phytoalexins. Evidence indicates that these mechanisms act in a complementary manner. Regulatory loci are also postulated to control resistance because they may determine the expression of defense reactions following infection. A thorough knowledge of elm defense mechanisms may be helpful in developing new varieties of elms through tissue culture manipulation and genetic engineering. This knowledge may also be used for the protection of susceptible trees in the field.

Introduction

The mechanisms responsible for resistance to Dutch elm disease (DED) caused by *Ophiostoma ulmi* have been studied extensively over the past 60 years. The main purpose of these studies was to elucidate and manipulate the natural means of resistance to DED. The advantages of such knowledge are two-fold. First, it may enable forest practitioners and arboriculturists to protect already existing elm specimens through the manipulation of these resistance mechanisms. Second, it may provide tree breeders with useful screening tools for resistance to DED. Although numerous investigations have emphasized the mechanisms of resistance to DED, this phenomenon remains poorly understood (Duchesne 1988; Elgersma 1982; MacHardy 1978).

The objective of this paper is to review some of the more recent advances regarding the mechanisms of resistance to DED. However, presenting an appropriate summary of the literature is difficult for three reasons. First, the various mechanisms of resistance to DED differ: (a) in their mode of action; (b) upon tree species: (c) upon the pathogen strain or isolate; (d) upon the timing of their action following infection by *O. ulmi*; and, (e) upon the environmental conditions which may alter their expression. Second, contradictory evidence is sometimes presented regarding the significance of certain mechanisms of resistance. Third, all evidence regarding the mechanisms of resistance to DED is circumstantial. Therefore, in reading this review it is important to acknowledge that generalizations will invariably introduce a certain level of uncertainty. Nevertheless, it is important to obtain an overview of the mechanisms of resistance to DED in order to emphasize areas of knowledge that need further investigation and, more importantly, to discuss new avenues of research and technology that may lead to the control of DED.

Localization of Infection

Host and nonhost resistance to DED is associated with the ability of elms to localize infection by limiting the movement of *O. ulmi* in the vascular system (Duchesne et al. 1990; Sinclair and Brener 1974; Sinclair and Larsen 1980; Sinclair et al. 1975). Moreover, the rapidity at which the pathogen is localized is presented as a resistance determinant; thus, a slow response by the plant will lead to a systemic spread of fungal propagules and, eventually, death of the plant resulting from a drastic impairment of water conductance of the functional xylem (MacHardy 1978). One may speculate that elms are particularly sensitive to vascular dysfunction because they are ring-porous trees wherein water is conducted mostly by the large earlywood vessels of the most recent growth ring (Newbanks et al. 1983). Because the pathogen relies mainly on the xylem flow for its spread within the tree, infection is mostly limited to the most recent growth ring. This emphasizes further the need for a rapid localization of the pathogen within the vascular system.

Shigo and Tippet (1981) have proposed that localization of *O. ulmi* by *Ulmus americana* follows the CODIT model (Compartmentalization of Decay In Trees) (Shigo 1984). Trees that are able to rapidly compartmentalize survive infection by *O. ulmi*, whereas trees with delayed compartmentalization are more severely affected by the fungus. This observation was further extended to other elm species and clones by Bonsen et al. (Bonsen et al. 1985) and was corroborated by analysis of nonhost resistance to DED (Rioux and Ouellette 1989; 1991a; 1991b). An easy indicator of localization of the pathogen in elms is the distribution of vascular browning, itself a marker of fungal spread. Pomerleau (1968) observed that vascular browning follows, by up to eight days, the spread of the pathogen in *U. americana*. Thus, one may speculate that the high levels of vascular browning found in *U. americana* results from a delayed compartmentalization process. *O. ulmi* may remain localized for long periods of time in asymptomatic elms (Banfield 1968; Jones et al. 1978). Three major hypotheses have been proposed to explain the localization of infection by *O. ulmi* in elms: vascular blockage, anatomical features, and formation of chemical barriers.

Vascular blockage

The rapidity at which vascular blockage takes place in the vascular system is a resistance determinant. DED being a vascular disease it is reasonable to expect that restriction in the xylem flow will impede the development of the disease for a short period of time. This sealing off response is thought to operate through the formation of tyloses, gels, the accumulation of fungal hyphae in the tracheids and vessels, and embolization (Newbanks et al.; 1983; Van Alfen and MacHardy 1978).

Tyloses are formed by the bulging of adjoining parenchyma cells through pit membranes into vessels (Beckman 1971). Elgersma (1973) observed that tyloses form faster in the resistant clone *U. hollandica* Mill. '390' than in the susceptible clone 'Belgica'. Further studies undertaken after the discovery of aggressive isolates of *O. ulmi* could not, however, correlate tylosis formation and resistance to DED (Elgersma and Heybroek 1979; Elgersma and Miller 1977; MacDonald and McNabb 1974). Tylosis formation was consequently described as a minor mechanism of resistance to DED (Elgersma 1982a; 1982b; MacDonald and McNabb 1974). Gels are mainly of a pectinaceous and hemicellulosic nature with a fibrillous structure (Ouellette, personal communication) and are deposited in vessels and fibers (Vandermolen et al. 1977). Ouellette (1981a; 1981b) reported the occurrence of gel-like deposits in the vessels and fibers of *U. americana* infected with *O. ulmi*. It is not known, however, whether these deposits are produced by the plant or the fungus, and whether they are implicated in resistance to DED for *U. americana*.

Anatomical features

Vessel width and length were first associated with resistance to DED by Elgersma (1970) who noted that clone '390' (resistant) had smaller and shorter vessels than the susceptible species *U. carpinifolia*. He suggested that these two factors may favor the action of tyloses and increase the number of obstacles to the transport of the fungus in the vascular system. The comparison of susceptible and putatively resistant clones of *U. americana* led to similar results (Sinclair et al. 1975), but the comparison of *U. americana* and *U. pumila* (resistant) showed little difference in vessel anatomy (Pope 1943). Therefore, vessel width and length cannot always be correlated with resistance to DED.

Pope (1943) characterized elm latewood by the absence of interconnections between vessels and groups of vessels. He further attributed the resistance of *U. pumila* to its greater proportion of latewood as compared to *U. americana*. In addition, McNabb et al. (1970) showed a trend of increased susceptibility with increasing average vessel size of different elm clones. In the latter study, size of vessel groups was more highly correlated with resistance than were vessel width and length.

Treatment of elms with growth-regulating factors such as 2,3,6-trichlorophenyl acetic acid (TCPA) "induced" resistance to DED. Trees showed smaller vessels and greater tylosis formation than controls after treatment with

TCPA (Brener and Beckman 1968). These results must be interpreted carefully because the effect of TCPA may be multiple (Vandermolen et al .1977). Thus, it is possible that the occurrence of small vessels was coincidentally associated with that of undetected mechanism(s).

Chemical barriers

Chemical, anatomical, and morphological studies have presented a considerable body of evidence for the presence of chemical barriers in elms. Chemical barriers may act directly through nonspecific oxidative stresses on living cells (Nebert et al. 1990) or indirectly through the formation of barriers that are impermeable to pathogens or their molecules.

Fungitoxic/fungistatic chemicals:

Extracts from seeds and seedlings of *U. americana* up to seven months-old contain capric acid, a substance inhibitory to spore germination of *O. ulmi* (Doskotch 1975; Schreiber 1970). Capric acid has not, however, been associated with resistance to DED in resistant elm species or in resistant specimens of *U. americana*.

The synthesis of phytoalexins by elms was the object of numerous investigations that sometimes present contradictory evidence, presumably because of different experimental conditions and methodologies. Phytoalexins are low molecular weight fungitoxic compounds accumulated by plants in response to several stress agents, notably in response to infection by microorganisms (Paxton 1981). Phytoalexin accumulation is associated with disease resistance. Fast accumulation of large amounts of phytoalexins is observed in incompatible host-pathogen interactions whereas delayed accumulation and/or accumulation of small amounts of phytoalexin is associated with compatible host-pathogen interactions (Darwill and Albersheim 1984; Deverall 1982; Keen 1981; Kuc 1976; Mansfield 1982).

Overeem and Elgersma (1970) were the first to investigate the production of phytoalexins in elms in response to infection by *O. ulmi*. Inoculation of *U. hollandica* with *O. ulmi* led to the accumulation of two naphthoquinones, the mansonones E and F, which had been originally isolated along with several closely related quinoid pigments from the West African tree *Mansonia altissima* Chev. (Galeffi et al. 1968; Marini-Bettolo et al. 1965; Tanaka et al. 1966). However, the presence of these compounds was not correlated with resistance to DED because cuttings of the resistant clone '390' did not accumulate more mansonones E and F than did the susceptible clone 'Belgica' (Elgersma and Overeem 1971). Bell and Mace (1981) found these results questionable because the extraction of mansonones in the latter experiment was conducted using boiling ethanol and, so, it is possible that the mansonones underwent heat degradation during extraction. It was also established that mansonone E and F accumulation in elms is also triggered by inoculation with *Verticillium albo-*

atrum, sulphuric acid, and ethanol (Elgersma and Overeem 1971; Overeem and Elgersma 1970).

New attention was given to mansonone accumulation in elms after it was observed that cross-protection against *O. ulmi* is associated with mansonone accumulation. Seedlings of *U. americana* that were cross-protected against an aggressive isolate by preinoculation with a non-aggressive isolate of *O. ulmi* accumulated larger quantities of mansonones E and F and ß-sitosterol than did seedlings inoculated with an aggressive isolate only (Hubbes and Jeng 1981; Jeng et al. 1983). These results presented the first circumstantial evidence implicating mansonone accumulation in resistance to DED. A thorough chemical analysis from wood of *U. americana* infected by *O. ulmi* revealed the presence of mansonones A, C, D, E, F, and G (Fig. 1), along with ß-sitosterol and scopoletin (Dumas et al. 1983).

Proctor and Smalley (1988) found that mansonone E and F concentrations were three times higher in a resistant variety of *U. americana* as compared to mansonone accumulation in a susceptible variety of *U. americana*. Inoculation of *U. americana* seedlings with different isolates of *O. ulmi* led to kinetics of mansonone accumulation that varied according to isolate aggressiveness (Duchesne et al. 1985; Duchesne et al. 1990). Accumulation of mansonones A, C, E, F, and G peaked four weeks after inoculation in seedlings inoculated with a nonaggressive isolate, whereas the accumulation of these substances was delayed to six weeks in seedlings inoculated with an aggressive isolate(Duchesne 1985; Duchesne et al. 1985). In another experiment, the inoculation of three aggressive and three nonaggressive isolates led to similar results, suggesting that the nonaggressive isolates induce more mansonone accumulation in U. americana than do the aggressive isolates (Duchesne et al. 1984; Duchesne et al. 1985; Duchesne et al. , 1990). It was then speculated that aggressiveness in *O. ulmi* is inversely correlated with mansonone accumulation (Duchesne et al. 1985, Fig. 2).

In *U. americana* and in *U. pumila* mansonone accumulation correlates with resistance to DED. The inoculation of *U. pumila* with an aggressive or a nonaggressive isolate of *O. ulmi* showed kinetics of mansonone accumulation that differed from those of *U. americana*. Mansonone accumulation reached its maximum two weeks after inoculation, and no difference was observed between the levels induced by the aggressive or by the nonaggressive isolate (Duchesne et al. 1986).

Investigation of the fungitoxic/fungistatic activity of mansonones suggests that the concentration of mansonone E extracted from *U. americana* and *U. pumila* is tenfold that required to inhibit 80% of the growth of *O. ulmi* in liquid culture using an enriched medium (Proctor and Smalley 1988; Wu 1985; Wu et al. 1989). Different conclusions were presented by Dumas et al. (1986) who investigated mansonone toxicity on linear growth of six isolates of *O. ulmi* using concentrations that were found in elm wood as reported from the chemical extraction of elm branches previously infected with *O. ulmi* (Dumas et al. 1983). The latter experiment suggests that mansonone accumulation cannot be solely responsible for the cross-protection of seedling against aggressive isolates of *O. ulmi* (Hubbes and Jeng 1981; Jeng et al. 1983). It is difficult, however, to assess the effect of the mansonones *in situ* because of factors such as differences between the physiological status of the fungus *in vitro* and *in situ*, actual concentrations of mansonones that fungi must encounter at the fungal-plant interface, and interactions between mansonones and other chemicals synthesized by elms or fungi. One interesting avenue of research is to compare anatomical features of the pathogen after localization in healthy elms with those of fungus exposed to mansonones *in vitro* (Wu et al. 1989).

Considerable evidence suggests that mansonone E and F accumulation in *U. pumila* and *U. americana* is analogous to the accumulation of phytoalexins in host-pathogen interactions and, so, mansonones act as phytoalexins in elms. However, four major questions must be resolved to ascertain the role of mansonones in disease resistance. First, they must be shown to accumulate in the immediate vicinity of the pathogen in the plant. Three investigations have addressed this question. Extraction of mansonones from infection pockets of a *U. americana* tree inoculated with *O. ulmi* shows that mansonones accumulated only in infection pockets but not in the surrounding sapwood. *O. ulmi* could not be recovered from asymptomatic sapwood whereas the fungus was isolated from infection pockets (Duchesne 1985). Moreover, mansonone accumulation correlates with vascular browning in *U. americana* (Duchesne et al. 1990). Proctor and Smalley (1988) showed that mansonone concentrations in discolored xylem of *U. pumila* seedlings three days after infection was more than 100 times greater than the mansonone concentration in the surrounding xylem. Although mansonone accumulation correlates with vascular browning in *U. americana* (Duchesne 1985), more documentation is required to demonstrate that mansonones are in contact with the fungus *in situ*. Third, mansonones must show fungitoxic/fungistatic activity *in situ* regardless of their biological activity *in vitro*. Fourth, mansonone accumulation must be concurrent to disease suppression. Resolving this problem is of paramount importance to determine the role of mansonone accumulation in relation to other mechanisms of resistance to DED. Moreover, the universality of mansonone accumulation as phytoalexins has to be established across the genus *Ulmus*. Although the heartwood, and sometimes the sapwood, of numerous elm species was shown to contain one or more mansonone (Burden and Kemp 1984; Duchesne 1985; Overeem and Elgersma 1970; Rowe et al. 1972; Suzuki et al. 1971; Yasuda et al. 1973), evidence has yet to be presented for a direct role of these substances in other species than *U. americana* and *U. pumila*.

Studies have indicated antagonism between aggressive isolates and nonaggressive isolates of *O. ulmi in situ* when co-inoculated in elms (Jeng et al. 1983; Scheffer et al. 1980). Co-inoculation of seedlings with both an aggressive and a nonaggressive isolate of *O. ulmi* suggests that aggressive isolates may have a suppressive effect on mansonone accumulation in *U. americana* (Duchesne et al. 1985). However, this suppressive influence of aggressive isolates of *O. ulmi* does not seem to operate in seedlings of *U. pumila* (Duchesne et al. 1984). It is, therefore, possible that the suppression of mansonone accumulation in elms is a virulence/pathogenicity factor of this fungus (Fig. 2).

Impermeable barriers:

Recent anatomical studies on the interaction between *O. ulmi* and host and nonhost trees give interesting results regarding the mechanisms of resistance (Rioux and Ouellette 1989; 1991a; 1991b). Particularly, suberin and lignin deposits are observed within barrier zones elicited by the inoculation of *U. americana* and other tree species with *O. ulmi* (Rioux and Ouellette 1991a). Both suberin and lignin are efficient barriers against pathogens because of their resistance to enzymatic degradation. These studies provide valuable insights into both host and nonhost resistance to DED.

Synergism and regulation of disease resistance

Fast-growing elms are more susceptible to DED than slow-growing ones (Elgersma 1981; Sinclair and Brener 1974). It was hypothesized that this phenomenon is associated with smaller vessel size and length in slow-growing trees as compared to fast-growing trees (Sinclair et al. 1975). Nevertheless, it is also possible that other metabolic factors associated with resistance to DED become favorably altered in slow-growing elms. It is important to emphasize that slow-growing elms are not truly resistant to DED, but this phenomenon warrants further investigation as it may provide clues regarding some mechanisms of resistance to DED.

Although several defense reactions against DED have been investigated, evidence for major resistance determinants has yet to be presented. There is a general consensus among scientists that the many minor resistance factors, either performed or inducible, act in a complementary manner (Duchesne et al. 1990; Elgersma 1982a; Elgersma 1982b; MacHardy 1978; Rioux and Ouellette 1991a). A synergistic interaction of defense reactions can also be hypothesized. This would render the elucidation of the mechanisms of resistance to DED even more difficult.

Anatomical features such as vessel diameter and the size of groups of vessels constitute a pre-formed line of defense. Following infection, stress responses that may be either anatomical or chemical evolve, and they constitute a second, inducible, line of defense. Tylosis formation as well as the development of barriers such as the ones described in the CODIT model have been associated with resistance to DED. Barrier formation may involve all of the inducible defense reactions and may be facilitated by pre-formed mechanisms of resistance. So far, the resistance mechanisms to DED were perceived as plant functions controlled by genes that are structurally encoded, such as the genes encoding for enzymes leading to mansonone synthesis, lignin synthesis, or vessel development. However, there is increasing evidence that resistance to DED may be associated with mechanisms that regulate the expression of the resistance factors rather than the resistance factors themselves.

The timing of expression of the different mechanisms of resistance to DED is critical for both anatomical and chemical means of defense to be effective in localizing the pathogen. This hypothesis is corroborated by three types of evidence. First, mansonone accumulation is faster in *U. pumila* than in *U. americana* (Duchesne et al. 1985). Second, mansonone accumulation is faster in *U. americana* inoculated with aggressive isolates than in *U. americana* inoculated with non-aggressive isolates. Third, barrier zone formation is faster in nonhost trees than in host trees inoculated with *O. ulmi* (Rioux and Ouellette 1991a). Because resistance is associated with the timely expression of the resistance mechanisms, their regulation becomes a critical determinant. One interesting hypothesis is that the regulation of resistance to DED is controlled by regulatory gene loci. Tepper and Anderson (1984) speculated that plant resistance to pathogens is associated with the presence of regulatory loci that control structural genes. Regulatory loci are genes that control the expression of other genes (Goff et al. 1990; Hedrick and McDonald 1980). At present we have no reports of attempts to identify such genes for the metabolic pathways encoding for enzymes leading to mansonone synthesis in elms. There are many cellular levels where the expression of mechanisms of resistance to DED may be regulated and future investigations should attempt to determine how the different mechanisms of disease resistance are regulated in elms.

Applications of Resistance Mechanisms

There are three general areas of DED research that would benefit greatly from a thorough knowledge of the mechanisms of resistance this disease: screening for selection of resistant elms varieties, genetic engineering of new elms through genomic transformation, and protection of trees in the field.

Screening elms for resistance to DED is time consuming and the results are often discouraging because symptom expression is influenced by age of trees, nutrition, vigor, and soil moisture, as well as month, mode, and locus of inoculation (Van Alfen and MacHardy 1978). The screening of elms in tissue culture is a promising alternative to large scale field trials. Numerous investigations have established that elms are amenable to cell cultivation and organogenesis (Duchesne 1985; Durzan and Lopushanski 1985; Fink et al. 1986; Redembaugh et al. 1981; Sticklen et al 1986; Sticklen et al. 1985). The interaction of *O. ulmi* and elms in tissue culture has received considerable attention (Krause et al. 1988; Pijut et al. 1990; Sticklen et al. 1991; Pijut et al. 1988; Rowe et al. 1972; Sinclair et al. 1975; Szczegola et al. 1987; Yang et al. 1989). It has been demonstrated that mansonone accumulation in tissue cultures of *U. americana* and *U. pumila* is similar to that of seedlings inoculated with the pathogen (Duchesne and Sticklen, unpublished results; Szczegola et al. 1987) suggesting that screening for mansonone accumulation may be done using cultivated cells. Tissue cultures can be used either in the screening of putatively resistant material or the isolation of new varieties generated by somaclonal variation (Diner and Karnosky 1987). However, it is important to realize that cell cultivation offers a limited means for *in vitro* selection and, so, a good screening program should also include extensive field analysis of promising material pre-screened *in vitro*.

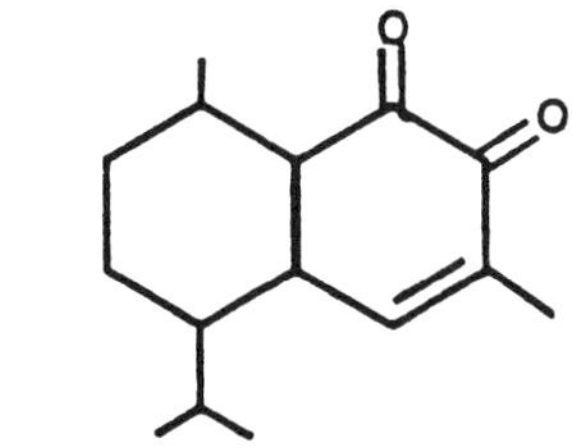

MANSONONE A

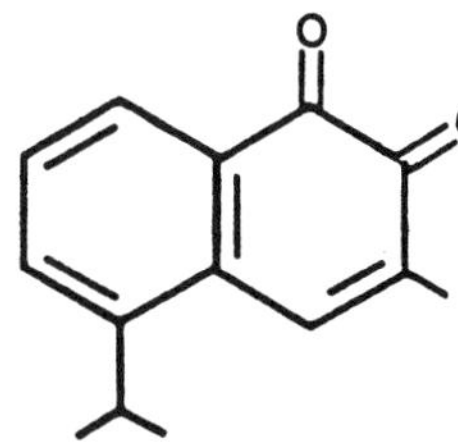

MANSONONE C

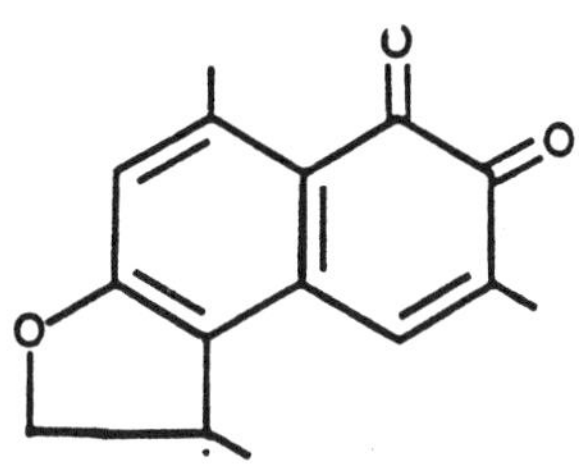

MANSONONE D

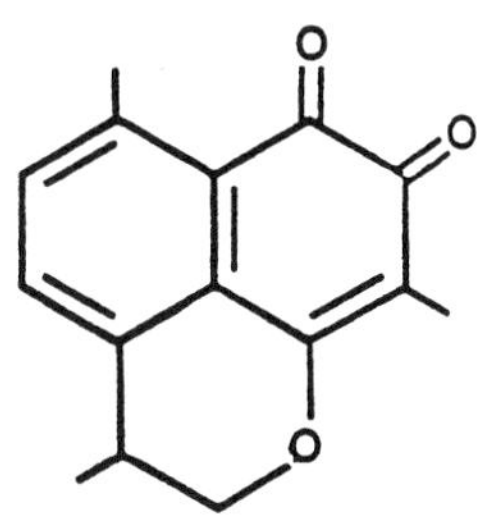

MANSONONE E

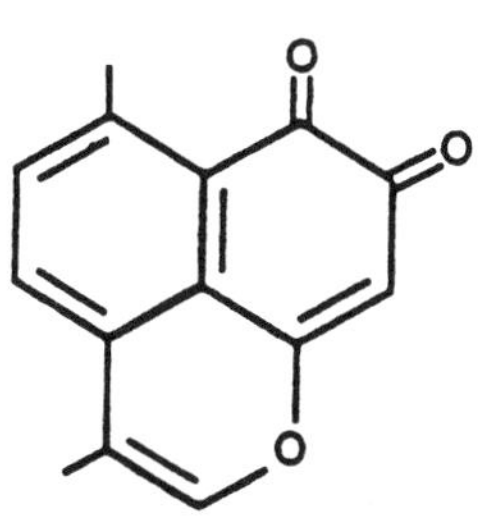

MANSONONE F

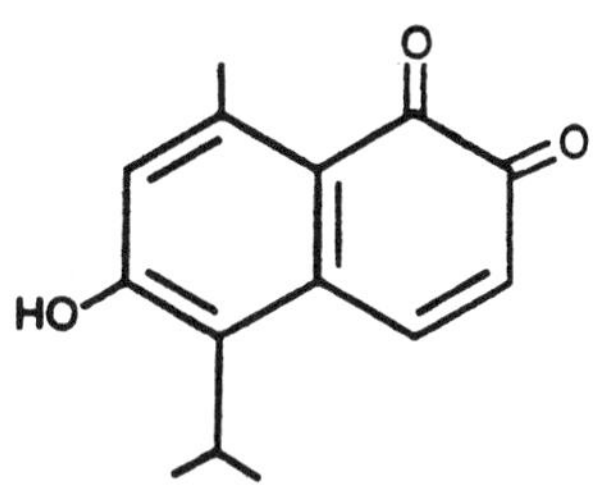

MANSONONE G

Figure 1. Mansonones from elms

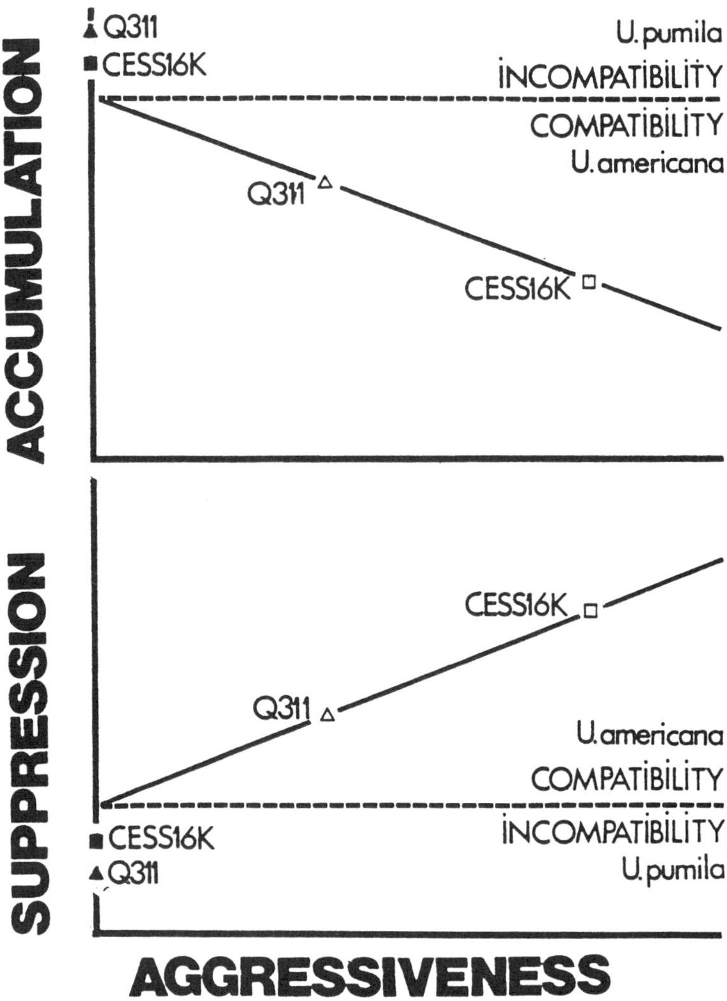

Figure 2. Model showing the relation between mansonone accumulation and compatibility between *U. americana*, *U. pumila*, and *O. ulmi*.

Genomic transformation of hardwoods was achieved for species such as *Juglans regia* (McGranaham et al. 1988), *Populus alba* x *grandidentata* (Fillati et al. 1987), and *Malus pumila* (James et al. 1989). This technology allows the introduction of foreign genes and their expression in the genome of trees. There are two general sources of genetic material that could be used for the development of transgenic elms. First, should the isolation of genes controlling resistance to DED be successful, then it will be possible to transform susceptible species with these genes. Such work is feasible only if resistance can be associated with a limited number of genes. Second, genes encoding for resistance mechanisms in other plants could be used for the genomic transformation of elms. For example such genes encode for the synthesis of hydroxyproline-rich glycoproteins (Collinge and Slusarenko 1987), protease inhibitors (Ryan 1990), fatty acid synthetase (Willemot et al. 1968), chitinase (Hedrick et al. 1988), pathogenicity-related proteins (Legrand et al. 1987), and glucanases (Kauffman et al. 1987). This is only possible if *O.* ulmi is sensitive to the actions of the products of those genes and also if they are expressed at times and levels that will interfere with pathogenesis. Transgenic resistance at the operational level may be difficult to maintain if it is multigenically rather than monogenically controlled (Duchesne 1992).

In the future, it may be possible to manipulate elm trees in order to protect them against DED. One field of research includes the injection of elms with substances or microorganisms that stimulate the onset of defense reactions. One example is the cross-protection of elms against an aggressive isolate of *O. ulmi* using a non-lethal inoculation with a non-aggressive isolate of this fungus to presumably induce mansonone accumulation (Hubbes and Jeng 1981; Jeng et al. 1983). However, for this to be effective, protective injections must be conducted shortly before infection by the pathogen or must have a lasting effect that protect trees over a long period of time. Another possible prophylaxis is the manipulation of environmental conditions that will favor the development and/or resistance of defense reactions. Little work has been conducted to document the link between environmental conditions and the expression and regulation of defense reactions in elms. Therefore, it is difficult to predict whether or not this will lead to the development of techniques that will help in the control of DED.

Conclusions

Considerable advances have been made in understanding the mechanisms of resistance to DED since its discovery in the 1920s. Pioneer research in this field emphasized single mechanisms of resistance in order to help breeders with the difficult task of developing disease-resistant trees. However, studies showed that resistance to DED is not based on a single mechanism of defense but rather on several defense mechanisms that interact with one another during the localization process. More recent evidence suggests that a combination of passive defense mechanisms along with the concerted and timely expression of active defense reactions is critical to localize the pathogen. Future research should thoroughly investigate the regulatory mechanisms controlling the expression of active defense reactions. Recent developments in molecular and cellular biology should contribute greatly toward this goal (Sticklen et al. 1991).

Acknowledgments

Thanks are addressed to A. Yapa for reviewing this manuscript.

References

Banfield, W.M. 1968. Dutch elm disease recurrence and recovery in American elm. Phytopath. Z. 62:21-30.

Beckman, C.H. 1971. The plasticizing of plant cell walls and tyloses formation. Physiol. Pl. Path. 1:1-10.

Bell, A.A., and Mace, M.E. 1981. Biochemistry and physiology of resistance. Pages 431-486 in Fungal wilt disease in plants. Academic Press, New York. M.E. Mace, A.A. Bell & C.H. Beckman,eds.

Bonsen, K.J.M., Scheffer, R.J., and Elgersma, D.M. 1985. Barrier zone formation as a resistance mechanism of elms to Dutch elm disease. IAWA Bull., n.s. 6:71-77.

Brener, W.D., and Beckman, C.H. 1968. A mechanism of enhanced resistance to *Ceratocystis ulmi* in American elms treated with sodium trichloro-phenylacetate. Phytopathology 58:555-561.

Burden, R.S., and Kemp, M.S. 1984. Sesquiterpene phytoalexins from *Ulmus glabra*. Phytochemistry 23:383-385.

Collinge, D.B., and Slusarenko, A.J. 1987. Plant gene expression in response to pathogens. Plant Mol. Biol. 9:389-410.

Darwill, A.G., and Albersheim, P. 1984. Phytoalexins and their elicitation - a defense against microbial infection in plants. Ann. Rev. Plant Physiol 35:243-275.

Deverall, B.J. 1982. Introduction. pp 1-20 in: Phytoalexins. J.A. Bailey and J.W. Mansfield, eds. Blackie and Son, London.

Diner, A.M., and Karnosky, D.F. 1987. Tissue culture application to forest pathology and pest control. pp. 351-373 in Cell and Tissue Culture in Forestry, vol.2, specific principles and methods: growth and developments. J.M. Bonga and D.J. Durzan, eds. Martinus Nijhoff Publishers, Boston.

Doskotch, R.W., Keely, S.L., and Schreiber, L.R. 1975. Isolation and identification of an antifungal agent from seeds of American elms. Phytopathology 65:634-635.

Duchesne, L.C. 1985. Phytoalexin accumulation and resistance to Dutch elm disease. M.Sc. thesis, University of Toronto, Toronto. 111 pp.

Duchesne, L.C. 1988. Resistance mechanisms to Dutch elm disease: a review. Nat. Can. (Rev. Ecol. Syst.). 115:163-167.

Duchesne, L.C. 1992. Ecological impact of biotechnology in Canadian Forests. For. Chronicle (in press).

Duchesne, L.C., Hubbes, M., and Jeng, R.S. 1986. Mansonone E and F accumulation in *Ulmus pumila*, resistant to Dutch elm disease. Can. J. Forest Res. 16:410-412.

Duchesne, L.C., Jeng, R.S., and Hubbes M. 1984. Effect of *Ceratocystis ulmi* train aggressiveness on phytoalexin accumulation in *Ulmus americana*. Can. J. Phytopath. 6:261.

Duchesne, L.C., Jeng, R.S., and Hubbes, M. 1985. Accumulation of phytoalexins in *Ulmus americana* in response to infection by a nonaggressive and an aggressive strain of *Ophiostoma ulmi*. Can. J. Bot. 63:678-680.

Duchesne, L.C., Jeng, R.S., Hubbes, M., and Sticklen, M.B. 1990. Accumulation of Mansonones E and F in seedlings of *Ulmus americana* in response to inoculation with *Ophiostoma ulmi*. Trees 4:187-190.

Dumas, M.T., Strunz, G.M., Hubbes, M., and Jeng, R.S. 1983. Isolation and identification of six mansonones from *Ulmus americana* infected with *Ceratocystis ulmi*. Experientia 39:1089-1090.

Dumas, M.T., Strunz, G.M., Hubbes, M., and Jeng, R.S. 1986. Inhibition of *Ceratocystis ulmi* by mansonones A,C, D, E, F, and G isolated from *Ulmus americana*. Eur. J. For. Pathol. 4:217-222.

Durzan, D.J., and Lopushanski, S.M. 1985. Propagation of American elm via cell suspension cultures. Can. J. For. Res. 5:273-277.

Elgersma, D.M. 1970. Length and diameter of xylem vessels as factors in resistance of elms to *Ceratocystis ulmi*. Neth. J. Pl. Path. 2: 161-172.

Elgersma, D.M. 1973. Tyloses formation in elms after inoculation with *Ceratocystis ulmi*, a possible resistance mechanism. Neth. J. Pl. Path. 79:218-220.

Elgersma, D.M. 1981. Host susceptibility and resistance. Pages 17-18 in Compendium of elm diseases. R.J. Stipes and R.J. Sinclair, eds. The American Phytopathological Society, StPaul, MN.

Elgersma, D.M. 1982a. Susceptibility and possible mechanisms of resistance to Dutch elm disease. pp. 169-177 in Proceedings of the Dutch elm disease symposium and workshop. E.S. Kondo, Y. Hiratsuaka and W.B.G. Denyer, eds. Manitoba Department of Natural Resources, Winnipeg.

Elgersma, D.M. 1982b. Resistance mechanisms in elms to Dutch elm disease. Pages 143-152 in Resistance of disease and pests in forest trees. Proceedings of the Third International Workshop on the Genetics of Host-parasite Interactions in forestry, 1980. H.M. Heybroek, B.R. Stephan and K. von Weissenberg, eds. Pudoc, Wageningen.

Elgersma, D.M., and Heybroek, H.M. 1979. Spread and survival of an aggressive and a nonaggressive strain of *Ophiostoma ulmi* in elms. Neth. J. Pl. Path. 85:235-240.

Elgersma, D.M., and Miller, H.J. 1977. Tylosis formation in elms after inoculation with an aggressive or a nonaggressive strain of *Ophiostoma ulmi* or with a non-pathogen in elms. Neth. J. Pl. Path. 83:241-243.

Elgersma, D.M. and Overeem, J.C. 1971. The relation of mansonones to resistance against Dutch elm disease and their accumulation, as induced by several agents. Neth. J. Pl. Path. 77:168-174.

Fillati, J.A., Sellmer, J., McCown, B., Haissig, B., and Comai, L. 1987. *Agrobacterium*-mediated transformation and regeneration of *Populus*. Mol. Gen. Genet 206:192-199.

Fink, C.V.M., Sticklen, M.B., Lineberger, R.D., and Domir, S.C. 1986. *In vitro* organogenesis from shoot tip, internode, and leaf explants of *Ulmus* x 'PIONEER'. Plant Cell Tissue Organ Cult. 7:237-245.

Galeffi, C., Casinova, C.G., Miranda Della Monache, M., and Marini-Bettolo, G.B. 1968. A new method for the isolation of the chloroform soluble constituents of *Mansonia* wood. Annali Ist. Sup. Sanità 4:305-316.

Goff, S.A., Klein, T.M., Roth, B.A., Fromm, M.E., Cone, K.C., Radicella, J.P., and Chandler, V.L. 1990. Transactivation of anthocyanin biosynthetic genes following transfer of B regulatory genes into maize tissues. EMBO J. 9:2517-2522.

Hedrick, P.W., and McDonald, J.F. 1980. Regulatory gene adaptation: an evolutionary model. Heredity 45:83-97.

Hedrick, S.A., Bell, J.N., Boller, T., and Lamb, C.J. 1988. Chitinase cDNA cloning and mRNA induction by fungal elicitor, wounding and infection. Plant Physiol. 86:182-186.

Hubbes, M., and Jeng, R.S. 1981. Aggressiveness of *Ceratocystis ulmi* trains and induction of resistance in *Ulmus americana*. Eur. J. Forest Pathol. 11:257-264.

James, D.J., Passey, A.J., Barbara, D.J., and Bevan, M. 1989. Genetic transformation of apple (*Malus pumila*) using a disarmed Ti-binary vector. Plant Cell Rep. 7:658-661.

Jeng, R.S., Alfenas, A.C., Hubbes, M., and Dumas, M.T. 1983. Presence and accumulation of fungitoxic substances against *Ceratocystis ulmi* in *Ulmus americana*: possible relation to induced resistance. Eur. J. Forest Pathol. 13:239-244.

Jones, R.K., Krass, J.J., and Sava, R.J. 1978. Isolation of *Ceratocystis ulmi* from 14-year-old rings of English elms in California. Plant Dis. Reptr. 62:994-996.

Kauffman, S.,Legrand, Geoffrey, P., and Lamb,C.. 1987. Biological functions of pathogenesis-related proteins-4PR proteins of tobacco have 1,3-beta-glucanase activity. EMBO J. 6:3209-3212.

Keen, N.T. 1981. Evaluation of the role of phytoalexins. pp. 155-177 in Plant Disease Control, R.C. Staple, ed. John Wiley and Sons, New York.

Krause, C.R. Ichida, J.M., Schreiber, L.R., and Domir, S.C. 1988. Host-parasite relationships of elm calli challenged with *Ceratocystis ulmi*. Phytopathology 78:1610.

Kuc, J. 1976. Phytoalexins. Pages 632-652 in Encyclopedia of plant physiology, vol. 4, R.H. Heitefuss and P.H. Williams, eds. Springer-Verlag, New York.

Legrand, M., Kauffman, S., Geoffrey, P., and Fritig, B. 1987. Biological function of pathogenesis-related proteins: four tobacco pathogenesis-related proteins are chitinases. Proc. Natl. Acad. Sci. USA 84:6750-6754.

MacDonald, W.L., and McNabb Jr., H.S. 1974. Electron microscope observations of *Ceratocystis ulmi* induced tyloses development in *Ulmus*. Eur. J. Forest Pathol. 4:2-10.

MacHardy, W.E. 1978. Mechanisms of resistance. pp. 25-27 in Dutch elm disease: perspective after 60 years. W.A. Sinclair, and R.J. Campana, eds. Cornell Univ. Agric. Expt. Stn. Search (Agriculture) 8.

Mansfield, J.W. 1982. The role of phytoalexins in disease resistance. pp. 106-132 in Phytoalexins, J.A. Bailey and J.W. Mansfield, ed. Blackie and Son, London.

Marini-Bettolo, G.B., Casinovi, G.G., and Galeffi, C. 1965. A new class of quinones: sesquiterpenoid quinones of *Mansonia altissima* Chev. Tetrahedron Lett. 52:4852-4864.

McGranaham, G.H., Leslie, C.A., Uratsu, S.L., Martin, L.A., and Dandekar, A.M. 1988. *Agrobacterium*-mediated transformation of walnut somatic embryos and regeneration of transgenic plants. Bio/Technology 6:800-804.

McNabb, H.S., Heybroek, H.M., and MacDonald, W.L. 1970. Anatomical factors in resistance to Dutch elm disease. Neth. J. Pl. Path. 76:196-204.

Nebert, D.W., Petterson, D.D., and Fornace, A.J. 1990. Cellular responses to oxidative stress: the [Ah] geen battery as a paradigm. Environmental Health Perspectives 88:13-25.

Newbanks, D., Bosch, A., and Zimmerman, M.H. 1983. Evidence for xylem dysfunction by embolization in Dutch elm disease. Phytopathology 73:1060-1063.

Ouellette, G.B. 1981a. Ultrastructural cell wall modifications in secondary xylem of American elms surviving the acute stage of Dutch elm disease: vessel members. Can. J. Bot. 59:2411-2424.

Ouellette, G.B. 1981b. Ultrastructural cell wall modifications in secondary xylem of American elms surviving the acute stage of Dutch elm disease: fibres. Can. J. Bot. 59:2425-2438.

Overeem, J.C., and Elgersma, D.M. 1970. Accumulation of mansonones E and F in *Ulmus hollandica* infected with *Ceratocystis ulmi*. Phytochemistry 9:1949-1952.

Paxton, J.D. 1981. Phytoalexins: A working redefinition. Phytopath. Z. 101:106-109.

Pijut, P.M., Domir, S.C., Lineberger, R.D., and Schreiber, L.R. 1990. Use of culture filtrates of *Ceratocystis ulmi* as a bioassay to screen for disease tolerant *Ulmus americana*. Plant Science 70:191-196.

Pijut, P.M., Lineberger, R.D., Domir, S.C., Ichida, J.M., and Krause, C.R. 1990. Ultrastructure of cells of *Ulmus americana* culltured *in vitro* and exposed to the culture filtrate of *Ceratocystis ulmi*. Phytopathology 23:787.

Pijut, P.M., Lineberger, R.D., Domir, S.C. and Schreiber, L.R. 1988. Response of elm callus to culture filtrate of *Ceratocystis ulmi* and correlation with whole plant disease reaction. HortScience 23:787.

Pomerleau, R. 1968. Progression and localisation de l'infection par le *Ceratocystis ulmi* dans l'orme d'amérique. Phytopathol. Z. 63:301-326.

Pope, S.A. 1943. Some studies on the Dutch elm disease and the causal organism. Ph. D. thesis, Cornell Univ., Ithaca, New York. 66 pp.

Proctor, R.H., and Smalley, E.B. 1988. Localized accumulation of mansonones E and F in elms following inoculation with *Ophiostoma ulmi*. Can. J. Pl. Pathol. 10:371.

Redembaugh, M.K., Westfall, R.D. and Karnosky, D.F. 1981. Diploid callus production from *Ulmus americana* anthers. Bot. Gaz. 142:19-26.

Rioux, D., and Ouellette, G.B. 1989. Light microscope observations of histological changes induced by *Ophiostoma ulmi* in various nonhost trees and shrubs. Can. J. Bot. 67:2335-2351.

Rioux, D., and Ouellette, G.B. 1991a. Barrier zone formation in host and nonhost trees inoculated with *Ophiostoma ulmi.* I. Anatomy and histochemistry. Can. J. Bot. 69:2055-2073.

Rioux, D., and Ouellette, G.B. 1991b. Barrier zone formation in host and nonhost trees inoculated with *Ophiostoma ulmi.* II. Ultrastructure. Can. J. Bot. 69:2074-2083.

Rowe, J.W., Seikel, M.K., Roy, D.N., and Jorgensen. 1972. Chemotaxonomy of *Ulmus.* Phytochemistry 11: 2523-2517.

Roy, G., and Neumann, P. 1989. Effect of *Ophiostoma ulmi* culture filtrate on *Ulmus americana* callus formation. Can. J. Pl. Pathol. 11:198.

Ryan, C.A. 1990. Protease inhibitors in plants: genes for improving defenses against insects and pathogens. Ann. Rev. Phytopathol. 28:425-449.

Scheffer, R.J., Heybroek, H.M., and Elgersma, D.M. 1980. Symptom expression in elms after inoculation with combinations of an aggressive and a nonaggressive strain of *Ophiostoma ulmi.* Neth. J. Pl. Path. 6:315-317.

Schreiber, L.R. 1970. Viability of *Ceratocystis ulmi* in young seedlings of American elms and the effects of extracts from their tissues on conidial germination. Phytopathology 60:31-35.

Schreiber, L.R., Domir, S.C., Ichida, J.M., and Krause, C.R. 1988. Factors affecting host-pathogen interactions between elm callus cultures and *Ceratocystis ulmi.* Phytopathology 78:1618.

Shigo, A.L. 1984. Compartmentalization: a conceptual framework for understanding how trees grow and defend themselves. A. Rev. Phytopath. 22:189-214.

Shigo, A.L., and Tippet, J.T. 1981. Compartmentalization of American elm tissues infected by *Ceratocystis ulmi.* Pl. Dis. 65:715-718.

Sinclair, W.A., and Brener, W.D. 1974. Dutch elm disease in clones from white elms resistant and susceptible to *Ceratocystis ulmi.* Phytopathology 64:675-679.

Sinclair, W.A., and Larsen, A.G. 1980. Localization of Dutch elm disease in 10-year-old white elm clones from resistant parents. Pl. Dis. 64:203-205.

Sinclair, W.A., Zahand, J.P., and Melching, J.B. 1975. Localization of infection in American elms elms resistant to *Ceratocystis ulmi.* Phytopathology 65:129-133.

Sticklen, M.B., Bolyard, M.G., Hajela, R.K., and Duchesne, L.C. 1991. Molecular and cellular aspects of Dutch elm disease. Phytoprotection 72:1-13.

Sticklen, M.B., Domir, S.C., and Lineberger, R.D. 1986. Isolation and culture of protoplasts from *Ulmus* x 'PIONEER'. Plant Science 47:29-34.

Sticklen, M.B., Lineberger, R.D., and Domir, S.C. 1985. Isolation and culture of protoplasts from *Ulmus* x 'HOMESTEAD'. Plant Science 41:117-120.

Suzuki, H., Yasuda, S., and Hanzama, M. 1971. The extractives of Ohyinire, *Ulmus laciniata* Mayr. The structure of lacinilene A and B, two new compounds from the heartwood. J. Hap. Wood. Res. Soc. 17:221-222.

Szczegola, M., Jeng, R.S., and Hubbes, M. 1987. Mansonone induction by *Ophiostoma ulmi* in callus cultures of elm species. Can. J. Pathol. 9:286.

Tanaka, N., Yasue, M., and Imamura, H. 1966. The quinoid pigments of *Mansonia altissima* wood. Tetrahedron Lett. 24:2767-2773.

Tepper, C.S., and Anderson, A.J. 1984. The genetic basis of plant-pathogen interaction. Phytopathology 74:1143-1145.

Van Alfen, N.K., and MacHardy, W.E. 1978. Dutch elm disease. Symptoms and host-pathogen interactions. pp. 20-25 in Dutch elm disease: perspective after 60 years. Cornell Univ. Agric. Exper. Stn Search (Agriculture) 8. W.A. Sinclair, and R.J. Campana eds.

Vandermolen, G.E., Beckman, C.H., and Rodehorst, E. 1977. Vascular gelation: A general response phenomenon following infection. Physiol. Pl. Path. 11:95-100.

Venn, K.O., Nair, V.M.G., and Kuntz, J.E. 1968. Effects of TCPA on oak sapwood formation in the incidence of development of oak wilt. Phytopathology 58:1071.

Willemot, C., Nair, V.M.G., and Kuntz, J.E. 1968. Effects of TCPA on oak sapwood formation in the incidence of development of oak wilt. Phytopathology 58:1071.

Wu, W.D. 1985. Toxic effects of elm phytoalexins mansonones on *Ophiostoma ulmi* Buis. Nannf., the causal agent of Dutch elm disease. M. Sc. thesis, University of Toronto, Toronto. 117 pp.

Wu, W.D, Jeng, R.S., and M. Hubbes. 1989. Toxic effects of elm phytoalexin mansonones on *Ophiostoma ulmi*, the causal agent of Dutch elm disease. Eur. J. For. Pathol. 19:343-357.

Yang, D., Jeng, R.S., and Hubbes, M. 1989. Manonone accumulation in elm callus cultures induced by elicitors of *Ophiostoma ulmi*, and general properties of elicitors. Can. J. Bot. 67:3490-3497.

Yasuda, S., Kinoshita, Y, and Hanzama, M. 1973. Phenolic constituents of Akinire, *Ulmus parvifolia* Jacq. Mokuzai Gakkaishi 19:593-598.

21
Alterations of Vessel Elements and Reactions of Surrounding Tissues in the DED Syndrome

G. B. Ouellette and Danny Rioux

Abstract

A summary of the principal observations accumulated by the senior author and colleagues since 1970, concerning the histopathology of DED, is presented. An overview of events occurring in disease development are exposed such as: invasion of secondary vessel elements and accumulation therein of extraneous material (including the alveolar network), attack of vessel walls (in association with masses of electron opaque material) and surrounding tissues up to the cambial area, pathogen action on proto- and metaxylem tissues, and similar observations of differentiating tissues, where cell malformation is often pronounced. Various types of cytoplasmic reactions in the axial parenchyma of secondary xylem, are mentioned. Some of these, considered detrimental, range from a high vacuolization of the cytoplasm, of pronounced organelle alterations up to complete cytoplasmic disintegration; other reactions may be considered of a defense type: increased opacity and density of the cytoplasm, accumulation of large amounts of opaque or more translucent matter in large vacuoles, and so on. These are ordinarily accompanied by wall modifications, such as an increased density and opacity of the protective layer. Intermediate situations exist. Wall modifications in parenchyma cells and fibers in normal wood or in barrier zones are thought to be associated with trees showing some degree of resistance or having partially recovered from acute symptoms. As a corollary, susceptibility can be associated with a lack of such reactions, particularly in shoot and twigs where tissue disturbances seem to be most prominent. It is stressed that all the disturbances observed are associated with the presence of masses or strands of opaque matter. These observations point to the necessity of characterizing further these opaque constituents, or at least recognizing that a problem exists that calls for the formulation of new concepts.

Introduction

Confronted with the widespread belief that the main action of plant wilt disease pathogens takes place in vessel lumina, and as a corollary, with the notion that the reactions of living cells would thus be directed against a "saprophyte" or be attributable solely to a secondary effect, one may consider it an anachronism to emphasize that such cell alterations may be germane in Dutch elm disease (DED) symptom development. In previous publications, the author and co-workers, using light (LM) and transmission electron (TEM) microscopes, have reported observations on reactions of xylem living cells, including those in the cambial area, that were interpreted as strong supportive evidence that such cells are primary sites of disease action. An overview of this evidence was presented in Ouellette and Rioux (1992) and in Rioux and Ouellette (1989). In these papers and in Rioux and Ouellette (1991a, 1991b) several types of reactions of invaded xylem tissues, including the formation of barrier zones, were described and interpreted as mechanisms that could impede xylem invasion and recurrence of the disease. These reactions were also considered indirect evidence that attack on living cells is of key importance in the DED syndrome. The present communication is an extension of the review article by Ouellette and Rioux (1992) and of the symposium paper by Ouellette (1982) given at the 1981 meeting on DED in Winnipeg, Canada. Although the object of the present paper is similar to that of papers mentioned previously, great care has been taken to obviate repeats. For this reason, an emphasis is now being placed on a possible sequence of events occurring from secondary xylem invasion radially to the cambial area and upwards to the shoots, including proto- and metaxylem and differentiating tissues. Features that may be similar to those already described are illustrated by new examples.

Results

I. Xylem invasion

A. *Secondary xylem*

1. *Vessel elements*

As the pathogen may rather quickly spread tangentially in branches following artificial inoculation, invasion of vessel elements at any sampling point some time after inoculation may have occurred over various periods of time. Being unable to precisely follow the upward movement of the pathogen in the vessel elements scored by the inoculation wound, our approximation of early to later vessel invasion rests on observations from samples collected over a period of a few to 30 hours or so after inoculation or at the extremities of or even shortly above advancing streaks of visible xylem discoloration. Establishing these early events is of utmost importance if one wants to follow the relation between vessel invasion and effect on adjoining tissues.

In light of these premises, early vessel invasion in inoculated elms is characterized by the rapid deposition of a thin opaque layer (Fig. 1), often initially in chambers of bordered as well as of half-bordered pits and around pit apertures (arrows) and then all along the vessel walls, sometimes in continuity with some figures constituting the alveolar network; the pit membrane of the progressively occluded pits can in general be easily differentiated from this opaque material. Fungal cells can be observed to be in continuity with or included in this material. This layer will be hereafter referred to as coating, although it may not have the same connotation or be as homogeneous (see below) as described or thought to be by other workers (Ouellette and Rioux 1992; Shi et al. 1992). The coating material is often continuous through the pit membrane with similar material present in the periplasmic areas of paratracheal parenchyma cells and distinct from their cytoplasm (see Fig. 13-2 in Ouellette (1978a)). As observed at the TEM level, this material was found to be very dense and opaque (Fig. 2) but not homogeneous, containing fibrillar and more opaque particles and at times vesicular and membranous structures (Figs. 2A and 3). The greater the number of fungal cells present in vessel lumina, the thicker the coating material is even early after inoculation (Fig. 3). Although the vessel element at the left (Fig. 1) illustrates a recent invasion, the large amount of amorphous homogeneous material including some fungal cells and at times covering or being circumscribed by coating-like material in the adjacent vessel, likely indicates an older invasion. It has always been difficult to evaluate the proportion of vessel elements that become completely occluded with such fibrillar material; there are indications that it may become partially or completely dislodged or broken down in the presence of the pathogen.

An important point concerning the opaque coating material is its presence within vessel walls (Figs. 2, 6) and its association with arrays of opaque strands that extend perpendicularly from it into the wall (Figs. 2A, 4 arrow heads). Opaque strands pervading the vessel wall have also been observed to be in continuity with contents of fungal cells through their walls (Fig. 4, all along between arrow heads). The opaque material is also associated with erosion of the vessel walls and presence of cavities therein (Fig. 5). Such a chiseling action may lead to the liberation of chips of wall material that can often be observed in vessel lumina (Ouellette 1978a). The use of a gold labeled exoglucanase as a probe for cellulose (Benhamou 1987) has helped confirm the origin of such wall fragments in vessel lumina and other associated alterations (Fig. 6), indicating that these disorders were not attributable to damage during the preparation of samples. Similarly, the link with the presence of the opaque material in vessel walls and an apparent alteration of the cellulose constituents was also disclosed, even in non-host trees inoculated with the pathogen (Fig. 7).

Membranes of bordered and half-bordered pits that are soon covered by the coating material are initially not visibly altered. Membranes of bordered pits eventually become thicker as if they were impregnated with a diffuse or aggregated translucent or more opaque matter, these membranes nevertheless retain their identity and are rarely visibly eroded. As confirmed by a gold complexed exoglucanase probe, the coating layer that becomes increasingly thicker with time over the pit membrane is clearly demarcated from it (Fig. 8).

The presence of strands of opaque material that are located across and parallel into the outside layer of pit membranes of half-bordered pits is the first noticeable indication of alteration of these membranes (Figs. 9, 10). As for the secondary wall of the vessel, these strands are continuous with the coating material covering the membranes (small arrows). As described elsewhere (Ouellette 1978b; Ouellette and Rioux 1992), a variety of alterations may subsequently occur, ranging from increasing accumulation of the opaque material in the pit membrane and the formation of small to larger cavities in the membrane, to partial to complete disruption of its outer portion. Distention to disruption of the membrane towards the adjoining parenchyma side also occurs frequently. Figures 11-15 illustrate a few of these potential alterations. Some of the observations indicate that such attacks on pit membranes may not only be progressive, but repetitive (Figs. 12-14). In the latter figure, portions of the pit membrane, likely previously impregnated with other type of material similar to that shown in Figure 15, are broken. At times, however, pit membranes are covered by a layer of fine fibrillar material that appears to be structurally different (Figs. 16-19) from the coating material just described. Continuity of these fibrils with similarly fine fibrils that extend perpendicular across the pit membrane (Fig. 16A, arrow) and even within the underlying protective layer, may be established. Somewhat analogous, but more opaque, material can at times be seen partially to completely filling the pit chambers of these cells, with very little of it being present along the vessel wall (Fig. 17). Traces of similar material may also be present between the secondary wall and the protective layer (Fig. 17, double arrows) and in the periplasmic areas next to the pits (Figs. 18, 19). Note that in the latter, vesicular structures are present within this fibrillar material and in the pit membrane (double arrows).

Later stages of infection are characterized by a thickening and apparent transformation of the coating which may comprise several layers of diverse opacities (Figs. 6, 8, 20, 24). The coating may contain parallel or contorted strata of structures that appear lamellar when thin (Fig. 21) but that become much thicker with time through the apparent addition of fine fibrillar opaque material or fusion of these strata (Figs. 20, 22), giving them a wall-like structural appearance. When not compressed between tylosis and vessel walls (Fig. 8), these structures expand into globose configurations of various sizes or intricacy, being continuous with the coating except over pit areas (Figs. 1, 22) where they expand likely through the action of gases liberated from the paratracheal parenchyma cells (Ouellette and Rioux 1992). Such configurations in the latter have been referred to "the alveolar network", which is generally present in a large proportion of vessel elements of the invaded xylem area. Vesicular structures are also present within these alveolar forms and within the pit membrane as well as in the periplasmic area, suggesting their liberation from the host. However, similar vesicular structures are also liberated from the pathogen cells. Initially, the alveolar forms may contain various amounts of fibrillo-granular material; with time, however, these forms generally become empty, but similar material may be retained in their labyrinths when they are compressed between tylosis and vessel walls. As already mentioned, tyloses form scarcely after inoculation in the elm species studied during the highest period of tree susceptibility (Ouellette and Rioux 1992).

The alveolar delimiting structures may also encompass or extend through the more homogeneous fibrillar material described above (Figs. 14, 18), apparently associated with an apparent transformation of the latter, some masses exhibiting a heterogeneous appearance. A similar relationship was observed in studies of infection of non-host trees inoculated with the DED pathogens, where these alveolar networks formed only in presence of these pathogens (Fig. 23), in comparison with wounded controls in which only the homogeneous material accumulated in vessel lumina near the wound.

Legends to figures

List of abbreviations

An, alveolar network; Ap, papilla-like apposition; AWL, additional wall layer; cg, coating material; F, fungal cells; Fm, fibrillar material; ML, middle lamella; Om, opaque material; P, parenchyma cell; Pc, pit chamber; Pm, pit membrane; PW, protective layer; S, secondary wall layer; St, secondary thickening; T, tylosis; V, vessel element; Va, vacuole; Vw, vessel wall; WI, wall ingrowth.
Figs. 1-62. TEM observations, fixation with glutaraldehyde and post-fixed with OsO_4, cross section, except where otherwise stated. LM observations, radial or tangential sections.

Legends to Figures 1-15

Figs. 1-5. Early stages of colonization of vessel elements in the secondary xylem. Fig. 1. LM observation. In the vessel element at the left, the coating material accumulates around and in pit chambers (arrows) and elsewhere continuously along the vessel wall. Some fungal cells linked with this coating. At one point the coating is continuous with the alveolar network. The vessel at the right is clogged with apparently homogeneous fibrillar material in which are embedded some fungal cells. The coating is perceptible at points, apparently in continuity with an alveolar form (above). Approx. 1000X. Fig. 2. The coating material appears to comprise fine fibrillo-granular material, more opaque material, and at times tubular-like structures. It is often continuous with similar material chiseling through the secondary vessel wall, encompassing some portions thereof (arrow). Note: absence of wall rupture where the opaque material (Om) is thin or discontinuous. 36800X. Fig. 2A. Enlarged portion near area labeled cg, showing greater detail of the Om, and fibrillar strands extending from it into the vessel cell wall. 57500X. Fig. 3. Coating material containing in addition to Om, as described in Fig. 2, vesicular structures and more translucent fibrillar material. At points, it is bound by a single membrane-like structure. Numerous fungal cells included in or surrounding the coating material. Pit membranes are slightly distended. Note: large bodies of fibrillo-reticulate matter present beneath the pits in the adjacent parenchyma cell, one of which is bound by a unit-like membrane (arrow, enlarged in Fig. 3A). Similar bodies are also observable in vessel lumina (see Fig. 20). Fig. 3, 20000X; Fig. 3A, 31250X. Fig. 4. A hypha appressed to a vessel wall; over almost its length (between arrow heads) numerous fibrillar strands and tubular-like structures extend from the fungal cytoplasm, through the fungal wall, into the host wall beneath. Erosion of the latter is visible at some locations (top part). Similar strands also extend from the fungal cell into the vessel lumen. 98750X. Fig. 5. A chiseled vessel wall portion and associated coating material. 36800X.

Figs. 6-8. Later stages of vessel colonization. From sections treated with a purified exoglucanase binding with cellulose, and complexed with colloidal gold. Fig. 6. Between labeled tylosis and host vessel wall, presence of a piece of cellulose labeled material (large arrow), surrounded by opaque matter that pertains to an alveolar network, some of it in continuity with the coating. Other portions of vessel wall eroded or traversed by opaque material (arrow head) extending into the adjacent parenchyma cell. Only remnants of the plasmalemma are visible in the latter. 16680X. Fig. 7. Development of *O. ulmi* in a vessel element of a non-host tree (*Sorbus americana*). Note: difference of labeling for cellulose in wall portions containing bands of opaque material. Neat labeling of fungal wall for cellulose. 28800X. Fig. 8. Coating material and alveolar constituents present between a tylosis and a vessel wall. Note: absence of labeling of the first and of the adjoining fibrillar layer; labeling of outer portions of the host vessel wall more pronounced than inner portion; distinct and regular labeling of the pit membrane; demarcation of tylosis wall, composed of three layers, the middle one, more translucent, showing no labeling. 21760X.

Figs. 9-15. Degrees of alterations of pit membranes of half-bordered pits, in association with opaque material similar to that forming the coating and often in continuity with it. Figs. 9-10. Presence of opaque matter extending as perpendicular bands or tubular-like structures into the outside portion of the pit membrane, even over the area abutting on the secondary wall of the parenchyma cell; continuity of these bands with the coating are often evident (arrows). In Fig. 9, strands of opaque material are pervading the vessel wall over the pit chamber (arrow head). In Fig. 10, the inner layer (protective layer) in the adjacent parenchyma cell is also not altered; it is even slightly thicker and more opaque than normal. Fig. 9, 27200X; Fig. 9A, 42500X; Fig. 10, 20000X; Fig. 10A, 31250X.

Figs. 11- 15. More advanced stages of alteration indicated by exfoliation of the outer portion of the half-bordered pit membrane (open arrows) with the formation of cavities therein, always in association with opaque material (thick short arrows, enlarged areas). The latter also contains lamellar-like structures (near open arrow, Fig. 12); these are similar to structures present in the vessel lumen, that are thought to correspond to components of the alveolar network (see below). Figs. 14 and 15 illustrate pit membranes that appear to have become impregnated before or after their becoming cavitated with other more translucent type material. Portions of these membranes are broken or chipped off in association with Om (Fig. 14, and enlarged portion near arrow). Alterations of the protoplasm in the adjacent parenchyma cells may be various and likely successive, as in the case of the pit membranes (Fig. 13). The secluded cytoplasmic portion (Fig. 14) contains numerous small vacuoles and vesicles, similar to those present in the inner portion of the pit membrane, and opaque intermixed matter. The bordering portion of the remaining cell contents is also impregnated with Om, as in Fig. 15 (arrow); see also Figs. 27, 28. Fig. 11, 27200X; Fig. 12, 20000; Fig. 12A, 31250X; Fig. 13, 272000; Fig. 14, 20000X; Fig. 14A, 31250X; Fig. 15, 27200X.

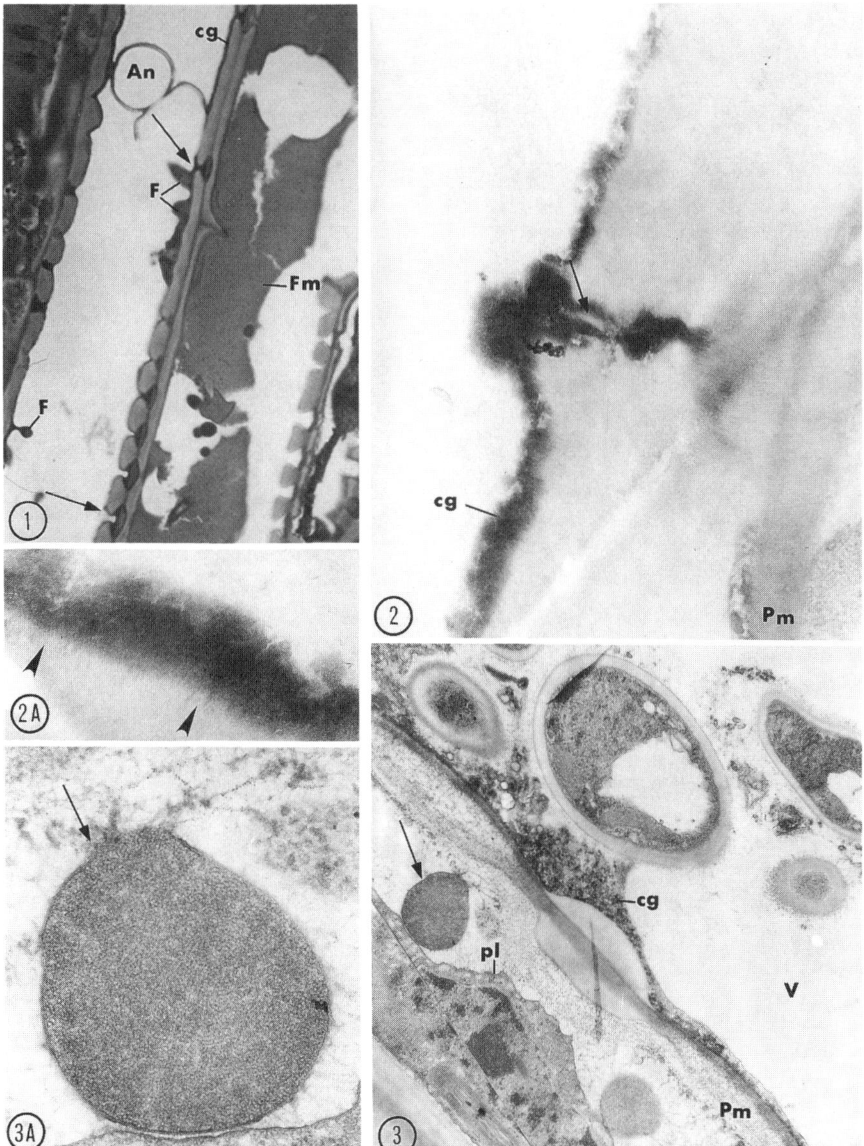

Figures 1-3. *See legend p.259.*

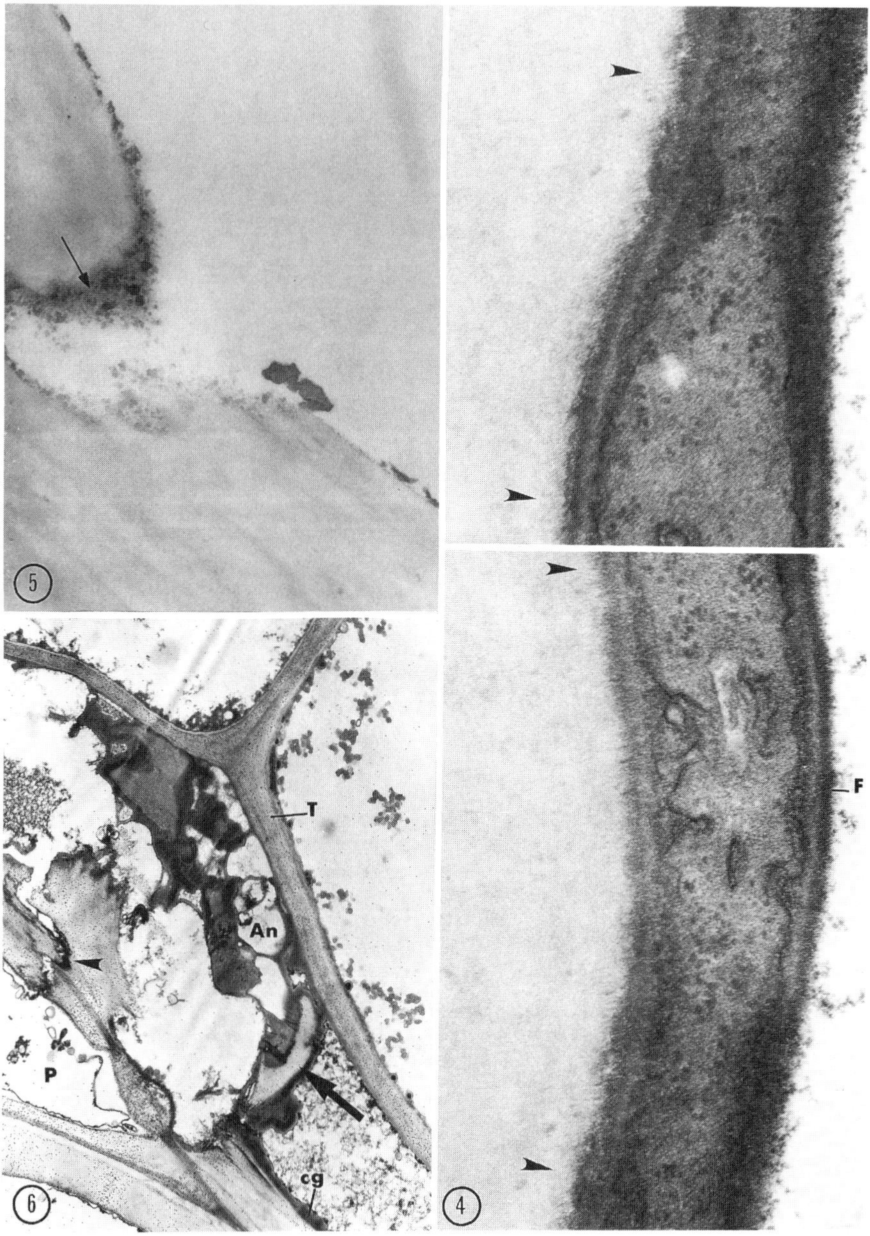

Figures 4-6. *See legend pp. 259-260.*

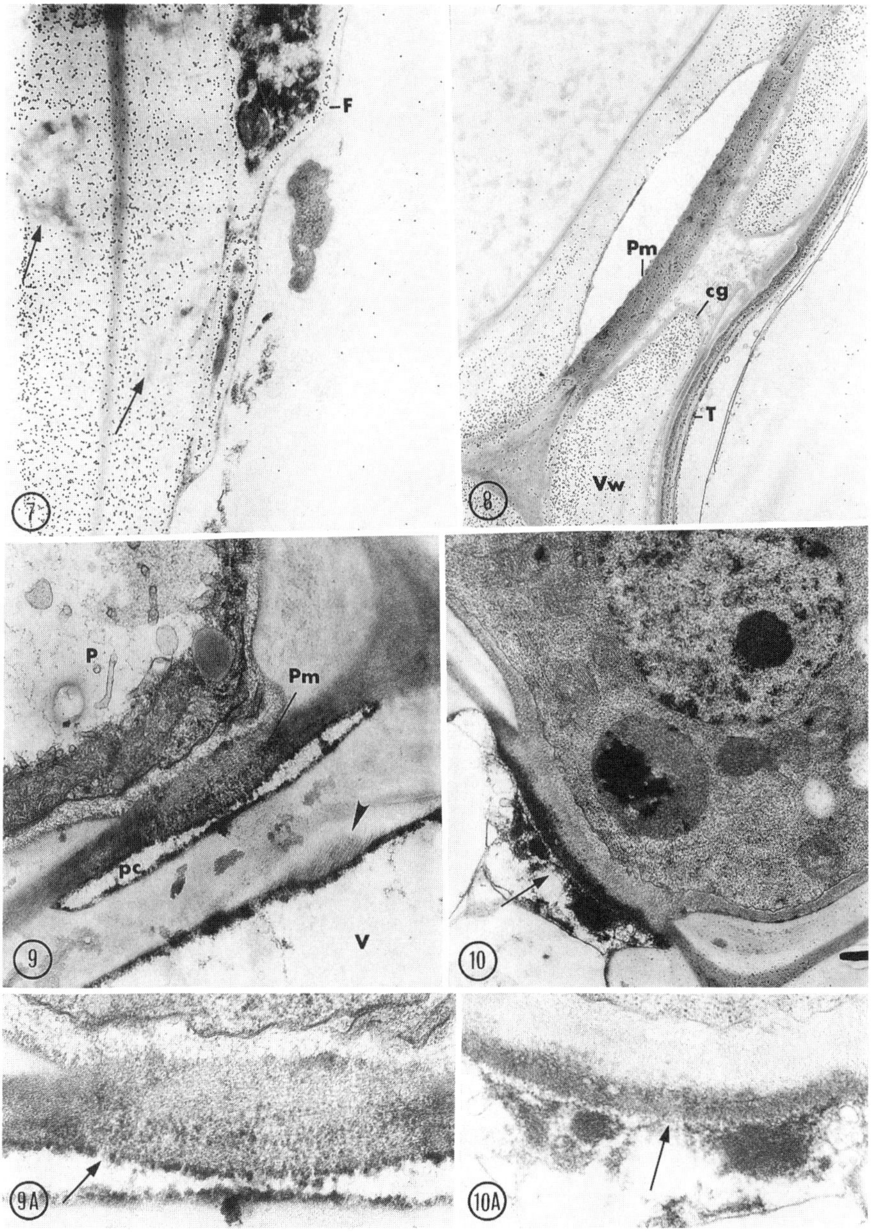

Figures 7-10. *See legend p. 260.*

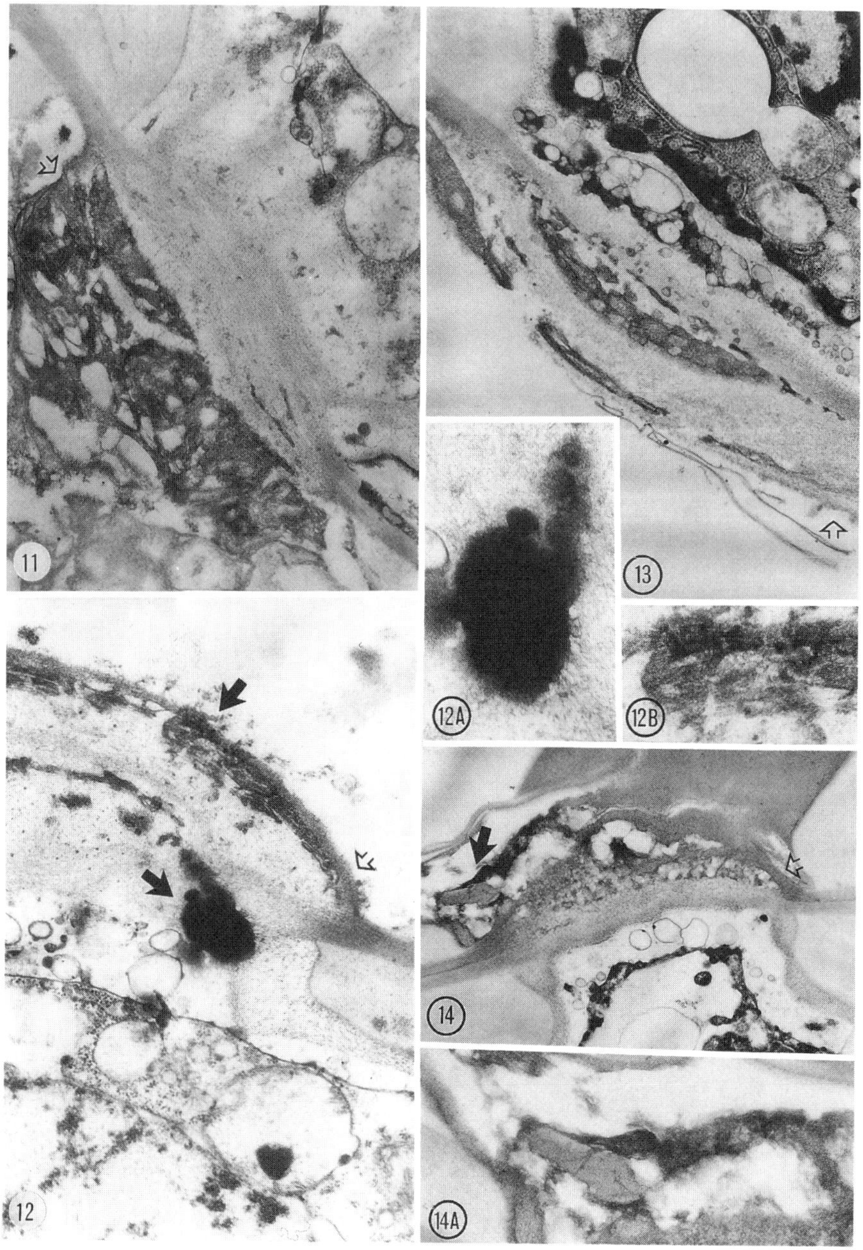

Figures 11-14. *See legend p. 260.*

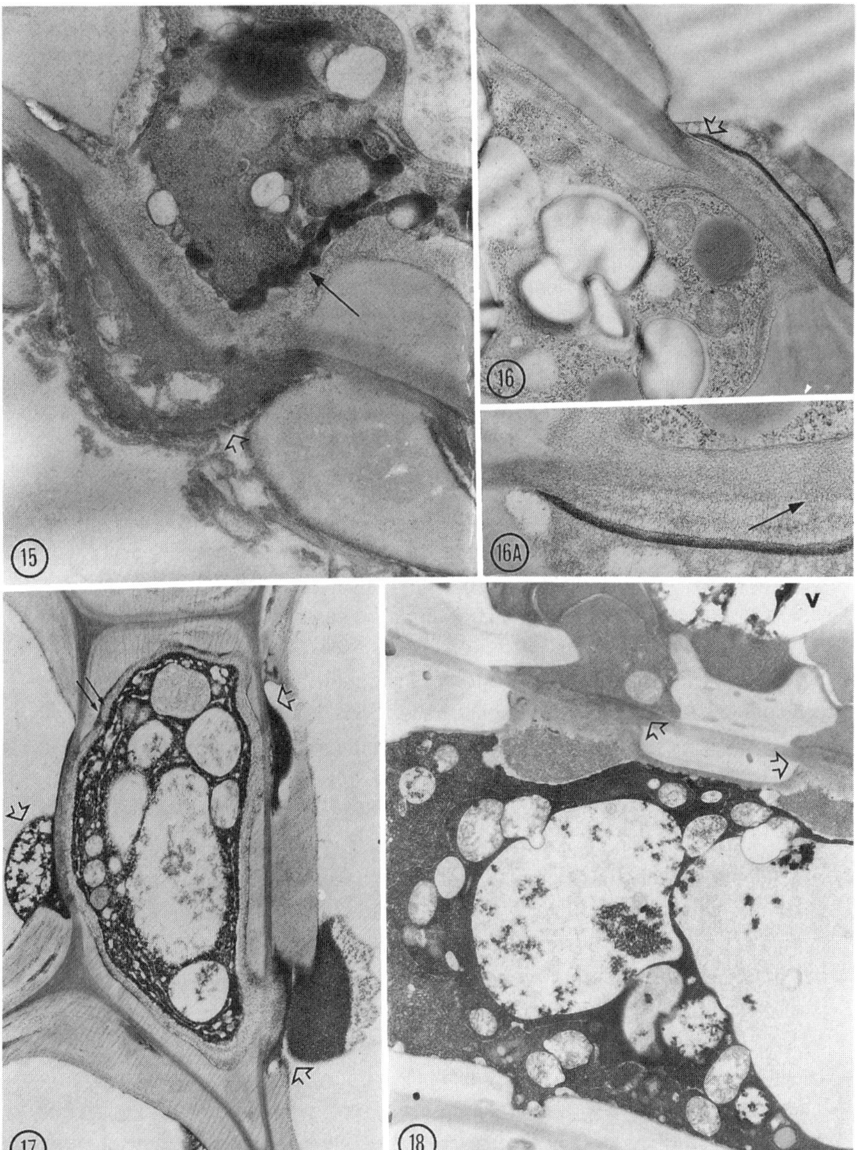

Figures 15-18. *See legend pp. 260, 267.*

2. *Effect on parenchyma cells*

Although typical pathogen cells have occasionally been observed in parenchyma cells shortly after inoculation, the great variety of cytoplasmic alterations observed in the majority of cells bordering invaded vessel elements cannot be traced to their direct penetration by fungal cells (Ouellette 1982; Ouellette and Rioux 1992). Sporadic observations of fungal cells attempting to penetrate papilla-like structures apposed to half-bordered pit membranes (Figs. 24, 25) indicate that such appositions may be responsible for excluding fungal invasion; papillae, however, are not abundantly formed. Therefore, alteration of such cells is to be sought through substances penetrating pit membranes. The degree of alteration of the first is apparently related to that of the latter and the end result is the complete degradation of the cytoplasm with various types of cell reactions in the course of events (Fig. 26, and other observations described below). These alterations may follow pronounced or slight retraction of the plasmalemma (Fig. 3). A peculiarity in this figure is the presence beneath the two half-bordered pits in the periplasmic area, of two large bodies of compact fibrillo-reticulate-like matter. One of these bodies is bound by a unit-like membrane (Fig. 3A) except at the top part where it appears to be continuous with material extending up to and even through the pit membrane. Similar bodies unbound (Fig. 20) or bound (Fig. 46) have been observed in vessel elements. One may refer to previous publications for a detailed description of the many alterations of parenchyma cell contents (Ouellette 1982; Ouellette and Rioux 1992).

A large proportion of cells exhibit some features, however, that have not yet been thoroughly described, for example that their cytoplasm is becoming cleaved or traversed by meandering bands of opaque material (OM) (Figs. 27, 28A). At first glance, this OM seems to be present in the periplasmic area (Figs. 27, 31), but in many cases it appears to be located inside the plasmalemma and to extend within the cytoplasm (arrows). The most pronounced cases of such anomalies have been observed in primary xylem (described under section B below). Considering that very few pathogen cells can be observed in paratracheal parenchyma, the alteration in several rows of cells, including fibers away from fungal elements and often close to the cambium (9,12), may be considered problematic. Such a phenomenon may be due to a chain reaction. The pronounced host wall degradation and associated reactions in these cells, particularly in fibers (Ouellette 1981b), can hardly be explained in this manner. In most samples examined a few days after inoculation, some distance above the point of inoculation or at the extremities of developing streaks, bands of OM which are irregular in contour and width, bifurcating or branching in various configurations to tightly encompass host wall material, have regularly been observed extending between, or even through, the secondary wall of several cells (Figs. 29-32). As shown in these examples and in previous works (Ouellette 1978a; Ouellette and Rioux 1992), this material is not homogeneous, and, as shown by its discontinuities at some location in a series of sections, its presence in the wall cannot be explained by its flowing into existing ruptures. The material can eventually lead to wall ruptures (Fig. 30). When such ruptures begin or end from a host cell containing cytoplasm, the latter may also flow into the ruptures. In such cases, however, host organelles are recognizable, if ruptures of any kind have occurred shortly before sample preparation, or the host cytoplasmic matter is completely disintegrated. In the cases just illustrated and others to be described below, it was judged that mechanical ruptures were not involved; also, the opaque material present in pit chambers can be differentiated from host cytoplasm (Fig. 31). Similar material (Fig. 30, as for Fig. 27) seems even to be related to pronounced vesiculation of the cytoplasm (arrow). Although the presence of typical hyphae is only rarely observed spreading radially towards the cambium, microhyphae often occur directly penetrating cell walls; such microhyphae are particularly noticeable in samples collected at the end of the growing season or early in the spring following the year of inoculation in surviving branches (Ouellette and Rioux 1992). In this situation, their passage is not related with noticeable wood discoloration.

Legends to Figures 16-32

Figs. 16-19. Fine fibrillar homogeneous material, distinguishable from the Om, covering almost intact pit membranes (open arrows) of half-bordered pits, and filling partially to completely the corresponding pit chambers. Some of the fine fibrils are visible extending in a parallel and perpendicular manner from the inside to the outside of the pit membranes and to the material overlaying these (Fig. 16A, arrow). Note: the absence of Om in the adjoining parenchyma cell, except perhaps for traces of opaque material between the protective layer and the secondary wall (Fig. 17, double arrows); the integrity of this protective layer; and the absence of coating material along the vessel wall nearby. The fibrillar material filling the pit chambers in Fig. 18 is obviously circumscribed or traversed by components of the alveolar network. Masses of similar fibrillar material have accumulated in the periplasmic area beneath the pits of the adjoining parenchyma cells. In some cases, vesicular structures are also present in the material filling the pit chamber as well as in the pit membrane (Fig. 19, double arrows and open arrow, respectively). A similar association is observable in the periplasmic area of that cell. Note that the adjacent parenchyma cell contains opaque cytoplasm and large vacuoles filled with near translucent fibrillar contents and masses of more opaque material. Fig. 16, 15840X; Fig. 16A, 24750X; Fig. 17, 16416X; Fig. 18, 12000X; Fig. 19, 16416X. Fig. 20. Various extraneous entities in a vessel lumen: the thick coating is covered by a body of material (arrow) analogous to that illustrated in Fig. 3; next to it stand a ragged translucent mass of wall-like material, and a broken piece of an alveolar network containing some finer translucent matter; and a layer of opaque matter, similar to that forming the coating in early colonization of vessels extends over these other constituents. The adjoining parenchyma cell is not appreciably altered. 15840X. Figs. 21-22. Stages in the formation of the alveolar network in vessel lumina, flattened as part of the coating in Fig. 21, and expanded over a pit aperture in Fig. 22, the delimiting layers of which are in continuity with the coating. Pit membranes in both cases have been affected as described previously. The parenchyma cell (Fig. 21) has reacted by forming a papilla-like structure; that in Figure 22 contains a large tannin-like body, the margin of which seems to encroach on the surrounding cytoplasm and to reach at points the plasmalemma. Increased wall deposition beneath the pit seems to have occurred. Fig. 21, 12000X; Fig. 22, 12000X.

Fig. 23. In *Prunus pensylvanica* inoculated with *O. ulmi,* an alveolar network developing through masses of fine fibrillar material in a vessel lumen. Some of the fine fibrillar material can be seen traversing perpendicular the pit membrane down to the periplasmic area of the adjoining parenchyma cell. 21760X. Figs. 24-25. Attempted penetrations of papilla-like structures in paratracheal parenchyma cells. In Fig. 24, the pit membrane of the half-bordered pit, in the vessel containing a tylosis and alveolar network, is covered with a layer a fine fibrillar material. Fig. 24, 20000X; Fig. 25, 15840X. Fig. 26. An empty paratracheal cell, the interior of which is bound by a wall-like layer; the latter is separated from the much altered protective layer, particularly in areas beneath the pits, by strands and masses of moderately opaque matter. 15840X.

Figs. 27-28. Presence of masses and bands of opaque material in the periplasmic area within the host cytoplasm (arrows) in paratracheal parenchyma cells comprised between two vessel elements. Fig. 28A, enlarged portion of each side of the area indicated by the arrow.

Fig. 27, 20000X; Fig. 28, 6800X; Fig. 28A, 15840X.

Figs. 29-32. Bands of opaque material (Om) extending intercellularly sometimes between several rows of axial parenchyma cells or fibers, or through secondary walls of some cells.

Fig. 29. Arrow heads point to smaller bands of Om extending through or encircling portions of host wall material. Configurations of Om indicate very close contact with host wall. 20000X.

Fig. 30. Om extending in small rupture of host wall. Presence of Om in periplasmic area of adjacent parenchyma cell or extending within the cytoplasm. Om is associated with pronounced cytoplasm vesiculation. 20000X.

Fig. 31. Band of Om, apparently bound at one end (thick arrow, and enlargement

Fig. 31B) by a wall-like layer that cuts across the host wall, and extends through a middle lamella to the pit chamber of a parenchyma cell (left); this band is also bridged (arrow head, and enlargement Fig. 31A) through the middle lamella with another band extending through the secondary wall layer of the adjacent parenchyma to the pit chamber and likely infringing upon the cytoplasm of that cell. Fig. 31, 15840X; Fig. 31A and 31B, 24750X.

Fig. 32. A knob-like mass of Om present in the pit membrane of the bordered pit that is continuous with the band of Om present in the middle lamella and in the outside portion of an adjacent pit (thick arrow). Thinner bands also linked with the knob extend into the secondary wall of the obliquely cut pit chamber. Note: the thickened protective layer in the parenchyma cell. 20000X.

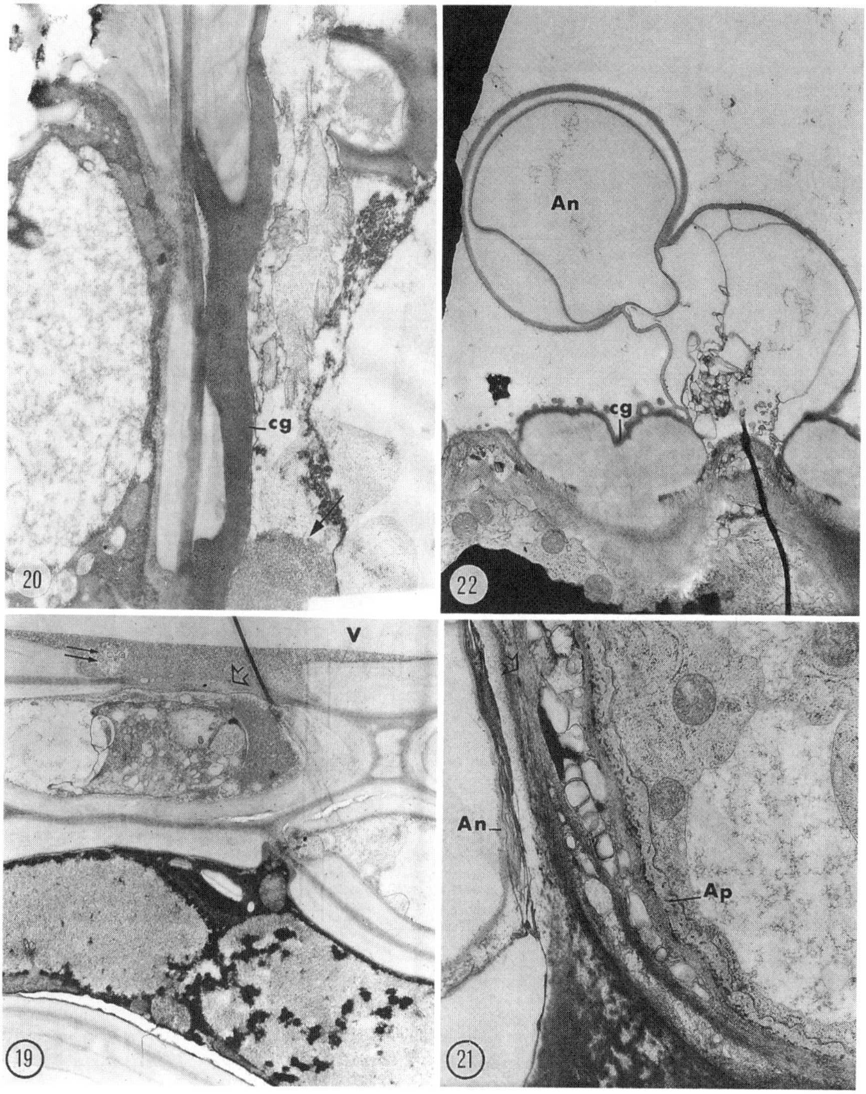

Figures 19-22. *See legend p. 267.*

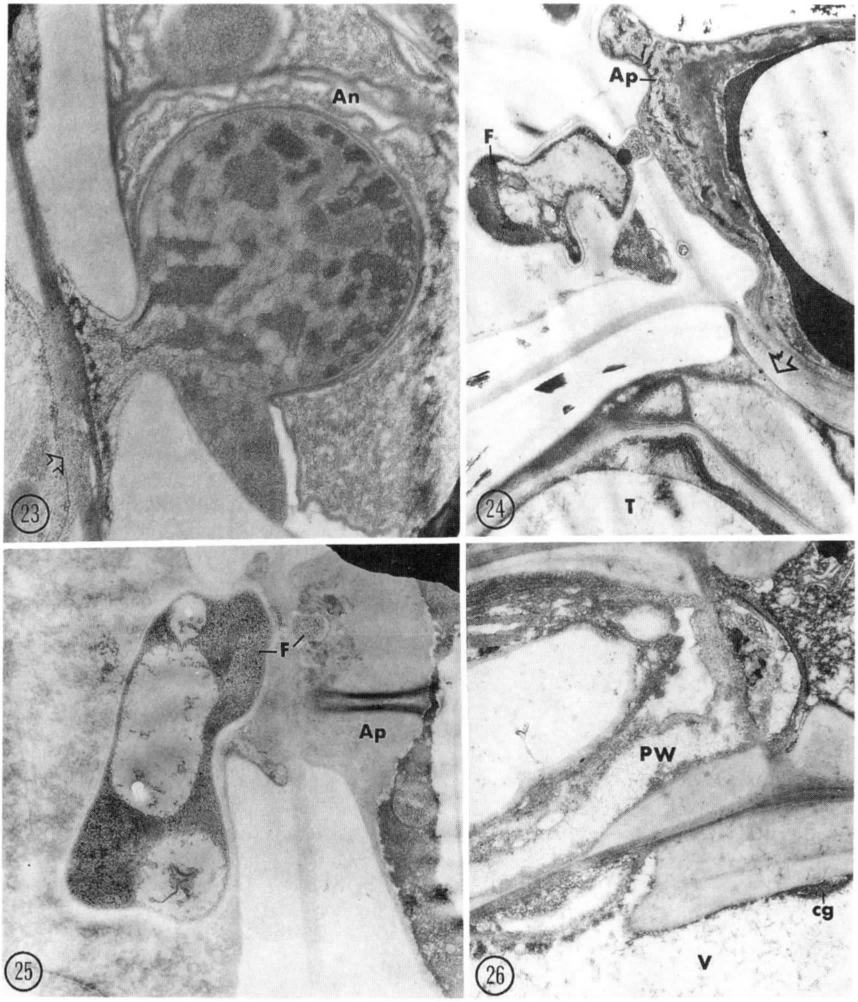

Figures 23-26. *See legend pp. 267-268.*

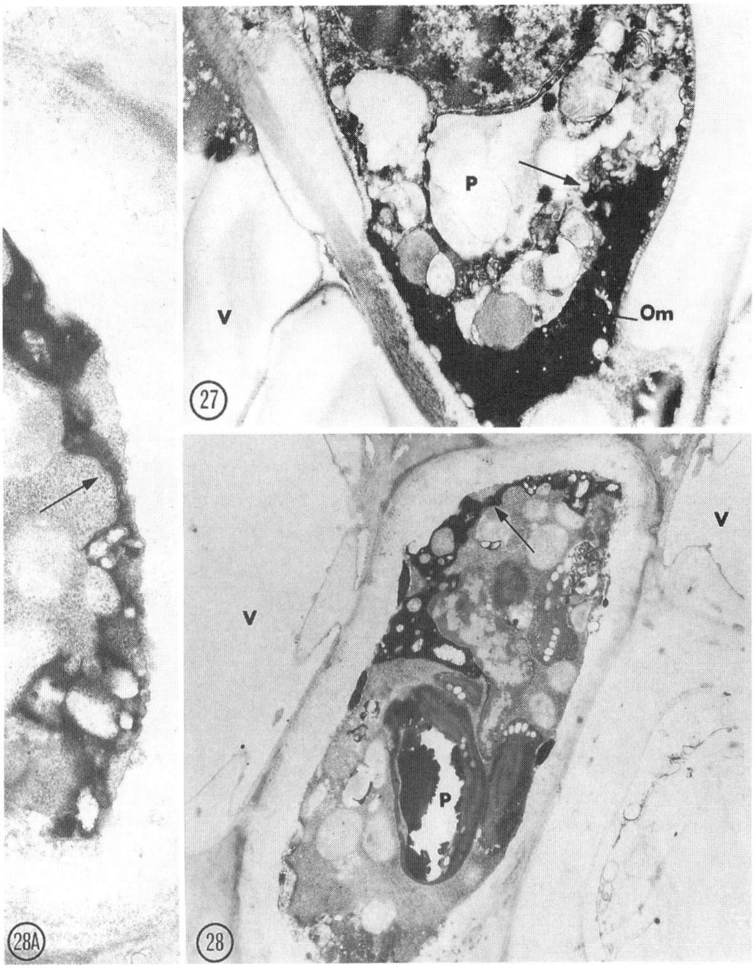

Figures 27-28. *See legend p. 268.*

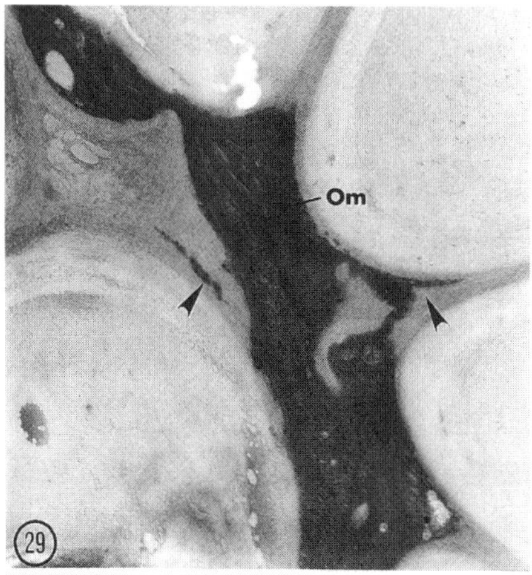

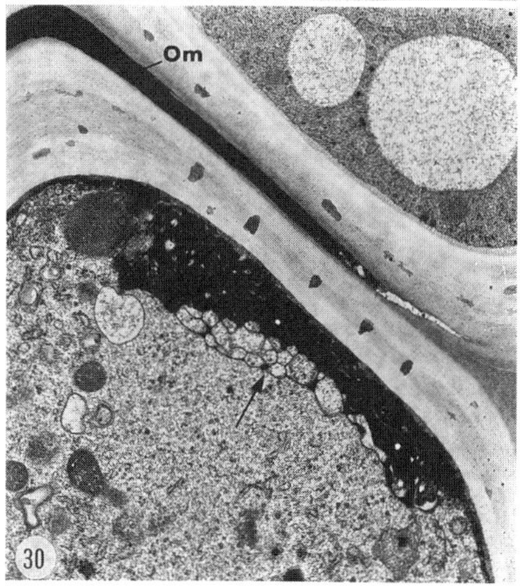

Figures 29-30. *See legend p. 268.*

Legends to Figures 33-37

Figs. 33-37. Colonization of proto- and metaxylem tissue. Figs. 33-34. LM observations. Two sections in a series of a few, with the central portion of altered parenchyma cells as a point of reference, showing the presence of small to larger hyphae or conidia in the vessel lumina or appressed to the secondary thickenings. Some of the fungal cells are linked with small hyphae crossing the pit membranes (lower F, Fig. 33, higher F, Fig. 34). Small opaque bands extend at several locations along and through the secondary thickenings, that are linked with small spore-like structures (arrows); approx. 800X. Fig. 35. A view of invaded tissues that could be comparable to the central portion of the Fig. 33 where altered parenchyma cells are abutting on clogged vessel ends, containing a variety of components: fungal cells, opaque and more translucent matter, degraded fungal cells, new wall material deposited by parenchyma cells (see below) and so on. 12000X. Fig. 36. A tylosis developing from parenchyma cell into a partially plugged protoxylem vessel. The pit membranes of the other protoxylem vessel are impregnated with an opaque substance. The secondary thickenings of this vessel which is clogged with homogeneous moderately opaque matter, are lined with a coating-like material. 5760X. Fig. 37. Alteration of half-bordered pit membranes in an early secondary xylem vessel, is associated with components similar to those described in preceding figures. A mass of opaque material is associated with an apparent degradation of the pit membrane (arrow and enlargement, Fig. 37A). Note: the high fragmentation of the cell cytoplasm. Fig. 37, 9900X; Fig. 37A, 24750X.

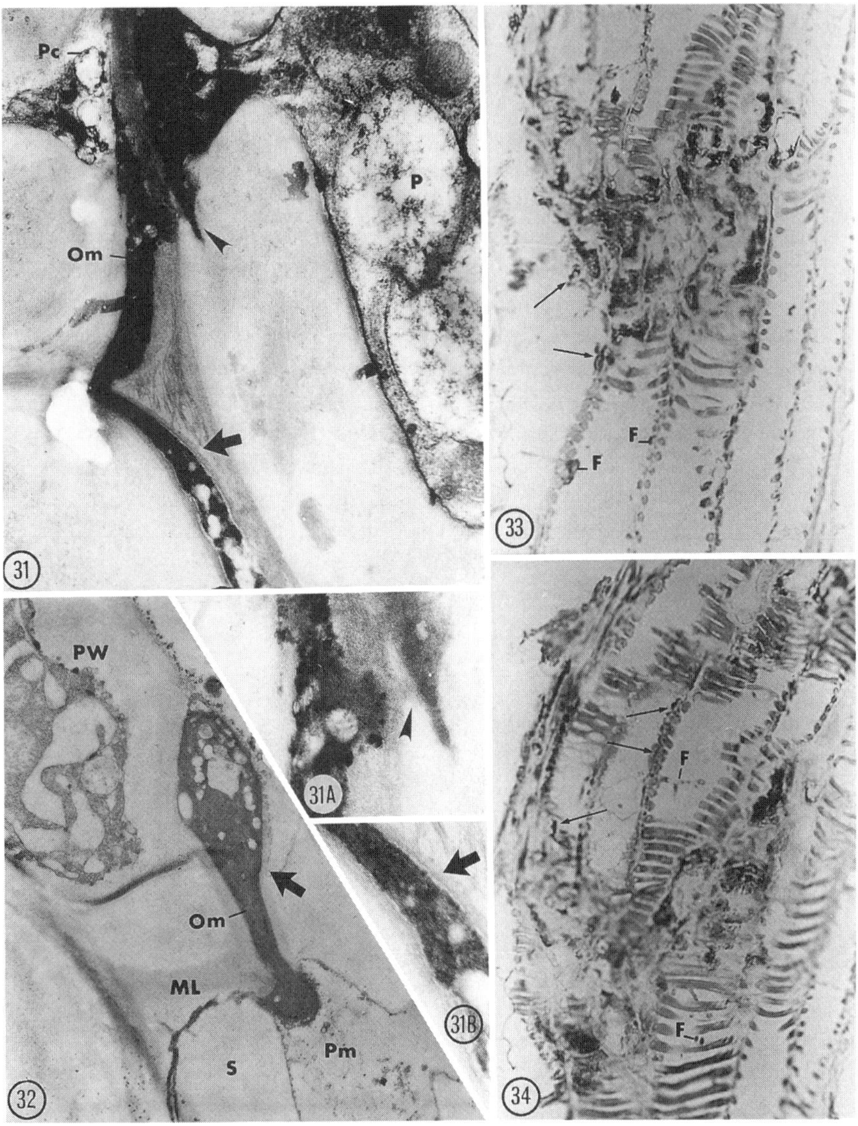

Figures 31a, 31b and 32. *See legend p. 268.*

Figures 33-34. *See legend p. 272.*

B. *Primary xylem colonization*

When the pathogen reaches the shoots, it may invade the major portion of the proto- and metaxylem vessel elements. The pathogen is recognizable as typical hyphae and conidia, or as microhyphae and small conidia (Figs. 33, 34). These small hyphae have often crossed pit membranes or penetrated the vessel's secondary thickenings. As described for secondary xylem invasion, coating material first accumulates around pit apertures and within pit chambers, and then as a fairly continuous layer along vessel walls (Ouellette and Rioux 1992). Pronounced alterations of these walls regularly occur. The vessel lumina may become at places completely occluded with various components, including fungal cells (Fig. 35). The latter obviously illustrates a very complex situation of an obliquely cut vessel abutting on altered and reactive parenchyma cells. Other vessel elements may contain more homogeneous fibrillo-granular material (Fig. 36). Tylosis-like structures may also form in these vessels (Fig. 36); pronounced swelling of the primary wall not subtending secondary thickenings also occur, almost sealing off the vessel lumina (Ouellette and Rioux 1992).

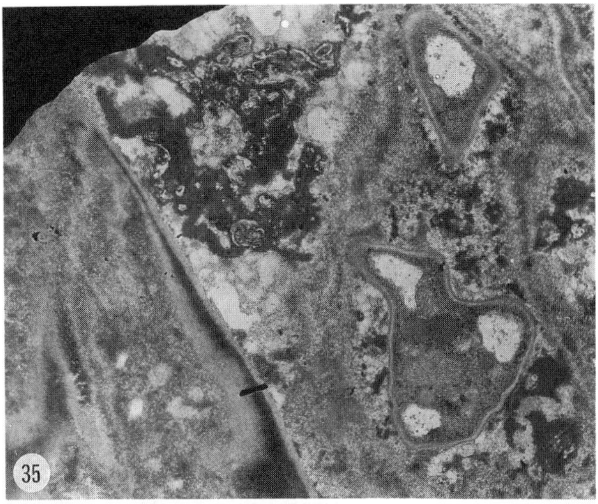

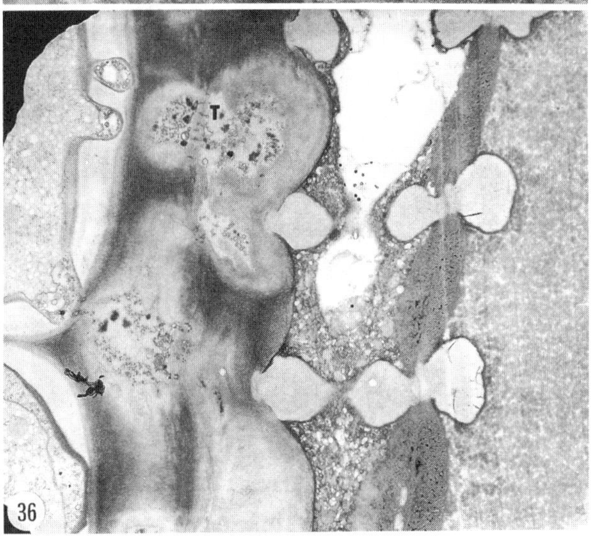

Figures 35-36. *See legend p. 272.*

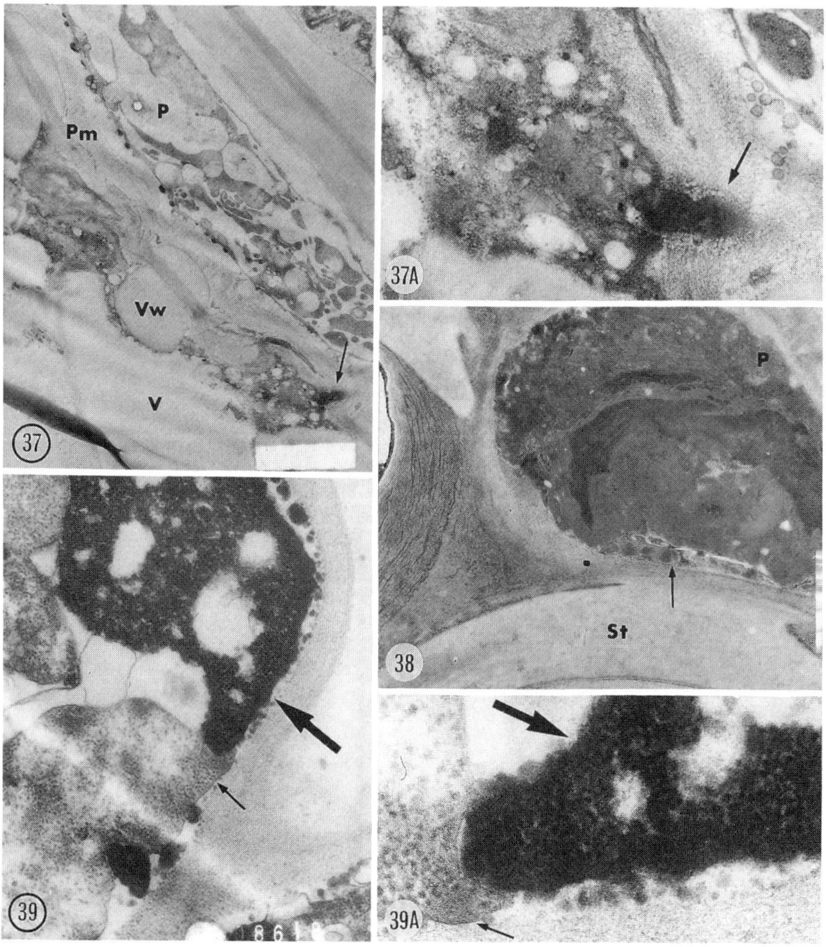

Figure 37. *See legend p. 272.*
Figures 38-39 *See legend p. 279.*

Paratracheal parenchyma cells adjoining these vessels are also affected (Figs. 33, 34). These illustrations represent median sections in a series of a few, the marginal ones showing almost no visible alteration of parenchyma cells. Two types of cell alteration observed at the TEM level are shown in Figures 38 and 39A. In Figure 38, the cell is filled with heterogeneous material, partially composed of opaque bands and masses, some recalling the association illustrated in Figure 28. Similar masses are also present here in the inner cell wall layer (arrow). Many cells also accumulate large opaque masses in the periplasmic area; this material may infringe upon the plasmalemma, pervading the cytoplasm (Figs. 15, 28 and 31). Pronounced fragmentation of the cytoplasm may ensue. Whatever the origin of these more opaque masses, they suggest a cytoplasmic-like organization, i.e. containing small bodies of ribosomal appearance (Fig. 39, large arrow). These bodies are permeated by a fine fibrillar-like substance that extends somewhat freely at the edge of these masses. Other examples of a similar association are presented below.

Alteration of the primary wall in the primary xylem is often marked by a pronounced distention and rupture of the wall microfibrils and by its impregnation with opaque material (Ouellette and Rioux 1992, and unpublished information). Pit membranes of half-bordered pits in early secondary xylem may be significantly affected, becoming pervaded by opaque material, some of which is apparently linked with membrane digestion (Fig. 37, 37A, arrow).

C. *Cambial area and differentiating tissues*

In host, as in non-host trees, the pathogen action, whether direct or indirect, is evident in the cambial area a few days after inoculation (Rioux and Ouelette 1989; Rioux and Ouelette 1991a, 1991b; Ouelette and Rioux 1989), and hence before any external symptoms are visible. The effect of this action is particularly pronounced in meristematic and differentiating tissues, as implied in these publications and others (Ouellette 1981a). This aspect, which has not been considered fully previously in DED investigations, has been studied in more detail in other plant wilt diseases (Baayen 1988; Harling and Taylor 1985; Pennypacker and Nelson 1972). The present paper offers opportunity to add data to this important aspect of the DED syndrome.

To follow more closely in time and space the development of the pathogen in these tissues, inoculations with a hypodermic needle were made in surface sterilized ends or bases of shoots, in an oblique manner through the outside tissues so as to breach slightly the primary xylem tissues. Observations of samples taken from material inoculated in this manner are illustrated in Figures 40 to 42; Figure 43 represents a healthy control. The inoculation point shown in Figure 40 was in tissue corresponding to the procambium and adjacent tissues (Fig. 43, left portion). Hyphae can be traced back to conidia present in and surrounding the wounded area. Further away, however, only masses and bands of opaque matter can be observed in between the cells and within the cell walls up to approximately 15 cm from the wound (Figs. 41, 42). This material may also extend to points within adjoining cells (Fig. 42, arrow, as an example). The cytoplasm in cells adjoining the path of the opaque material is altered, but obviously does not become much opaque. An example of corresponding tissue examined in TEM is presented in Fig. 45, showing the opaque material in the primary wall of a parenchyma cell that seems to have expanded to partially fill the enlarged space present between the two adjacent cells; whether this space corresponds to a small dead cell or to an intercellular area is difficult to establish. Numerous collapsed cells (Fig. 47) in this and other similar samples could, however, often be observed in association with the opaque material in the cell walls (Fig. 44, arrows). In many other situations, the origin of the labyrinths of wall-like layers present between still intact cells could not be determined, as some of these layers could have corresponded to new wall material produced by the cells to exclude some of the opaque material, as suggested by the presence of a few wall-like strata in the lower part of Figure 42, where no collapsed cells are detectable (see below, and illustrations in Ouellette and Rioux 1992). As viewed in the cross sections of samples taken from shoots recently inoculated or in a state of disease recurrence, indications of cell attacks are often visible throughout the vessel bundles (Fig. 44, arrows) from the pith outside to the cambial area (left to right side of the photograph). The intercalary meristem can also be affected at places. Typical fungal cells could be observed at one point or another in vessel elements bordering the affected cambial or differentiating tissue (Ouellette and Rioux 1992). Alveolar networks are also present in the young, often malformed vessel elements near the cambium (Fig. 47, and illustrations in Ouellette and Rioux 1992).

II. Reactions to infection

A. *Secondary xylem*

Numerous types of reactions occurring in parenchyma cells have been observed (Ouellette 1982; Ouellette and Rioux 1992). The present illustrations show but a few variations of these reactions, particularly some of those that appear to reflect a defense reaction such as: an increased density and opacity of the cytoplasm (Figs. 17, 18, 47, 53), papilla-like formation as already mentioned, and the accumulation of translucent to opaque material in large vacuoles in cells containing either opaque or more translucent cytoplasm (Figs. 19, 46, 48, 49). Wall modifications are significant changes that can occur in such cells. One of the first changes is an increased opacity of the protective layer (Figs. 48, 49), sometimes concomitant with a thickening of that layer (Fig. 48). These cells may also form lamellate, suberin-like layers (Fig. 49, arrow) (Ouellette 1981b; Rioux and Ouellette 1991b). Other wall modifications in parenchyma cells are thick wall-like deposits in some parts of the cell, apparently to exclude masses of opaque material (Figs. 50, 50A) as described above, types of wall ingrowths separating disintegrating cytoplasm from cytoplasm of other appearance (Fig. 51) (Ouellette 1981b), and the formation of additional layers of other types (ibid).

Legends to Figures 38-47

Figs. 38-45 and 47. Disease effect on meristematic and differentiating tissues. Fig. 38. A parenchyma cell located between metaxylem vessel elements near the pith of a twig. This cell, which lacks a secondary wall, is filled up with heterogeneous matter, some standing out as more opaque bands, in the cell lumen or apparently even in the periplasm; the imperfectly formed inner wall layer of the cell is impregnated with similar material (arrow). 9900X. Fig. 39. A typical case of an affected parenchyma cell in its early stage of differentiation, containing fragmented cytoplasm with small to large masses of Om located in the periplasmic area (large arrow) or at times permeating the cytoplasm (see Figs. 27-28). This material is similar to that already described: opaque bodies the size of ribosomes (compare with host cytoplasm nearby) held in a matrix of fine fibrillar material of the same opacity. Whereas the cell cytoplasm is bound by a membrane (small arrow), Om is delimited by that diffuse component. Fig. 39, 50000X; Fig. 39A,90000X (enlarged portion near large arrow).

Figs. 40-43. Development of the pathogen in young shoots. Figs. 40-42. From three samples taken near the point of inoculation (Fig. 40) to the recognizable tip (Fig. 41) of tissue invasion. These are also from a series of a few sections. Fig. 43. From water inoculated control. The tissues shown in Figs. 40-42 would correspond, at a higher magnification, to the tissues shown at the left of the protoxylem in Figure 43. In Fig. 40, note the presence in the wound area of typical conidia linked to a normal hypha (right) or to a more opaque thin band (bottom side). In Figs. 41 and 42, many similar bands are present intercellularly at times giving the impression of being circumscribed by several more or less continuous host wall-ike layers (Fig. 42, below Om label). Some of the material also seems to extend within some of the cells (Fig. 41, cell at the middle left and below arrow, and Fig. 42, arrow). Figs. 40-42 Approx. 1100X; Fig. 43, approx. 600X.

Fig. 44. Various cellular alterations in and near a vessel bundle in a shoot. These alterations are recognizable throughout the bundle from the medullary sheath to the cambium as collapsed cells, likely in association with bands of Om (arrows), or as cells lacking secondary walls. Some cell reactions are also visible, such as an increased wall opacity and the presence of large opaque bodies in several of the cells. In the twig below where this sample was taken, the cambial area was also affected, following an apparent deposition of healthy tissues over the invaded protoxylem. Approx. 650X.

Fig. 45. Presence of Om in the wall of a cell located near the cambium, and also in the area above, corresponding either to a dead cell, or to an enlarged intercellular space. Cytoplasm in these cells is altered to various intensities. 20000X.

Fig. 46. Accumulation of nearly translucent fibrillar material in a large vacuole bound by a band of opaque cytoplasm in a paratracheal parenchyma cell. The pit membrane of the half-bordered pit, covered by a layer of Om (empty arrow), similar to that along the rim nearby, is altered as described above. Note: the presence of bound bodies of fibrillo-reticulate matter analogous to that shown in Fig. 3A, in the periplasmic area and vessel lumen (curved arrows). 27200X. Fig. 47. An affected zone of the cambial area, showing some collapsed cells next to hypertrophied cells in the cambium (right). New cell divisions are also visible (bottom right). The two larger vessels (top) are separated from two smaller vessels by a cell lacking a secondary wall, except perhaps for the small isthmus comprised between the recurved secondary wall of the adjoining vessel; an alveolar network is present in the latter. Various reactions are exhibited in the other parenchyma cells. 7296X.

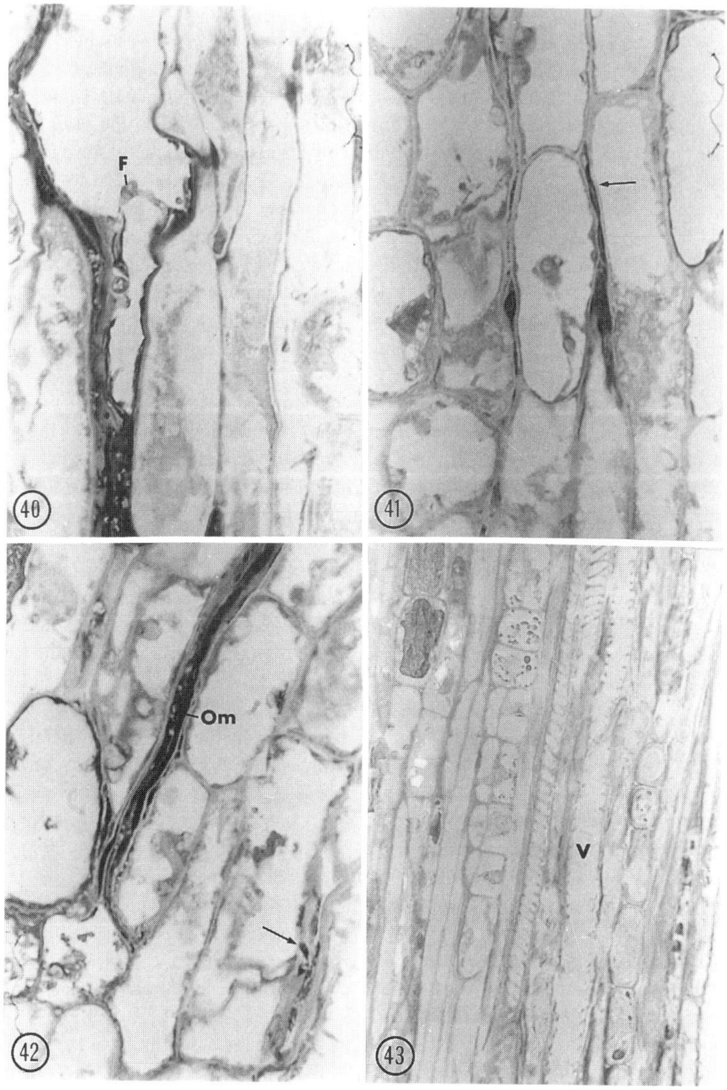

Figures 40-43. *See legend p. 279.*

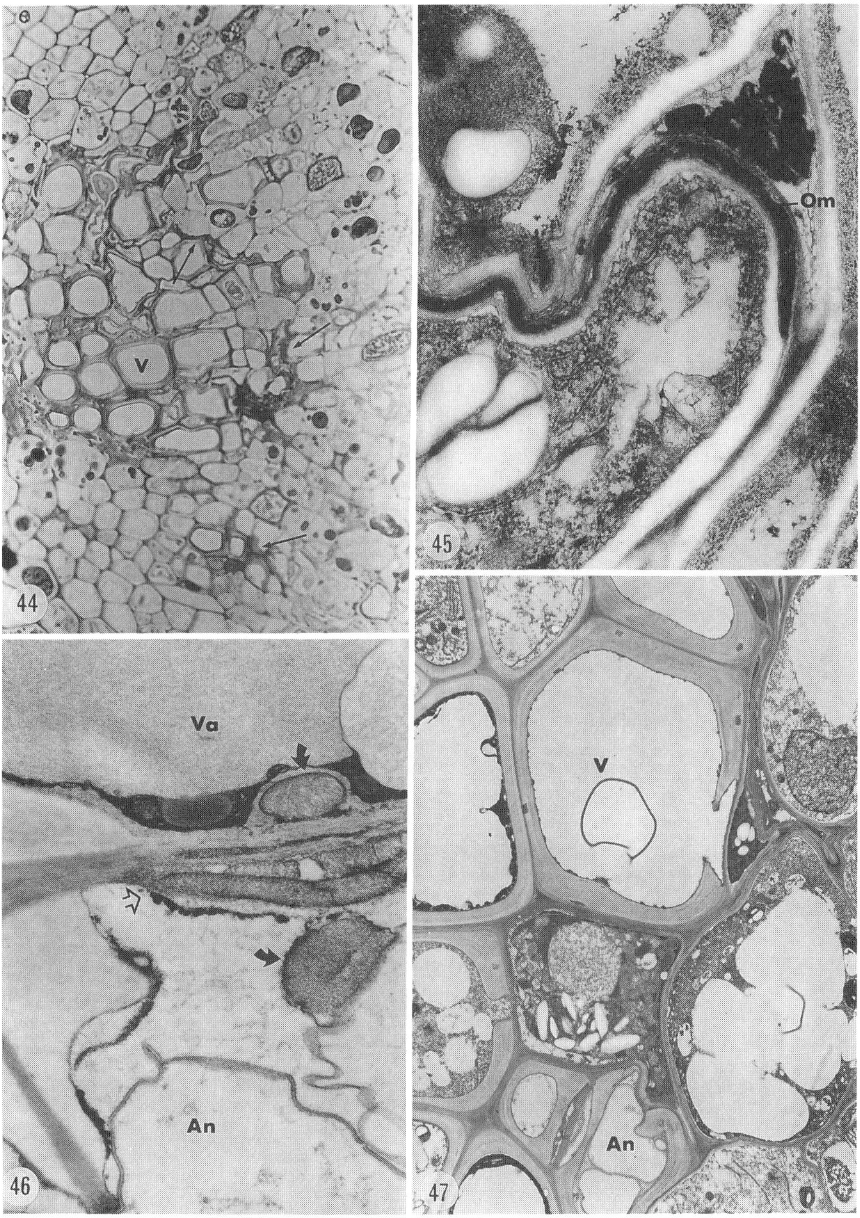

Figures 44-47. *See legend p. 279.*

Legends to Figures 48-52

Figs. 48-51. Various cell reactions in the secondary xylem.

Fig. 48. Thickening and increased opacity of the protective layer; large opaque body (arrow) surrounded by normal cytoplasm. 12000X.

Fig. 49. A suberin-like layer laid over the opaque protective layer. A very thin layer of cytoplasm separates it from a large vacuole filled with fine translucent matter. 12000X.

Fig. 50. An apposition of material structurally similar to wall components of the gelatinous lay formed in fibers, laid on one side of the cell over masses of Om similar to that already describ above (enlarged portion Fig. 50A). Fig. 50, 36800X; Fig. 50A, 57500X.

Fig. 51. A wall ingrowth separating a mass of opaque cytoplasm from some that is appreciab altered. An extra secondary wall layer is present in the cell below. 15840X.

Fig. 52. An illustration from inoculated *Ulmus pumila* (already published in Ouellette (198 showing one of the typical reactions of paratracheal parenchyma cells. In addition to an increas wall deposition in the unexpanded tylosis, the wall has obviously produced bands of new w material to seal off affected cytoplasm (arrow, and below the half-bordered pit). The alveol network visible over the pit contains large body of near translucent matter similar to those nearby the vessel. 15840X.

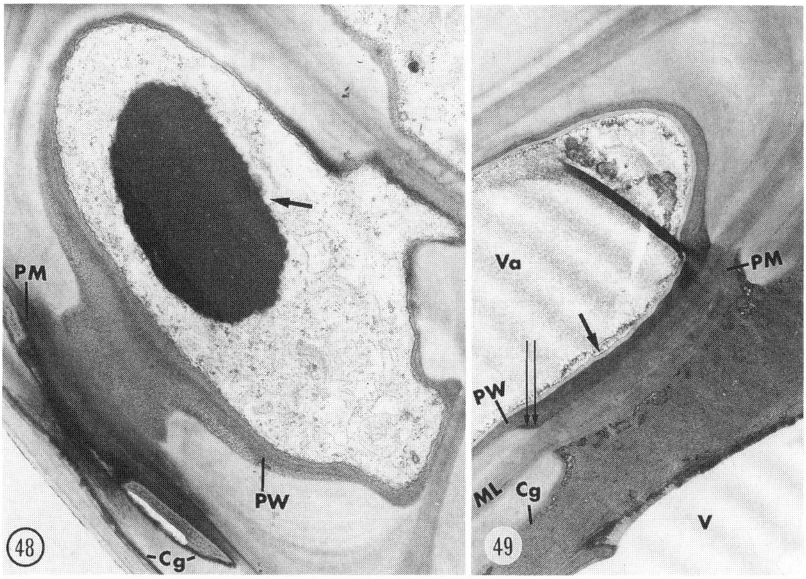

Figures 48-49. *See legend above.*

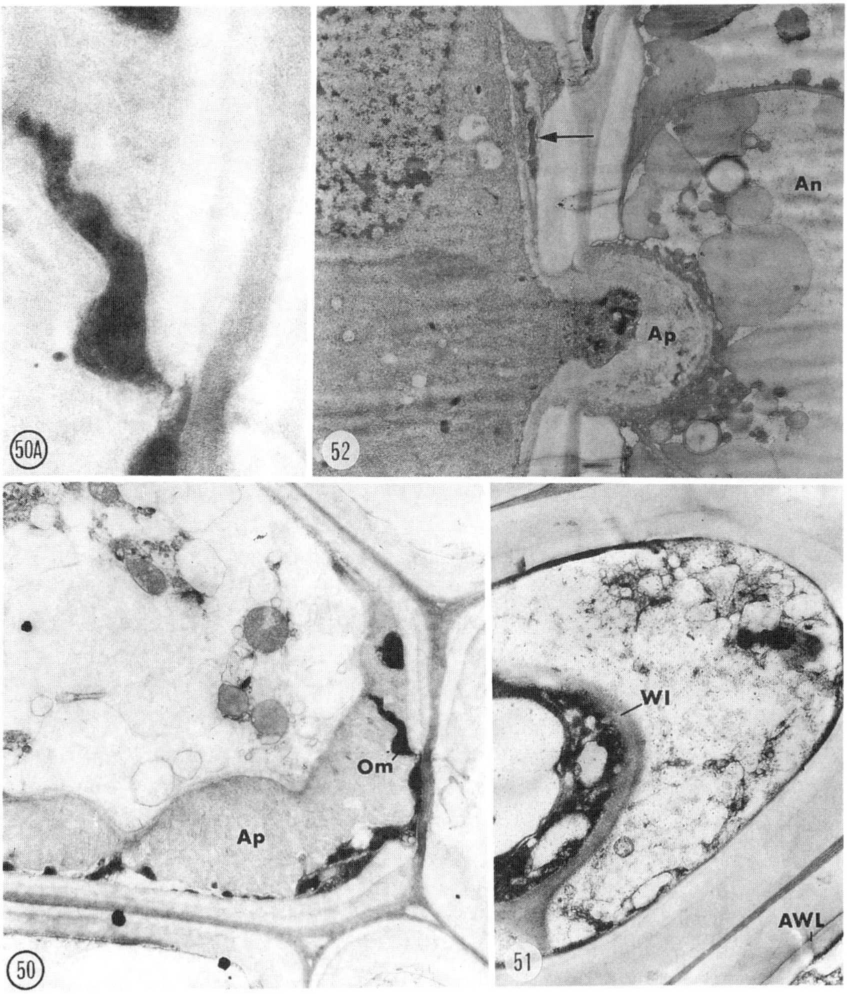

Figures 50-52. *See legend p. 282.*

Of the numerous reactions also observed in parenchyma cells in *Ulmus pumila,* the complete exclusion of punctually-altered cytoplasm (Fig. 52) seems to be one of the most efficient defense reactions. Several types of wall modifications in fibers also occur that have been thoroughly described, culminating generally in the formation of one or more suberin-like layers (Ouelette 1981b; Rioux and Ouelette 1991b).

B. *Primary xylem and adjacent tissues*

As cells of these tissues are attacked when they lack a secondary wall, complex situations are encountered where either intact, collapsed, disintegrating, or reacting cells are intermixed. Many of the cytoplasmic reactions in the primary xylem and differentiating tissues are analogous to those occurring in the secondary xylem. Only two of these will be mentioned which have a more direct bearing on the main object of the present communication. One significant reaction of young cells is the pronounced punctual or general thickening of the primary wall with the deposition of opaque matter (Fig. 54, bottom part). Masses of unorganized wall-like material may also form throughout the labyrinth of the numerous vacuoles that could develop in these cells (not illustrated here). Wall ingrowths may also form in cells delimiting normal appearing cytoplasm (Fig. 55) or portions of altered cytoplasm. New wall material is also often deposited to exclude masses of OM at certain points (Fig. 54) or almost completely around the cell (Fig. 53). This opaque material is structurally similar to that already described. Reactions in the cambial area may also be diverse. One significant feature is the differentiation of vessel elements that are abnormal in shape and wall thickness (Fig. 56). Square-shaped vessels may even differentiate within rows of cells that correspond to ray cells (not illustrated here). Surrounding these and crushed cells in the same region are cells containing large tannin bodies. Many of these surviving cells will eventually divide to generate barrier zone cells and then more normal tissues if the attack is not too pronounced (Rioux and Ouellette 1991a; Rioux and Ouellette 1991b). Many cells laid down in the xylem portion following cambial attack lack a secondary wall.

III. Considerations concerning the opaque material

In the previous sections, it was shown that many of the disturbances occurring in tissues affected by DED are associated with the presence of an opaque material (OM) constituted of more than one component. The following illustrations summarize the evidence that the OM does not result from artefacts of fixation or of other kinds, a fact which is necessary to understand the problem of the origin and nature of this OM. These illustrations supplement those already published in Ouellette (1978a) and Ouellette and Rioux (1992). Firstly, OM is present in vessel lumina or along vessel walls, tightly intermixed with other types of material that is considered to belong to advanced vessel colonization (Fig. 57). In this situation as in others (Fig. 58A), it is linked with host wall erosion and disintegration (arrow heads). In the latter, it is not conceivable that movement of the OM was from this cell with a still fairly intact plasmalemma, and through a such narrow channel to a mass of much larger diameter in the intercellular area. Secondly, cell contents may completely disintegrate, except for such matter which is present in the lumen as well as in the periplasmic area (Figs. 59, 59A). When shoots are injected with tritiated thymidine 24 h or so after inoculation and then sampled at various time intervals, similar, unbound matter having penetrated host cell walls is highly labeled (Figs. 62, 62A); the host tissues in these examples were mature and did not show labeling higher than background, except for the occasional mitochondrion and plastid. In searching for other indications that OM originates from the pathogen and not the plant, although this and other possibilities are not automatically excluded, the following may be invoked: connections with the free OM and fungal cell contents have often been discernible (Fig. 60). Also, the portions of fungal cells that are in close contact with host wall seem to lack a wall, whereas the free portion is bound by a wall. These fungal cell contents may also be more opaque. Similarly, microhyphae penetrating host wall often appear to lack a delimiting wall (Ouellette and Rioux 1992, and unpublished observations). Finally, as shown in a study of samples of the fungus growing in sterilized elm sections fixed under high pressure freezing, many of the fungal cells (of various configurations) were filled with components (Fig. 61) structurally analogous to the OM under study.

Legends to Figures 53-62

Figs. 53-56. Cell reactions in differentiating tissues. Fig. 53. A group of affected ray cells (tangential section). Cells A and B have no secondary wall, as well as some located right of these, next to other cells that possess such a wall. Cell B contains masses of Om between the initial wall and an another internal wall layer which is apparently continuous at points with the outside layer (arrow heads) and which arches inside the cell (near "cell B" label) over opaque matter. The internal wall layer appears to be discontinuous at both ends of this arch. 12000X. Fig. 54. An opaque mass of matter similar to Om, secluded within a mound of primary wall-like fibrils. Om also present in the periplasm. The wall of the adjacent cell contains also a large amount of opaque bodies. Fig. 54, 36800X; Fig. 54A, 57500X. Fig. 55. Wall ingrowths in a moderately affected cell which bordered at the right a number of more severely affected cells. 7200X. Fig. 56. A row of vessel elements bordered on each side by several hypertrophied cells in the cambial region. Although these vessel elements were obviously affected while they were differentiating, as indicated by the opaque contents that could correspond to their cytoplasmic remnants (top left), they nevertheless have formed a thick secondary wall. Many of the surrounding cells contain large tannin-like bodies. 10944X.

Figs. 57-61. Recapitulation of observations concerning the opaque material (Om).

Fig. 57. A band of Om similar to that shown in Fig. 39 extending freely in a vessel lumen. Om is associated here with vessel wall breakdown (arrow heads). It is permeating or is tightly bound to the configurations of the alveolar network typical here of advanced infection; the large arrow points to an area of discontinuity of Om occupied by a portion of the network component. The small wormy opaque strands located above the area were also often observed in vessel lumina. 20000X.

Fig. 58. A mass of Om extending between cells having extra secondary wall layers, ending into a cell lacking a secondary wall. A neat digested path is visible in the wall layer of that cell (arrow head, enlargement 58A). Note: the presence of Om in the periplasm nearby, as well as in the cell below. The channel leading to the intercellular space and the latter are lined with a thin opaque band similar to that bordering the gap in Om above, and which extends at points into the still intact wall. Fig. 58, 9440X; Fig. 58A, 31250X.

Fig. 59. Large bodies of Om, but less opaque than usual but with recognizable bodies of ribosomal size (Fig. 58A, enlarged portion indicated by arrow) that are likely located in a periplasmic area of this affected paratracheal parenchyma cell. Whereas this cell contains completely disintegrated cytoplasm, the other cells nearby show two types of reactions as illustrated in some of the preceding figures. Fig. 59, 15840X; Fig. 59A, 57500X.

Fig. 60. Om comprised between two fungal cells, appearing to be embedded in their walls (above upper arrow head) with possible areas of continuity with the cytoplasm. The fungal wall below these masses is not discernible, or is completely permeated (as in Fig. 4) by opaque strands. 57500X.

Fig. 61. A tiny bifurcating hypha, next to a normal appearing cell, having opaque contents in which can be recognized a few opaque bodies interpersed within arrays of parallelly oriented fibrils (Fig. 61A, enlarged portion). This material, from inoculated sterilized *Ulmus glabra* sections, was fixed by high pressure freezing. Fig. 61, 29600X; Fig. 61A, 46250X.

Fig. 62. Radioautographic microphotographs from samples injected with tritiated thymidine, 1 day after inoculation with the pathogen. Note high labeling of fungal cells and of a band of unbound cytoplasmic-like matter (Fig. 62A, enlarged portion near curved arrow) cutting across a pit membrane of a metaxylem vessel. Fig. 62, 36800X; Fig. 62A, 57500X.

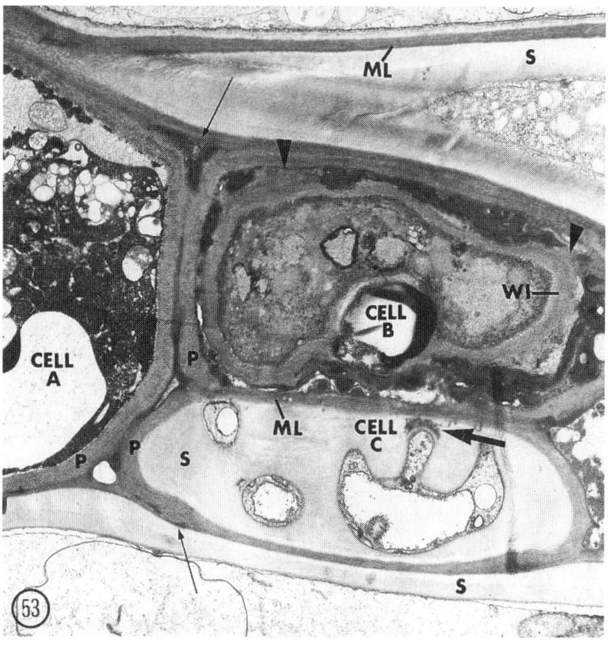

Figure 53. *See legend p. 285.*

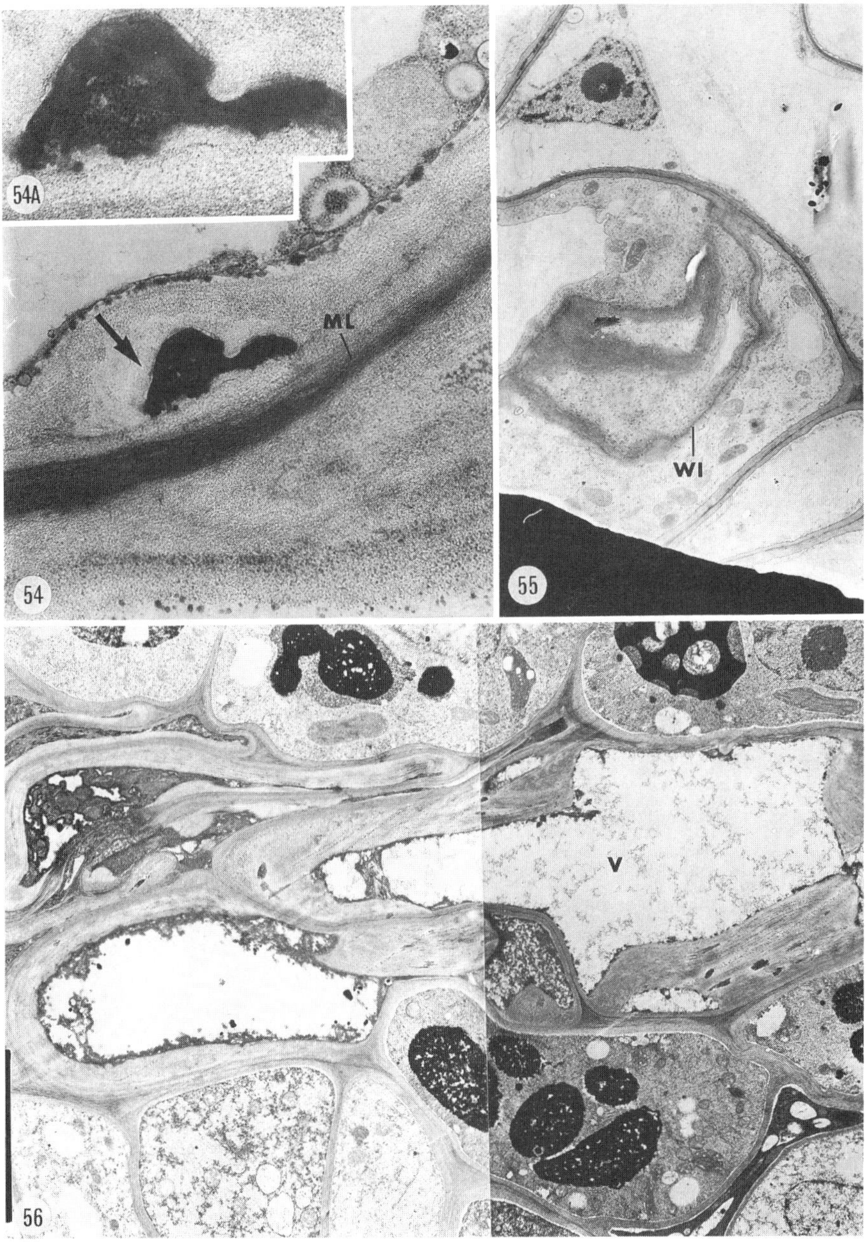

Figures 54-56. *See legend pp. 285-286.*

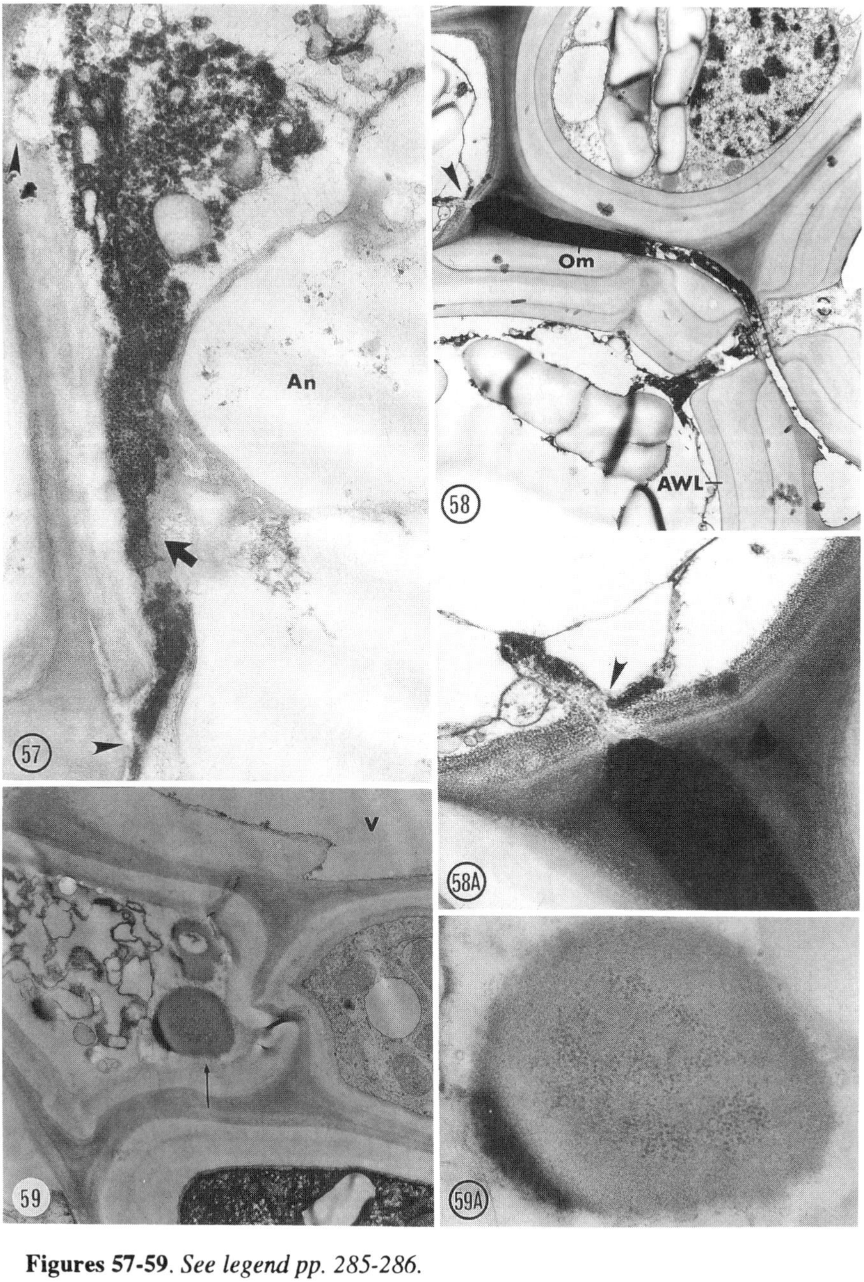

Figures 57-59. *See legend pp. 285-286.*

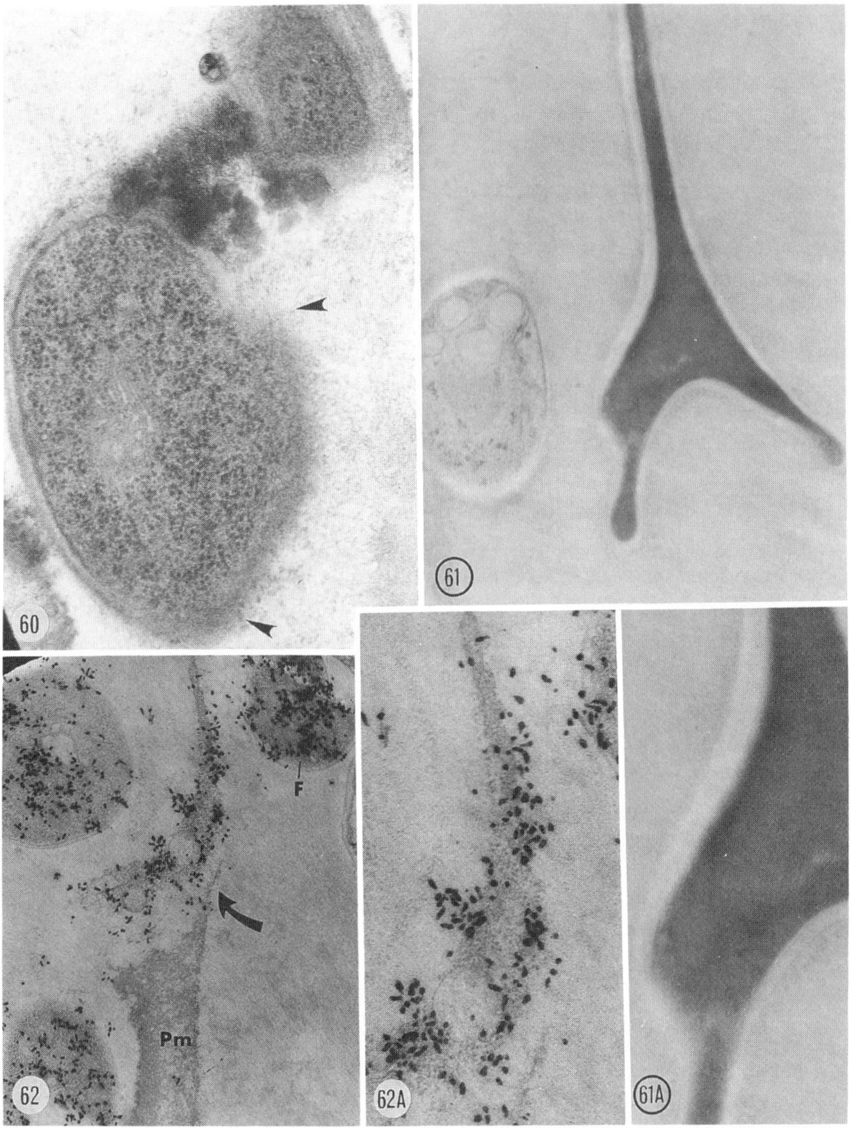

Figures 60-62. *See legend p. 286.*

Discussion

The following points should be recalled regarding the relative importance of disturbances occurring in vessel elements and surrounding tissues in DED development: 1) the material that accumulates along the walls of vessel elements is heterogeneous; 2) one of the components constituting the coating is associated with penetration and alteration of these walls as well as of pit membranes of half-bordered pits; 3) this component is analogous with masses and/or strands of opaque matter that accumulates in periplasmic areas or extends intercellularly over long distances and across host cell wall layers; 4) the alterations of axial parenchyma cells are concomitant with, and not subsequent to, vessel invasion by the pathogen; 5) independently of the degree of vessel occlusion, the external symptoms are mild or absent when the parenchyma cells are slightly altered, and as a corollary, when the cell reactions considered to be a defense type are more pronounced; 6) no extensive wilting occurs before the pathogen reaches the current shoot growths, the intensity of external symptoms appearing to be proportional to the degree of alterations of these tissues; 7) the detrimental effect of the disease on tissue differentiation may be significant in preventing recovery; 8) tylosis formation does not seem to have a bearing on early disease development.

When these points are weighed against each other concerning their relative importance in the DED syndrome, the effect on living cells, whether in the primary or secondary xylem, and whether they are in a state of differentiation or not, appears to be determinant at least in the chronic stage of the disease. As already mentioned (Ouellette and Rioux 1992), therefore, effective reactions of defense have to be sought in those cells, so that radial and even tangential spread of the pathogen in stems and branches can be delayed or prevented. This does not exclude that a mechanism liable to prevent fungus growth and vertical spread of the pathogen in vessel elements would not be ideal. Such a wall 1 type of resistance (Shigo and Marx 1977) does not seem to occur to any appreciable extent in elms, including *U. pumila,* and finding means of enhancing such a resistance is very difficult. Present investigations have shown that the opaque material (OM) can be observed as masses and strands of variable sizes and is associated with various types of host cell alterations. A main objective in research investigations of this disease would then be to characterize to a greater extent the origin and nature of this OM. In the authors' opinion, no great breakthroughs can be made in understanding the disease syndrome unless this problem is solved. An essential step in this direction would be to recognize simply that this problem exists, i.e. that it cannot be explained by current knowledge and that it may be necessary to formulate new concepts, or to re-establish already expressed but overlooked postulates. One possibility is that it may arise from the fungus; however, any other reasonable possibility is not excluded. Better tools now exist to determine the nature of the components of this OM. In this respect, a number of probes are now being tested at the light and confocal microscopes, and at the TEM level. Whatever the issue of these investigations is, it will not exclude the possibility that this OM carries some of the toxins that are now being studied, in addition to a possible complex of enzymes.

Acknowledgments

The authors express their gratitude to Nicole Lecours and Claude Moffet for their technical assistance.

References

Baayen, R.P. 1988. Fusarium wilt of carnation. Disease development, resistance mechanisms of the host and taxonomy of the pathogen. Ph. D. thesis, Univ. Amsterdam, 165 pp.

Benhamou, N., Chamberland, H., Ouellette, G.B., and Pauze, F.J. 1987. Ultrastructural localization of B-(1→4)-D-glucans in two pathogenic fungi and in their host tissues by means of an exoglucanase-gold complex. Can. J. Microbiol. 33:405-417.

Harling, R., and Taylor, G.S. 1985. A light microscope study of resistant and susceptible carnations infected with *Fusarium oxysporum* f. sp. *dianthi.* Can. J. Bot. 63: 638-646.

Ouellette, G.B. 1978a. Ultrastructural observations on pit membrane alterations and associated effects in elm xylem tissues infected by *Ceratocystis ulmi.* Can. J. Bot. 56:2567-2588.

Ouellette, G.B. 1978b. Light and electron microscopic studies on cell wall breakdown in American elm xylem tissues infected with Dutch elm disease. Can. J. Bot. 56:2666-2693.

Ouellette, G.B. 1981a. Ultrastructural cell wall modifications in secondary xylem of American elm surviving the acute stage of Dutch elm disease: vessel members. Can. J. Bot. 59:2411-2424.

Ouellette, G.B. 1981b. Ultrastructural cell wall modifications in secondary xylem of American elm surviving the acute stage of Dutch elm disease: fibres. Can. J. Bot. 59:2425-2438.

Ouellette, G.B. 1982. Comparative ultrastructural observations of reactions of xylem parenchyma cells to *Ceratocystis ulmi* infection in *Ulmus* spp. Pages 267-297 in Proceedings of Dutch elm disease symposium and workshop, Winnipeg, Man., October 5-9, 1981. E.S. Kondo, Y. Hiratsuka, and W.B.G. Denyer, eds.

Ouellette, G.B., and Rioux, D. 1992. Anatomical and physiological aspects of resistance to Dutch elm disease. In Defense mechanisms of woody plants against fungi. R. A. Blanchette and A. R. Biggs, eds., Springer-Verlag, New York. (In Press).

Pennypacker, B.W., and Nelson, P.E. 1972. Histopathology of carnation infected with *Fusarium oxysporum* f sp. *dianthi.* Phytopathology 62:1318-1326.

Rioux, D., and Ouellette, G. B. 1989. Light microscope observations of histological changes induced by *Ophiostoma ulmi* in various nonhost trees and shrubs. Can. J. Bot. 67:2335-2351.

Rioux, D., and Ouellette, G. B. 1991a. Barrier zone formation in host and nonhost trees inoculated with *Ophiostoma ulmi.* I. Anatomy and histochemistry. Can. J. Bot. 69:2055-2073.

Rioux, D., and Ouellette, G. B. 1991b. Barrier zone formation in host and nonhost trees inoculated with *Ophiostoma ulmi.* II. Ultrastructure. Can. J. Bot. 2074-2083.

Shi, J., Mueller, W.C. and Beckman, C.H. 1992. Vessel occlusion and secretory activities of vessel contact cells in resistant or susceptible cotton plants infected with *Fusarium oxysporum* f. sp. *vasinfectum.* Physiol. Mol. Plant Pathol. 40:133-147.

Shigo, A.L., and Marx, H.G. 1977. Compartmentalization of decay in trees. USDA Forest Service, Washington, DC.

22
Conventional and Molecular Genetic Approaches to the Study of Pathogenicity in *Ophiostoma ulmi sensu lato*

Louis Bernier

Abstract

One of the enduring questions about Dutch elm disease in general, and the biology of *Ophiostoma ulmi sensu lato* in particular, is how the fungus kills its host. Over the last two decades, the search for pathogenicity determinants has been stimulated by the discovery that two subgroups differing in pathogenicity occur within natural populations of *O. ulmi.* The non-aggressive (NAG) subgroup is characterized by strains with a lower level of pathogenicity, whereas the aggressive (AG) subgroup comprises highly pathogenic strains. The difference in pathogenicity between the NAG and AG subgroups has been correlated with differences for various attributes such as asexual sporulation, mycelial growth rate and morphology, pigmentation, intramycelial protein and enzyme electrophoretic patterns, DNA restriction fragment length polymorphisms, toxin production, and secretion of cell-wall degrading enzymes. Few genetic analyses have been carried out on the inheritance of the above-mentioned traits. Results from these analyses indicate that pathogenicity is under a predominantly additive nuclear gene control. Only one trait, growth-temperature response, could be correlated with pathogenicity in the progeny of controlled crosses between AG and NAG parents. This presentation will examine and discuss various strategies which have been tested or which appear promising for the study of pathogenicity in *O. ulmi.* Emphasis will be placed on genetic analysis of the pathogen, with both traditional and molecular approaches being discussed.

Introduction

Dutch elm disease has drastically altered the rural and urban landscape of various parts of Europe and North America. For many years, intensive research efforts and vast sums of money were devoted to understanding the mechanisms underlying pathogenesis and to finding ways to stop the spread of the disease in individual trees and in elm populations. Unfortunately, all the theories and control methods which have been proposed and tested so far have not prevented elms from dying in large numbers. Today, Dutch elm disease can claim the dubious honor of being a plant pathology textbook example of a tree disease for which no real cure has been found.

This grim preamble is not intended, however, to suggest that the challenge presented by the disease cannot be solved. On the contrary, Dutch elm disease may in fact be one of the few tree diseases for which research may soon have answers. It is, of course, a complex system but, unlike several other tree diseases, the components of the system are highly amenable to investigation and manipulation. This is particularly true of the pathogen, *Ophiostoma ulmi* (Buism.) Nannf. *sensu lato,* although researchers interested in Dutch elm disease have not yet taken full advantage of this fact. This will change, however, with the rapid development of increasingly powerful analytical methods, as well as with the involvement of scientists other than plant pathologists who are sharing an interest in various aspects of the biology of the fungus.

The following sections will present an overview of what is known about the genetics and physiology of *O. ulmi,* with particular attention to its pathogenicity. Work on some putative pathogenicity genes will be discussed, linkage maps will be presented, and some molecular strategies for the mapping and cloning of pathogenicity genes, or other genes of interest, will be reviewed.

Investigation and manipulation of *Ophiostoma ulmi* in the laboratory

From the investigator's standpoint, one of the most exciting features of *O. ulmi* is the ease with which it can be studied and manipulated in the laboratory. The fungus grows well and rapidly on a variety of synthetic and semidefined media, both solid and liquid (Brasier 1981; Bernier and Hubbes 1990a). It exhibits a yeast-mycelial dimorphism which can be controlled in liquid media by choosing the culture conditions (static vs. agitated) and the nitrogen source (inorganic vs. organic) (Kulkarni and Nickerson 1981; Muthukumar and Nickerson, 1984). Long term conservation of cultures of interest is easily accomplished (and highly recommended!) by storage of yeast cells in 15% glycerol at -80 C, or cryopreservation of either yeast cells or mycelium in liquid nitrogen. Unlike many other fungal plant pathogens, sexually compatible isolates of *O. ulmi* can usually be crossed easily on a variety of substrates (Holmes 1977; Asina and Hubbes 1980; Brasier 1981; Marshall et al. 1982; Bernier and Hubbes, 1990a). Genetic analysis of *O. ulmi* is facilitated by the fact that it is heterothallic and that hyphae and spores are predominantly uninucleate and the somatic nuclei are haploid (Brasier 1988).

The above-mentioned characteristics of *O. ulmi* have already led to a certain level of "domestication" of the fungus for laboratory investigations. For example, spontaneous drug-tolerant mutants are easily obtained (Webber et al. 1986; Mitchell 1987). Protocols for the recovery and analysis of mutants through the use of chemical and physical mutagenesis, enrichment techniques, and replica-plating on selective media have been described (Bernier 1988; Bernier and Hubbes 1990a). Since sexual matings can give rise to fruiting bodies (perithecia) in less than three weeks, near-isogenic lines differing only for selected markers can be obtained within a reasonable time frame.

In recent years, it has become increasingly clear that *O. ulmi* is an excellent candidate for studies at the molecular level. Both deoxyribonucleic and ribonucleic acids (DNA and RNA, respectively) can be purified by standard procedures. As a consequence, diagnostic and taxonomic studies formerly relying on morphological, proteinic and isoenzymatic markers (Stipes 1970; Bernier et al. 1983; Jeng and Hubbes, 1983; Jeng et al. 1988) are now performed routinely using DNA restriction fragment length polymorphisms (Bates et al. 1991; Jeng et al. 1991; Hintz et al. 1991). The more spectacular applications of molecular biology techniques, however, will be the cloning of genes of interest, and the *in situ* mutagenesis of these genes in order to assess directly their importance. The availability of both genomic and expression DNA libraries, and the development of a very efficient transformation system for *O. ulmi* are now putting some of these developments within close reach (Dewar 1990; Royer et al. 1991).

Investigations on the pathogenicity of *O. ulmi,* however, suffer from some drawbacks. Inoculation of elm seedlings and saplings, and accurate quantification of the host response are time-consuming and are not carried out as easily as in other plant-parasite interactions. In addition, elm breeding programs are not as advanced as those for agriculturally important crops. Therefore, investigators have access to a rather limited number of elm species and cultivars. Fortunately, several new elm varieties may be shortly available to researchers (Smalley and Guries 1993; and see elsewhere in this volume), whereas alternatives such as the use of elm calli have already shown their potential for *in vitro* screening of various molecules secreted by *O. ulmi* (Szczegola et al. 1987; Yang et al. 1989; Pijut et al. 1990). Our laboratory is currently testing a very simple bioassay based on the inoculation of Golden Delicious apples (Fulbright 1984). Preliminary results suggest a good correlation between pathogenicity on apples and on *Ulmus americana* seedlings. Such artificial systems cannot, of course, replace elm trees but they offer the possibility of accelerating some aspects of the work.

Natural variation of pathogenicity in *Ophiostoma ulmi*

Since *O. ulmi is* a parasite, the study of its pathogenicity has understandably been at the forefront of research on this organism. Little attention, however, has been given to the study of the genetic control of pathogenicity, in comparison with the efforts put on the clarification of the biochemical and physiological processes involved. Following reports by Walter (1937), Went (1938), and Tyler and Parker (1945) on the occurrence of variability in the pathogenicity of *O. ulmi,* Holmes (1965) was the first investigator to attempt a genetic study of this trait. He analyzed the pathogenicity of single-ascospore cultures towards a resistant elm species, *U. carpinifolia* "Christine Buisman", and a susceptible species, *U. hollandica* "Belgica". The progeny from parents collected in the same geographic area were reported to show little variation, whereas the progeny from parents collected in different areas were more variable and contained individuals able to overcome the resistance of the *U. carpinifolia* clone used in the study.

During the following decade, British researchers firmly established the occurrence, within natural populations of *O. ulmi,* of biological races differing in pathogenicity. Gibbs and collaborators (1972) observed the dynamics of a very severe epidemic of Dutch elm disease in England and reported that isolates collected from outbreak regions were more pathogenic than isolates from areas where the disease was endemic. The more pathogenic isolates, termed "aggressive" (AG), were originally described as fast-growing and fluffy, whereas less pathogenic isolates were slow-growing and waxy in culture, and were termed "non-aggressive" (NAG). Interestingly, the aggressive isolates were pathogenic to clones of *U. hollandica* previously considered to be highly resistant. Soon, it became obvious that isolates from other parts of Europe and from North America could also be assigned to either subgroup (Gibbs and Brasier 1973).

The observations reported by Gibbs et al. (1972) and Gibbs and Brasier (1973) prompted a systematic search for other traits that could be associated with pathogenicity. Over the following two decades, various morphological, physiological, biochemical and molecular markers were thus correlated with increased pathogenicity (for a review, see Brasier 1991). It is important, however, to stress that most of these observations were done on isolates from the field and were rarely verified in Fl progeny from controlled crosses between aggressive and non-aggressive parents. If the AG and NAG subgroups are indeed reproductively isolated in nature, as suggested by Brasier (1987), many if not most of the above-mentioned observations may thus just reflect the extent of evolutionary divergence between the subgroups and may not be correlated with pathogenicity.

Genetic analysis of pathogenicity in *Ophiostoma ulmi*

Laboratory crosses between aggressive and non-aggressive isolates usually yield progeny with decreased pathogenicity (Bernier 1983; Brasier 1987; Kile and Brasier 1990). In addition, none of the markers studied until now (growth rate, colony morphology, isoenzymes, pigmentation, cerato-ulmin production), with the sole exception of growth-temperature response, has been correlated conclusively with pathogenicity in the F_1 progeny of such crosses (Brasier and Gibbs 1976; Bernier 1983; Brasier 1987; Kile and Brasier 1990). There is no evidence of a maternal effect, indicating that no cytoplasmic genetic elements are involved. Therefore, data suggest strongly that pathogenicity in *O. ulmi* is under a predominantly additive nuclear gene control (Brasier 1987; 1988).

Although the analysis of progeny from crosses between aggressive and non-aggressive parents has already contributed most of our current knowledge on the genetic basis of pathogenicity in *O. ulmi,* this type of approach has serious limitations. Firstly, it makes it difficult to dissect the polygenic system into its individual, simpler, components. Secondly, the extent of the genetic differences between the subgroups raises the possibility that various additional factors may affect the results and further complicate the analysis. The analysis of F_1 progeny from crosses between Eurasian (EAN) aggressive isolates sheds maybe a more direct light on the control of pathogenicity in *O. ulmi* (Brasier 1987). Isolates belonging to the EAN race exhibit a wide range of pathogenicity on *U. procera.* When isolates with either low or moderate pathogenicity are crossed with highly pathogenic isolates, the progeny segregate in a nearly 1:1 ratio, suggesting the implication of a single gene with two alleles. This may well be the most convincing evidence so far for the implication of a specific pathogenicity gene, even though the product of this gene is unknown.

Construction of genetic maps

The analysis of traits controlled by a single gene is a relatively easy task, provided that individuals carrying mutant alleles at this locus are available and that their phenotype is easily detectable. When the phenotype of interest is difficult to assess, it is common to look for a closely linked marker which is easier to score. In the case of a complex phenotype, such as pathogenicity in *O. ulmi*, the use of several additional, well-characterized markers is a more complex, but still attractive, enterprise resulting in the construction of a linkage map on which pathogenicity genes may be identified and localized. This approach requires that the sexual (or parasexual) cycle of the organism studied can be completed in the laboratory so that linkages can be established. As stated before, *O. ulmi* is an ideal organism in this respect, since sexual matings are easily conducted on relatively simple media and mature perithecia can often be harvested in less than three weeks.

Attempts to use morphological markers such as growth rate or mycelial habit have not been successful since these traits are also under polygenic control (Brasier and Gibbs 1976). Tolerance to the fungicides benomyl and iprodione, however, was found to be controlled, in each case, by a single nuclear gene linked to the mating-type *(MAT)* locus (Brasier and Gibbs 1975a, 1975b; Webber et al. 1986; Mitchell 1987). In a systematic search for additional markers, mutants were induced by exposure to the chemical mutagen N-methyl-N'-nitro-N-nitrosoguanidine (MNNG) and recovered through various enrichment methods (Bernier and Hubbes 1990a). Meiotic analysis of monogenically controlled nuclear mutations in random ascospore progeny first allowed the mapping of eight of these mutations on three linkage groups (Bernier and Hubbes 1990b). More recent data suggest the occurrence of at least four linkage groups in *O. ulmi* (Fig. 1). No pathogenicity gene, however, has yet been assigned a location on this map.

In spite of their usefulness, induced mutations provide us with relatively few markers and mutants may require a fair amount of "genetic cleaning" in order to eliminate additional, hidden mutations which may bring undesirable effects. In recent years, two molecular strategies have become increasingly popular in genome mapping programs. The first one is based on the analysis of DNA restriction fragment length polymorphisms (RFLPs). These naturally occurring markers result from a variety of genetic changes, such as base substitutions, insertions or deletions. Polymorphisms are made visible by the digestion of the DNA with restriction enzymes, followed by the electrophoretic separation of the resulting fragments and their transfer onto a nitrocellulose or nylon membrane, and the molecular hybridization of the immobilized fragments with labeled cloned sequences used as probes. RFLPs are often co-dominant but are restricted, for practical reasons, to DNA regions with low or single copy sequences. A marked advantage of RFLP analysis is the possibility to map several loci from the analysis of a single cross, if the parent isolates have been chosen carefully. A recent report by Tzeng et al. (1992) on RFLP mapping of the maize pathogen *Cochliobolus heterostrophus* provides an impressive account of the possibilities offered by this type of analysis.

A major drawback of the traditional RFLP technique, however, is its reliance on the availability of suitable probes and on the need to perform Southern hybridization for the detection of polymorphisms. These factors make the technique tedious, especially when large sample sizes are involved, and potentially unsafe if radiolabeled probes are used. The development of methods to amplify DNA sequences through the polymerase chain reaction, or PCR, (Saiki et al. 1985; Mullis et al. 1986) is currently revolutionizing the way linkage maps are constructed. One particularly exciting application of PCR technology is the obtention of random amplified polymorphic DNAs (RAPDs). RAPD markers result from the amplification of genomic DNA with single primers of arbitrary nucleotide sequence (Williams et al. 1990). They are usually dominant and segregate in Mendelian fashion. Their use for mapping genomes offers the advantages already provided by RFLPs, plus additional benefits: hundreds of synthetic oligodeoxynucleotide primers are available commercially, and polymorphisms are detected directly on ethidium bromide-stained agarose gels, or silver-stained polyacrylamide gels.

So far, published uses of RFLPs for the analysis of *O. ulmi* have been restricted to general comparisons of the nuclear and mitochondrial genomes of the AG and NAG subgroups (Bates et al. 1991; Jeng et al. 1991; Hintz et al. 1991). Work is currently under way in our laboratory in which both induced mutations and RAPD markers are used to produce a linkage map. We are also attempting to establish a direct relationship between linkage groups and the chromosomes they represent. To this end, *O. ulmi* chromosomes must first be identified directly, either cytologically or electrophoretically. We have chosen the latter approach, in light of the encouraging results obtained by Royer and co-workers (1991) who have reported the electrophoretic separation of six chromosomes in the North American (NAN) aggressive isolate MH75 (TOR).

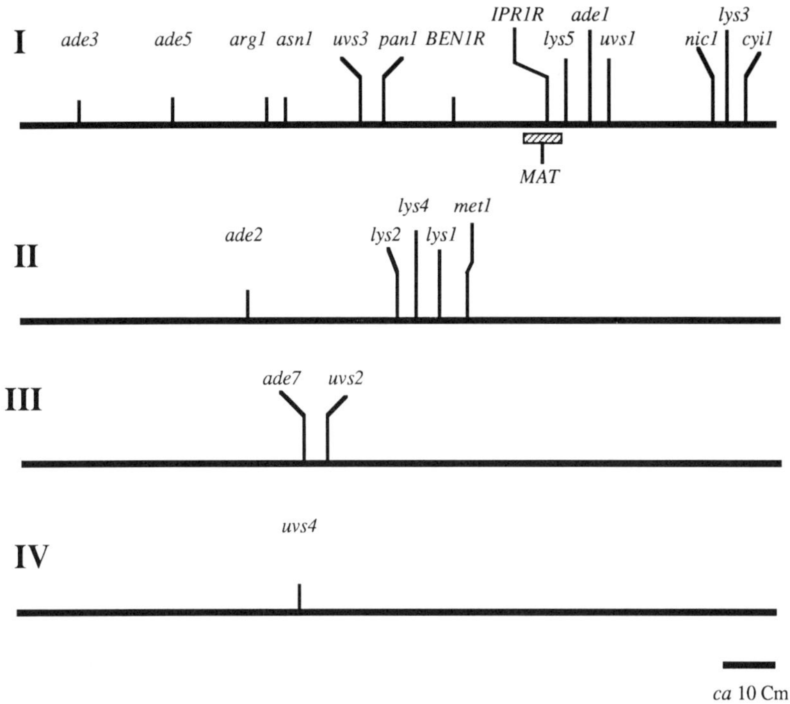

Figure 1. A simplified linkage map of *Ophiostoma ulmi*.
The auxotrophic markers are *ade* for adenine, *arg* for arginine, *asn* for asparagine, *cyi* for cystine, *lys* for lysine, *met* for methionine, *nic* for nicotinamide, and *pan* for pantothenic acid. Drug resistance markers are *BENIR* for benomyl and *IPRI* for iprodione. Other markers are *uvs* for ultraviolet-sensitive, and *MAT* for mating-type. The hatched area under linkage group I represents the region where the *MAT* locus is thought to reside. The order of closely linked genes, such as the *lys2, lys4, lysl,* and *met1* loci, is still tentative. Distances (in Centimorgans) are approximate.

We have analyzed chromosomal DNAs from 12 isolates representing the NAN, EAN and NAG biotypes, using three different pulse-field gel electrophoresis (PFGE) systems. Seven of these isolates, including MH75, shared a common chromosomal DNA banding pattern, whereas each of the other five had unique polymorphisms. Differences in electrophoretic karyotypes were not associated with race or subgroup (Dewar and Bernier, 1993a, 1993b). Work is currently under way to assign specific markers to chromosomes separated by PFGE techniques.

Identification of putative pathogenicity genes

Various explanations have been offered to explain pathogenicity in *O. ulmi.* It has been observed that the fungus produces both pectic and cellulolytic enzymes *in vitro* (Elgersma 1976). Scanning electron microscope studies of cell walls in infected wood tissue suggest that these enzymes may be important *in vivo* (Scheffer and Elgersma 1982). Moreover, aggressive isolates display higher cell-wall degrading activity than their non-aggressive counterparts (Scheffer and Elgersma 1982; Svaldi and Elgersma 1982). While the role of cell-wall degrading enzymes is fairly well documented in some fungal pathogen-plant interactions, the information available in this case is too scant to draw solid conclusions. There is clearly an opportunity here for some simple genetic work taking advantage of the existing variability in cell-wall degrading activity among isolates of *O. ulmi.* The analysis of crosses between aggressive and non-aggressive parents, or between low or moderate and highly aggressive isolates, could indicate whether these enzymes play a major or minor role in the expression of pathogenicity.

Much attention has been devoted to non-enzymatic, phytotoxic molecules extracted and purified from *O. ulmi* culture filtrates. Claydon and co-workers (1980) described several phenolic metabolites toxic to elm shoots. None of these putative toxins, however, induced Dutch elm disease symptoms when injected into seedlings (Scheffer 1983). Strobel et al. (1978) recovered a glycopeptide fraction which was later partially characterized as a polydisperse peptidorhamnomannan with a molecular weight ranging from 105 to 120 kilodaltons (Nordin and Strobel 1981). Although the production of this glycoprotein was confirmed *in vivo* (Scheffer and Elgersma 1981), its high molecular weight suggests that it may interfere with water flow by physical plugging and cavitation and that it may thus be an aggressin rather than a true toxin (Van Alfen and McMillan 1982). Scheffer and co-workers (1987) reported the occurrence of variations in the production of peptidorhamnomannan by isolates *in vitro,* but no differences between the mean values for the AG and NAG subgroups. No attempts were made to cross low toxin producing and high toxin producing isolates.

The *O. ulmi* toxin which has been the most thoroughly investigated is the small protein cerato ulmin (CU), initially described by Takai (1974). There is ample biochemical and physiological evidence (reviewed elsewhere in this volume) that CU possesses various attributes of a true toxin. Work by Russo and co-workers (1982) suggests that the molecule may also act as an aggressin. So far, however, there is no genetic evidence that CU is a major determinant of pathogenicity. As mentioned earlier, CU production *in vitro* and pathogenicity are not correlated in the Fl progeny from crosses between aggressive and non-aggressive isolates (Bernier 1983; Kile and Brasier 1990). On the other hand, the possibility that the protein plays a minor role has not yet been explored. This could likely be done by crossing EAN and NAN aggressive isolates differing both in CU production and in pathogenicity (Brasier 1987; Brasier et al. 1990), and analyzing the segregation of both traits in the Fl progeny. Obviously, the status of CU will soon become clearer since its complete amino acid sequence is known, low CU-producing mutants are available, and cloned CU genes about to be sequenced (Jeng et al. 1987; Bernier 1988; and see elsewhere in this volume).

Besides the more traditional models offered by enzymes and toxins, additional candidates are now being considered. The production of phytoalexins belonging to the group of mansonones by elm seedlings and calli in response to infection by *O. ulmi* is well documented (Duchesne 1988; Sticklen et al. 1991; and see elsewhere in this volume). Once again, the observation that non-aggressive isolates induce mansonones in higher concentrations than the aggressive isolates opens the door for simple genetic tests on the inheritance of this trait. Further comparison among isolates belonging to the EAN and NAN races within the aggressive subgroup could also be attempted. It could be possible to quantify and compare the production, by Fl progeny, of the glycopeptide elicitor described by Yang and co-workers (1989). Likewise, the susceptibility of *O. ulmi* isolates to specific mansonones *in vitro* is variable (Dumas et al. 1986) but the genetic basis of this variability has not been studied. Finally, the use of *in vitro* bioassays may offer the opportunity to identify genes controlling recognition between the pathogen and its host.

New approaches to the genetic study of pathogenicity in *O. ulmi*

Traditional genetic analysis offers correlative evidence for the implication of one or several genes in the determination of a given phenotype. With the advent of molecular biology, an ever increasing number of avenues is available for the direct study of regulatory and structural genes. Molecular techniques have already allowed the identification and manipulation of various genes controlling plant-pathogen interactions. Several fungal pathogens of agricultural crops are currently the object of intense scrutiny. Tree pathogenic fungi, however, have elicited considerably less interest.

Molecular genetic techniques offer much potential for the study of pathogenicity in *O. ulmi*. It is likely, however, that several distinct strategies are needed for the elucidation of the various attack mechanisms which seem to operate in the fungus. Fortunately, an array of techniques is now available for cloning fungal genes (for a recent review, see Timberlake 1991). Some approaches could be simple. *Ophiostoma ulmi* obviously possesses some pathogenicity genes which are also found in other fungi, such as genes coding for cellulases and pectinases. Attempts to clone such genes could proceed from computer searches in sequence data banks, obtention of clones from related species or genera, and heterologous Southern or dot blot hybridizations under conditions of low stringency.

Ophiostoma ulmi, on the other hand, also possesses genes which appear to be unique, such as the structural gene coding for CU. In this case, a direct cloning approach also appears feasible since the product of the gene is known, well-characterized, and easily detected and quantified. The gene can either be identified by screening a genomic DNA bank with a synthetic oligodeoxynucleotide probe, or by screening an expression library with antibodies against the toxin. In fact, it appears that a CU gene has already been cloned (see elsewhere in this volume). Once the gene is cloned, it can be mutated *in vitro* and reintroduced into a wild-type isolate, through genetic transformation, to selectively inactivate or replace the resident wild-type allele. It then becomes possible to create specific mutants and to assess directly the importance of CU for pathogenicity. It is, of course, necessary to verify that only the CU structural locus has been inactivated. An efficient transformation system allowing site-specific integration of the mutated gene into the appropriate host chromosome is thus required. The successful cloning of both a CU gene and its promoter will, in turn, allow studies on the regulation of toxin production.

In the case of pathogenicity genes for which no specific product has been identified, indirect cloning techniques are also available. If the gene confers a phenotype that is scored easily, it can be cloned by complementation of mutant strains with pools of cosmid vectors carrying fragments from a genomic DNA library from a wild-type isolate. After successive rounds of transformation with subpools which are progressively decreasing in size, it is possible to isolate a single complementing clone. This technique, known as "sib selection", was devised by Vollmer and Yanofsky (1986) to clone several genes in *Neurospora crassa* by transformation with banks constructed in the highly efficient pSV50 cosmid vector. Unfortunately, it has not been possible to transform *O. ulmi* with this vector (Royer et al. 1990).

Some fungal genes of interest, however, do not produce a phenotype which can be detected easily and rapidly. Genes coding for avirulence and for mating-type factors often fall into this category. In such cases, it may be possible to simplify the screening procedure by using chromosome walking techniques. In order to do so, a marker closely linked to the gene of interest is cloned first and will serve as a starting point for the walk through a genomic DNA library. This marker can be another gene, or a RFLP or RAPD locus, and must be easily scored. Overlapping clones are isolated, and the clone carrying the gene of interest is identified by transformation and expression of the desired phenotype in the recipient strain. The walk can be shortened by making a library which is specific to the chromosome carrying the markers which are being studied, and by constructing the library in vectors capable of accepting very large inserts, such as yeast artificial (YAC) chromosomes (Burke et al. 1987). It is, however, currently difficult to prepare chromosome-specific libraries in *O. ulmi,* since good electrophoretic separation of the chromosomes is still hard to obtain (Royer et al. 1990; Dewar and Bernier 1993a, 1993b).

As can be seen from the above, there are several prerequisites to be met before a chromosome walk is launched. An alternative, simpler technique termed "genomic substraction" has been proposed by Straus and Ausubel (1990). This procedure identifies DNA sequences which are present in a wild-type individual but are missing in a homozygous deletion mutant. This is achieved by submitting wild-type DNA and an excess of labeled mutant DNA to several cycles of hybridization, by which sequences common to both individuals are eliminated. The DNA sample collected at the end of the procedure is thus enriched in sequences that are absent in the deletion mutant. These sequences can be linked to oligonucleotide adaptors and amplified directly by PCR. *Ophiostoma ulmi* deletion mutants with decreased pathogenicity could be analyzed by this technique. Field isolates differing in pathogenicity (such as AG and NAG, or EAN and NAN) are less suitable candidates for genomic substraction, since they are also likely to be heterozygous at several other loci. Several rounds of controlled backcrosses to a fixed parent would first be required for producing the near-isogenic lines differing in pathogenicity which could be analyzed.

Finally, a method based on PCR amplification of RAPD markers, developed by Michelmore and collaborators (1991), may prove very useful for the analysis of pathogenicity and other traits in *O. ulmi.* In this procedure, two pooled DNA samples of individuals from a 1:1 segregating population originating from a single cross are compared. Each pool, or bulk, thus contains individuals which are identical (homozygous) for the trait of interest, but are arbitrary (heterozygous) for all other genes. The bulked DNAs are then screened for differences using RAPD markers. Most RAPD markers will be monomorphic between the bulks, but any polymorphic marker will be genetically linked to the locus that controls the trait used to construct the pools in the first place. The polymorphic RAPD markers can then be used for walking the chromosome towards the gene of interest. As mentioned earlier, RAPD primers required for screening the bulks are available by the hundreds. Another advantage of this procedure is that field isolates can be studied and compared readily, without any need for mutational work or preparation of near-isogenic lines, as long as they exhibit differences that segregate in a simple Mendelian fashion.

Conclusion

Our knowledge of the genomic architecture and gene regulation in *O. ulmi* is obviously still very superficial. Previous work has focused mainly on accounting for overall differences among the EAN, NAN, and NAG biotypes. We are still far from understanding the mechanisms and pathways which are activated when the pathogen comes in contact with its host. The availability of several new molecular techniques, in conjunction with the continued use of classical inheritance studies, is creating exciting opportunities for investigations on this front. It seems reasonable to predict that at least a few pathogenicity genes, as well as other important genes, will be identified, mapped, and cloned in the short future. Likewise, molecular techniques are also being used to understand susceptibility and resistance in the host. The Dutch elm disease may thus earn the symbol status of a "model system". Hopefully, this will also translate into viable and long-term control methods against the disease.

Acknowledgments

Unpublished experiments from the author's laboratory are supported by grants from the Natural Sciences and Engineering Research Council (Canada), Fonds FCAR (Quebec), and Ministere des Forets du Quebec.

References

Asina, S. and Hubbes, M. 1980. Compatibility in *Ceratocystis ulmi*. Eur. J. For. Pathol. 10:20-1208.

Bates, M.R., Buck, K.W., and Brasier, C.M. 1991. Molecular variation in the Dutch elm disease fungus. *In* Molecular Evolution. M.T. Clegg and S.J. O'Brian, eds. UCLA Symposia on Molecular and Cellular Biology New Series 122: 171-178.

Bernier, L. 1983. Relationships between isoenzymes, cultural characters, and pathogenicity in *Ceratocystis ulmi* (Buism.) C. Moreau. M.Sc.F. Thesis, University of Toronto, Toronto.

Bernier, L. 1988. Induction, characterization and mapping of mutations in *Ophiostoma ulmi,* the causal agent of Dutch elm disease. Ph.D. Thesis, University of Toronto, Toronto.

Bernier, L. and Hubbes, M. 1990a. Mutations in *Ophiostoma ulmi* induced by N-methyl-N'-nitro-N-nitrosoguanidine. Can. J. Bot. 68:225-231.

Bernier, L. and Hubbes, M. 1990b. Meiotic analysis of induced mutations in *Ophiostoma ulmi* Can. J. Bot. 68:232-235.

Bernier, L., Jeng, R.S., and Hubbes, M. 1983. Differentiation of aggressive and non-aggressive strains of *Ceratocystis ulmi* by polyacrylamide gel electrophoresis of intramycelial enzymes. Mycotaxon 17:456-472.

Brasier, C.M. 1981. Laboratory investigation of *Ceratocystis ulmi. In* Compendium of elm diseases. R.J. Stipes and R.J. Campana, eds. American Phytopathological Soc., St. Paul, MN. pp. 76-79.

Brasier, C.M. 1987. Some genetical aspects of necrotrophy with special reference to *Ophiostoma ulmi. In* Genetics and Plant Pathogenesis. P.R. Day and G.J. Jellis, eds. Blackwell Scientific Publications, Oxford. pp. 297-310.

Brasier, C.M. 1988. *Ophiostoma ulmi,* cause of Dutch elm disease. *In* Advances in plant pathology. Vol. 5. G.S. Sidhu, ed.. Academic Press, New York. pp. 207-223.

Brasier, C.M. 1991. *Ophiostoma novo-ulmi sp.* nov., causative agent of current Dutch elm disease pandemics. Mycopathologia 115:151-161.

Brasier, C.M. and Gibbs, J.N. 1975a. MBC tolerance in aggressive and nonaggressive isolates of *Ceratocystis ulmi.* Ann. Appl. Biol. 80:231-235.

Brasier, C.M. and Gibbs, J.N. 1975b. Highly fertile form of the aggressive strain of *Ceratocystis ulmi.* Nature 257:128-131.

Brasier, C.M., Takai, S., Nordin, J.H., and Richards, W.C. 1990. Differences in cerato-ulmin production between the EAN, NAN and non-aggressive subgroups of *Ophiostoma ulmi.* Plant Pathol. 39:231-236.

Burke, D.T., Carle, G.F., and Olson, M.V. 1987. Cloning of large segments of exogenous DNA into yeast by means of artificial chromosome vectors. Science 236:806-812.

Claydon, N., Elgersma, D.M., and Grove, J.F. 1980. The phytotoxicity of some phenolic products of *O. ulmi* to *Ulmus* sp. Neth. J. Plant Pathol. 86:229-238.

Dewar, K. 1990. Generation and screening of an *Ophiostoma ulmi* cDNA library. M.Sc.F. Thesis, University of Toronto, Toronto.

Dewar, K. and Bernier, L. 1993a. Electrophoretic karyotypes of the elm tree pathogen *Ophiostoma ulmi (sensu lato).* Submitted to Mol. Gen. Genet. (In press).

Dewar, K. and Bernier, L. 1993b. Variabilité du caryotype électrophorétique chez *Ophiostoma ulmi,* agent de la maladie hollandaise de l'orme. Phytoprotection 74 (Abstr. In press).

Duchesne, L.C. 1988. Resistance mechanisms to Dutch elm disease: A review. Nat. Can. 115:163-167.

Dumas, M.T., Strunz, G.M., Hubbes, M., and Jeng, R.S. 1986. Inhibition of *Ceratocystis ulmi* by mansonones A, C, D, E, F, and G isolated from *Ulmus americana.* Eur. J. For. Pathol. 16:217-222.

Elgersma, D.M. 1976. Production of pectic and cellulolytic enzymes by aggressive and non-aggressive strains of *Ophiostoma ulmi.* Neth. J. Plant Pathol. 82:161-172.

Fulbright, D.W. 1984. Effect of eliminating dsRNA in hypovirulent *Endothia parasitica.* Phytopathology 74:722-724.

Gibbs, J.N. and Brasier, C.M. 1973. Correlation between cultural characters and pathogenicity in *Ceratocystis ulmi* from Europe and North America. Nature 241:381-383.

Gibbs, J.N., Heybroek, H.M., and Holmes, F.W. 1972. Aggressive strain of *Ceratocystis ulmi* in Britain. Nature 236:121-122.

Hintz, W.E., Jeng, R.S., Hubbes, M., and Horgen, P.A. 1991. Identification of three populations of *Ophiostoma ulmi* (aggressive subgroup) by mitochondrial DNA restriction-site mapping and nuclear DNA fingerprinting. Exp. Mycol. 15:316-325.

Holmes, F.W. 1965. Virulence in *Ceratocystis ulmi.* Neth. J. Plant Pathol. 71:97- 112.

Holmes, F.W. 1977. Distinction between sex and compatibility in *Ceratocystis ulmi* as shown by unisexual male cultures. Mycologia 69:1149-1161.

Jeng, R.S. and Hubbes, M. 1983. Identification of aggressive and non-aggressive strains of *Ceratocystis ulmi* by polyacrylamide gradient gel electrophoresis of intramycelial proteins. Mycotaxon 17:445-455.

Jeng, R.S., Bernier, L., and Brasier, C.M. 1988. A comparative study of cultural and electrophoretic characteristics of the Eurasian and North American races of *Ophiostoma ulmi.* Can. J. Bot. 66:1325-1333.

Jeng, R.S., Bernier, L., and Svircev, A.M. 1987. Low cerato-ulmin producing mutants from an aggressive isolate of *Ophiostoma ulmi,* the causal agent of Dutch elm disease. Can. J. Plant Pathol. 9:280 (Abstr.).

Jeng, R.S., Duchesne, L.C., Sabourin, M., and Hubbes, M. 1991. Mitochondrial DNA restriction fragment length polymorphisms of aggressive and non-aggressive isolates of *Ophiostoma ulmi.* Mycol. Res. 95:537-542.

Kile, G.A., and Brasier, C.M. 1990. Inheritance and inter-relationship of fitness characters in progeny of an aggressive x non-aggressive cross of *Ophiostoma ulmi.* Mycol. Res. 94:514-522.

Kulkarni, R.K. and Nickerson, K.W. 1981. Nutritional control of dimorphism in *Ceratocystis ulmi.* Exp. Mycol. 5:148-154.

Marshall, M.R., Hindal, D.F., and MacDonald, W.L. 1982. Production of perithecia in culture by *Ceratocystis ulmi.* Mycologia 74:376-381.

Michelmore, R.W., Paran, I., and Kessell, R.V. 1991. Identification of markers linked to disease resistance genes by bulked segregant analysis: a rapid method to detect markers in specific genomic regions using segregating populations. Proc. Natl. Acad. Sci. USA 88:9828-9832.

Mitchell, A.G. 1987. Characterization of iprodione tolerance in *Ophiostoma ulmi.* Trans. Br. Mycol. Soc. 88:283-288.

Mullis, K.B., Faloona, F.A., Scharf, S., Saiki, R.K., Horn, G.T., and Erlich, H.A. 1986. Specific enzymatic amplification of DNA *in vitro:* the polymerase chain reaction. Cold Spring Harbor Symp. Quant. Biol. 51: 63-273.

Muthukumar, G. and Nickerson, K.W. 1984. Ca(II)-calmodulin regulation of fungal dimorphism in *Ceratocystis ulmi.* J. Bacteriol. 159:390-392.

Nordin, J.H. and Strobel, G.A. 1981. Structural and immunochemical studies on the phytotoxic peptidorhamnomannan of *Ceratocystis ulmi*. Plant Physiol. 67:1208-1213

Pijut, P.M., Lineberger, R.D., Domir, S.C., and Schreiber, L.R. 1990. Use of culture filtrates of *Ceratocystis ulmi* as a bioassay to screen for disease tolerant *Ulmus americana.* Plant Sci. 70:191-196.

Royer, J.C., Dewar, K., Hubbes, M., and Horgen, P.A. 1991. Analysis of a high frequency transformation system of *Ophiostoma ulmi,* the causal agent of Dutch elm disease. Mol. Gen. Genet. 225:168-176.

Russo, P.S., Blum, F.D., Ipsen, J.D., Abul-Hajj, Y.J., and Miller, W.G. 1982. The surface activity of the phytotoxin cerato-ulmin. Can. J. Bot. 60:1414-1422.

Saiki, R.K., Scharf, S., Faloona, F., Mullis, K.B., Horn, G.T., Erlich, H.A., and Arnheim, N. 1985. Enzymatic amplification of B-globin genomic sequences and restriction site analysis for diagnosis of sickle cell anemia. Science 230:1350-1354.

Scheffer, R.J. 1983. Toxins in Dutch elm disease. *In* Research on Dutch elm disease in Europe. D.A. Burdekin, ed. Forestry Commission Bull. 60, HMSO, London. pp. 82-85.

Scheffer, R.J. and Elgersma, D.M. 1981. Detection of a phytotoxic glycopeptide produced by *Ophiostoma ulmi* in elm by enzyme-linked immunospecific assay (ELISA). Physiol. Plant Pathol. 18:27-32.

Scheffer, R.J. and Elgersma, D.M. 1982. A scanning electron microscope study of cell wall degradation in elm wood by aggressive and non-aggressive isolates of *Ophiostoma ulmi.* Eur. J. Forest Pathol. 12:25-28.

Scheffer, R.J., Liem, J.I., and Elgersma, D.M. 1987. Production *in vitro* of phytotoxic compounds by non-aggressive and aggressive isolates of *Ophiostoma ulmi,* the Dutch elm disease pathogen. Physiol. Molec. Plant Pathol. 30:321-335.

Smalley, E.B. and Guries, R.P. 1993. Breeding elms for Dutch elm disease resistance. Phytoprotection 74 (In press).

Sticklen, M.B., Bolyard, M.G., Hajela, R.K., and Duchesne, L.C. 1991. Molecular and cellular aspects of Dutch elm disease. Phytoprotection 72:1-13.

Stipes, R.J. 1970. Comparative mycelial protein patterns in four species of *Ceratocystis.* Mycologia 62:987-995.

Straus, D., and Ausubel, F.M. 1990. Genomic substraction for cloning DNA corresponding to deletion mutations. Proc. Natl. Acad. Sci. USA 87:1889-1893.

Strobel, G.A., Van Alfen, N., Hapner, K.D., McNeil, M., and Albersheim, P. 1978. Some phytotoxic glycopeptides from *Ceratocystis ulmi,* the Dutch elm disease pathogen. Biochim. Biophys. Acta 538:60-75.

Svaldi, R. and Elgersma, D.M. 1982. Further studies on the activity of cell wall degrading enzymes of aggressive and non-aggressive isolates of *Ophiostoma ulmi.* Eur. J. For. Pathol. 12:29-36.

Szczegola, M., Jeng, R.S., and Hubbes, M. 1987. Mansonone induction by *Ophiostoma ulmi* in callus cultures of elm species. Can. J. Plant Pathol. 9:286 (Abstr.).

Takai, S. 1974. Pathogenicity and cerato-ulmin production in *Ceratocystis ulmi.* Nature 252:124-126.

Timberlake, W.E. 1991. Cloning and analysis of fungal genes. *In* More gene manipulations in fungi. J.W. Bennett and L.L. Lasure, eds.. Academic Press, San Diego. pp. 51-85.

Tyler, L.J. and Parker, K.G. 1945. Pathogenicity in the Dutch elm disease fungus. Phytopathology 35:257-261.

Tzeng, T.H., Lyngholm, L.K., Ford, C.F., and Bronson, C.R. 1992. A restriction fragment length polymorphism map and electrophoretic karyotype of the fungal maize pathogen *Cochliobolus heterostrophus.* Genetics 130:81-96.

Van Alfen, N.K. and McMillan, B.T. 1982. Macromolecular plant-wilting toxins: artifacts of the bioessay method? Phytopathology 72:132-135.

Vollmer, S.J. and Yanofsky, C. 1986. Efficient cloning of genes of *Neurospora crassa.* Proc. Natl. Acad. Sci. USA 83:4869-4873.

Walter, J.M. 1937. Variation in mass isolates and monoconidium progenies of *Ceratostomella ulmi.* J. Agr. Res. 54:509-523.

Webber, J.F., Mitchell, A.G., and Smith, F. 1986. Linkage of the genes determining mating type and fungicide tolerance in *Ophiostoma ulmi.* Plant Pathol. 35:512-516.

Went, J.C. 1938. Compilation of the investigations of the susceptibility of different elms to *Ceratostomella ulmi* Buisman in the Netherlands. Phytopath. Z. 11:181-201.

Williams, J.G.K., Kubelik, A.R., Livak, K.J., Rafalski, J.A., and Tingey. S.V. 1990. DNA polymorphisms amplified by arbitrary primers are useful as genetic markers. Nucl. Acids Res. 18:6531-6535.

Yang, D., Jeng, R.S., and Hubbes, M. 1989. Mansonone accumulation in elm callus induced by elicitors of *Ophiostoma ulmi,* and general properties of elicitors. Can. J. Bot. 67:3490-3497.

23
DNA Polymorphism, Perithecial Size and Molecular Aspects of D Factors in *Ophiostoma-ulmi* and *O. novo-ulmi*

C.M. Brasier, M.R. Bates, N.W. Charter and K.W. Buck

Introduction

Since the early 1970s there has been considerable progress in our understanding of Dutch elm disease biology, from beetle feeding preferences to the efficiency of vector transmission, and from the role of fungal toxins to elm resistance mechanisms. Included in this progress is discovery that there have been not one but two pandemics of the disease in this century, and that these pandemics were caused by not one but two very different forms of fungal pathogen. We now know that the pathogens themselves have diseases, the so-called d-factors, which have the potential to break the cycle of Dutch elm disease and so influence epidemic outcome. Understanding why the pandemics arose and controlling the resulting elm disease must be two of the greatest challenges facing modern tree pathology. We review here the results of some of our own studies on the question of the origins of the Dutch elm disease pathogens, and on the mode of action of d-factors, with special emphasis on molecular approaches.

Ophiostoma ulmi and O. novo-ulmi.

In the early 1970's combined growth rate and pathogenicity tests showed that in both Britain and North America there were two distinct populations within the Dutch elm disease fungus *Ophiostoma* (*Ceratocystis*) *ulmi* (Gibbs and Brasier 1973). The two groups were known first as the aggressive and non-aggressive strains, then as more of their biological properties became known, as the aggressive and non-aggressive subgroups of *O. ulmi.* By the early 1980's it was clear that the two subgroups not only differed in a wide range of characters including colony form, growth rate, pathogenicity, cerato-ulmin toxin production, temperature-growth relationships and sexual behaviour but they were also strongly reproductively isolated at the pre- and post-zygotic levels. At this time it also became evident that the aggressive subgroup itself comprised two partially reproductively isolated population units, the Eurasian (or EAN) and the North American (or NAN) races (Brasier 1979; 1982). By the late 1980's the known biological discontinuities and other major differences between the aggressive and non-aggressive subgroups included perithecial size and a range of molecular characters (Table 1). It was increasingly evident that the two populations were behaviourally, morphologically and genetically distinct if closely related *Ophiostomas*. The aggressive subgroup was therefore redesignated *O. novo-ulmi* sp. nov. to distinguish it from the non-aggressive, which was equated to *O. ulmi* sensu Buisman (Brasier 1991). *O. ulmi* and *O. novo-ulmi* may comprise part of a complex of specific and sub-specific population units which includes two other sibling species recently identified within *O. piceae* (see Brasier 1993).

Table 1. Major differences between *Ophiostoma ulmi* and *Ophiostoma novo-ulmi*†

Complete discontinuities	Other major differences
Growth rate at 20°C	Upper temperature limits
Temperature optimum	Colony patterns
Pathogenicity in xylem	Perithecial sizes
Cerato-ulmin production	Mating systems
Pathogenicity in bark	Vegetative compatibility systems
Reproductive isolation	Synnematal production
	nDNA polymorphisms
	mtDNA polymorphisms
	mtDNA size
	Protein patterns
	Isozyme patterns
	Monoclonal antibodies

†For full detailes of and a reference list covering these diffeences see Brasier 1991.

Geographical origins and spread of O. ulmi *and* O. novo-ulmi.

Historical and geographical surveys of the two Dutch elm disease *Ophiostomas* indicate that they have rather different histories of spread (see Brasier 1990a). *O. ulmi is* associated with the first pandemic of Dutch elm disease which began in north west Europe in the early 1900's and spread rapidly eastwards across Europe to south west Asia and to westwards North America. *O. novo-ulmi is* associated with the current and much more destructive second pandemic of the disease. This species was first recorded in its NAN form in the UK during the early 1970s (Brasier and Gibbs 1973; Gibbs and Brasier 1973), but is thought probably to have originated in its EAN form near the Romania/Moldavia/Ukraine area, possibly during the 1940s. EAN *O. novo-ulmi* appears then to have migrated westwards across Europe and east to parts of Asia. It is suggested that the EAN was also imported into North America in the

Illinois-Indiana area at about this time, where it evolved into the biologically distinct NAN form while adapting to the highly susceptible American elm population (Brasier 1990a). There is further evidence that the NAN was introduced into Britain from North America in the 1960s (Brasier and Gibbs 1973), from whence it has spread into western Europe, overlapping in places with the westwardly migrating EAN (Brasier 1987). *O. novo-ulmi is* rapidly replacing *O. ulmi* as it migrates into the latter species' territory (Brasier 1987; 1990a).

So far as can be determined neither *O. ulmi* nor *O. novo-ulmi* is present in China (Brasier 1990a). Also, the two species may each have appeared in Europe for the first time only this century. Both species have spread like an introduced pathogen on a series of epidemic fronts. To account for their relatively sudden appearance, both introduction from the Himalayas (the one major elm region not yet surveyed for the occurrence of these pathogens), and recent unusual evolutionary events have been proposed (Brasier 1990a). The latter includes the further possibility that a secondary speciation event involving hybridisation between *O. ulmi* and another *Ophiostoma* species such as *O. piceae* may have given rise to *O. novo-ulmi.*

Perithecial Size Variation In *O. ulmi* and *O. novo-ulmi*

Perithecial morphology may throw some light on the evolutionary history and geographic spread of *O. ulmi* and *O. novo-ulmi.* Biometrical studies with nine different crosses of *O. ulmi* and ten of each of the EAN and NAN races of *O. novo-ulmi* show that EANs have significantly longer perithecial necks (Brasier 1991). Indeed at up to 680 (-1070) µm they are much longer than the ~380 µm traditionally attributed to *O. ulmi sensu lato.* The EAN also shows considerable variability in this character, the mean neck lengths of each of the ten different crosses studied ranging from 310 to 580 µm. The overall mean neck lengths for the three taxa were EAN, ~450 µm; NAN, ~296 µm; and *O. ulmi,* ~348 µm.

Interestingly, the perithecial bases of the EAN are actually smaller than those of the NAN, and those of the NAN are smaller than those of *O. ulmi.* (Overall means are EAN ~103 µm; NAN ~115 µm; and *O. ulmi* ~122 µm). As a consequence the perithecia of the EAN tend to be more elegant looking than the rather dumpy, short-necked perithecia characteristic of *O. ulmi.* The NAN often has short necked but sometimes has longer necked perithecia. This morphological difference in appearance is reflected in the mean neck length: bowl width ratios of these taxa. That of the EAN averages 4.4:1 compared with only 2.9:1 in *O. ulmi* and 2.5:1 in the NAN (Brasier 1991). Again the EAN exhibits considerable variation around this mean, and the NAN shows more variability than does *O. ulmi.*

One hypothesis for the origin of EAN *O. novo-ulmi* is that it has evolved only recently from *O. ulmi.* If so, this process has involved a substantial change in perithecial dimensions. One possible explanation is that the longer perithecial neck lengths, smaller bases and greater overall variability of EAN perithecia reflects a recent secondary speciation event, such as hybridisation between *O. ulmi* and another *Ophiostoma* with longer necks and smaller bases (cf. Brasier 1990a). As a consequence, an allele or pool of alleles for longer perithecial neck lengths may be present in parts of the EAN population. This raises a further possibility that this character might become less variable in the EAN in the future under continued natural selection. Alternatively, it may remain rather variable across the EANs geographical range. The smaller overall variability and more *O. ulmi*-like neck length of the NAN might reflect the initial introduction of shorter-necked EAN genotypes into North America; a reversion of the introduced EAN genotypes to a more *O. ulmi*-like perithecial type under natural selection; or even some local introgression of *O. ulmi* nuclear genes into the introduced EAN gene pool (though the latter is not supported by our recent molecular studies on the NAN). These studies are continuing, and may be extended to studies on the heritability of perithecial size.

Molecular Origins of *O. ulmi* and *O. novo-ulmi*

Nuclear DNA polymorphism studies

Traditional genetical studies reveal strong pre- and post-zygotic isolation between *O. ulmi* and *O. novo-ulmi*, and partial pre-zygotic isolation between the EAN and NAN in *O. novo-ulmi* (Brasier 1977, 1979, 1984; Brasier and Gibbs 1976; Kile and Brasier 1990). We are now conducting molecular studies of nuclear and mitrochondial DNA to gain further insights into the evolutionary history of these taxa, and to test the hypothesis of an evolutionary origin of *O. novo-ulmi* from *O. ulmi* and of NAN from EAN (Brasier 1990a).

In preliminary studies, nuclear DNA of 37 isolates of *O. ulmi* and *O. novo-ulmi* (11 *O. ulmi;* 12 EAN and 14 NAN) selected to cover a wide range of geographical locations was digested with restriction enzymes BamH1 or HindIII and probed with 86 cloned fragments of *O. ulmi* nuclear DNA to search for restriction fragment length polymorphisms (RFLPs). Ten enzyme-clone combinations (36%) did not distinguish between *O. ulmi* and *O. novo-ulmi,* consistent with previous evidence that these two species are related, and either share a common progenitor or *O. novo-ulmi* evolved from *O. ulmi.* Ten other combinations, however, distinguished all *O. novo-ulmi* isolates from all *O. ulmi* isolates, and a further five distinguished most *O. ulmi* isolates from most *O. novo-ulmi* isolates. This is consistent with the multiple biological differences that exist between the two species (Table 1). Most probes detected only one or two bands, and hence probably represented single copy sequences. If *O. novo-ulmi* has evolved from *O. ulmi,* therefore, this process has probably involved a very large number of mutations.

To search for enzyme-clone combinations that would distinguish between EAN and NAN in *O. novo-ulmi,* nuclear DNA of EAN isolate M3 (Moscow) and NAN isolate C112 (UK) was digested with BamH1 and HindIII and probed with all 86 cloned fragments from *O. ulmi.* Of 172 enzyme-probe combinations tested, fourteen (8.1%) revealed RFLPs between M3 and C112. Since no RFLP was revealed by both enzymes with a single clone it is likely that the RFLPs were due to point mutations rather than to substantial length polymorphisms. These results are clearly consistent with the observed biological properties of the EAN and NAN, and the conclusion that they are rather closely related but distinct forms of *O. novo-ulmi* (cf. Brasier 1979,1982, 1991).

The same 14 enzyme-probe combinations were then used to search for RFLPs in all 37 *O. ulmi* and *O. novo-ulmi* isolates. Of these, four (~28%) distinguished all 12 EAN from all 14 NAN isolates and also all 11 *O. ulmi* isolates. A further five (~35%) distinguished all EAN isolates from all *O. ulmi* isolates and from many NAN isolates, while other NANs showed EAN-like bands or novel bands. Four other enzyme-clone combinations distinguished most *O. ulmi* from most *O. novo-ulmi* isolates but not most EAN from most NAN isolates, while some *O. ulmi*-like bands or novel bands were exhibited by some *O. novo-ulmi* isolates. Another enzyme-clone combination gave identical bands for all *O. ulmi* and EAN isolates and for seven NANs, but a different band for seven other NANs.

Our study has revealed rather more molecular variability in *O. ulmi* and *O. novo-ulmi* than previously described. Duchesne et al. (personal communication) detected no variation in six *O. ulmi* isolates using a ribosomal probe. Similarly Duchesne et al. (personal communication), and also Hintz et al. (1991) using a ribosomal probe and a human minisatellite probe (five enzyme-probe combinations) distinguished only four nuclear DNA types in twelve *O. novo-ulmi* isolates (including some EAN isolates, some NAN, and some unspecified). In contrast, the 39 enzyme-probe combinations used by us revealed five nuclear types among eleven *O. ulmi* isolates, and eleven nuclear types among twelve EAN and fourteen NAN *O. novo-ulmi* isolates. The extent of diversity revealed clearly reflects the number of enzyme-clone combinations used and the number of isolates analysed.

In general, the eleven *O. ulmi* isolates studied appeared closely related, showing limited variation despite their diverse geographic origins. The *O. novo-ulnli* isolates however showed rather more variation. While nine EAN isolates showed the same RFLP profile despite origins from different countries, three other isolates showed a total of ten polymorphisms of which one isolate (R136, Romania) showed six *O. ulmi*-like RFLPs. The latter is consistent with the proposed evolution of EAN *O. novo-ulmi* from *O. ulmi* in the Romania-Moldavia-Ukraine region (see below). The NAN showed even more variation, with six RFLP types among fourteen isolates. The evidence also suggests that some of these RFLP types have been generated by sexual recombination between the two most molecularly different types. For full details of these results, see Bates et al. 1990 and 1993a.

Further evidence of an O. ulmi-*EAN* O. novo-ulmi *relationship from nuclear DNA studies*

The occurrence of *O. ulmi*-like RFLPs in Romanian EAN isolate R136 (see above) might be a product of the postulated evolutionary relationship between *O. ulmi* and EAN *O. novo-ulmi* and, in particular, evidence for the origin of the EAN within the Romania-Moldavia-Ukraine area (Brasier 1990a). To examine this possibility further a concentrated sample of 13 Romanian EAN isolates, all of different vc groups (Brasier 1984) from R136 and from different locations in north-east Romania, was compared with a wide geographical sample of 37 EAN isolates from the adjacent areas of Russia and from across Europe to south-west Asia. The isolates were screened for *O. ulmi*-like polymorphisms by digesting DNA with BamH1 and HindIII and probing with nDNA clones which had revealed *O. ulmi* polymorphisms in EAN isolates R136 and M3 in the previous tests.

The probes revealed *O. ulmi*-like polymorphisms in five of the 13 Romanian isolates, whereas only eight of the 37 'wider sample' isolates revealed either *O. ulmi*-like or novel polymorphisms. Interestingly, a majority of these eight isolates came from the Ukraine, Russia and the Caucasus. Thus in the Romania-Moldavia-Ukraine region the EAN may indeed exhibit more molecular diversity than in the rest of the EAN-colonised areas. This preliminary evidence is therefore consistent with the hypothesised origin of EAN *O. novo-ulmi* in this region. Again, one possibility is that a range of hybrid genotypes may have emerged in this region as a result of an interspecific hybridisation event involving *O. ulmi*, with some hybrids now containing more *O. ulmi* DNA than others. R136 and other EAN isolates carrying *O. ulmi* or novel polymorphisms might therefore be remnants of such a hybrid population, the commoner non-*O. ulmi* DNA carrying EAN RFLP type being a fitter more widely distributed genotype (cf. Brasier 1977). Another explanation could be that the *O. ulmi*-like RFLPs represent a degree of *O. ulmi*—EAN introgression, but this does not account for the striking geographical concentration of these isolates in the Romania-Ukraine region. These observations now need to be verified with a wider range of DNA probes.

Further evidence for the EAN-NAN relationship from nuclear DNA studies

It has been suggested that NAN *O. novo-ulmi* may have evolved from a founder population of EAN-infected elm material imported into the North American Illinois-Indiana area during the 1940s (refs. Brasier 1988b, 1990a and C.M. Brasier & A.G. Mitchell, unpublished observations). One way of genetically defining *O. novo ulmi* populations is by their vegetative compatibility (vc) group structure (Brasier 1984; 1988a). A striking and unusual feature of the present day NAN population in North America is that it comprises only three dominant vc groups. These are the North American NAN vc supergroup, the North

American NAN second vc group which is related to the former group by its vc genes (carrying the same w-allele), and the European NAN vc supergroup, which is probably not closely related to the former groups. These two vc groups comprised ~60%, ~20% and ~9% respectively of a North American NAN *O. novo-ulmi* population sample of 110 isolates. The remaining ~10% comprised a small but genetically highly heterogeneous vc component (see Brasier and Kirk 1991). The possible significance of such dominant vc clones and of the associated heterogeneous population components has been discussed elsewhere (Brasier 1988a; 1990b; Brasier and Kirk 1991). The European NAN vc supergroup is presumed to have originated in North America but is also now the dominant vc group in the European NAN population, though numerous other vc groups in the NAN exist in Europe (Brasier 1988a; 1990b).

Many EAN isolates are genetically related in vc terms to the North American NAN vc supergroup in that they carry the same vc w-allele (C.M. Brasier, unpublished). This includes the EAN vc supergroup, the dominant vc group in the EAN across Europe and areas of Asia (Brasier 1988a). It also includes a number of the EAN isolates described above which carry *O. ulmi* or novel DNA polymorphisms. The fact that such EAN isolates share their w-allele with the dominant American NAN vc supergroup may be evidence for the postulated emergence of the NAN from the EAN outlined above.

To investigate these relationships further, a sample of 34 North American NAN isolates was compared molecularly with 6 European NAN isolates and 17 European and central Asian EAN isolates (their nuclear DNA was digested with HindIII or BamH1 and probed with a number of DNA clones). The study included 21 American NAN vc supergroup isolates, which all showed an identical RFLP profile as also did a single European representative of this vc group. Twelve European NAN vc supergroup isolates tested (8 USA and 4 European) also showed a common RFLP profile. This would appear to represent considerable genetic stability and a strong selective advantage of these two vc 'supergroups' (cf. Brasier 1988a; 1990b).

However, the European NAN vc supergroup isolates also share some of their polymorphisms with EAN isolates, suggesting a possible evolutionary sequence of EAN → European NAN vc supergroup → American NAN vc supergroup. One group of 12 EAN isolates also exhibits a polymorphism found in American NAN vc supergroup isolates, suggesting that the hypothesised initial EAN founder population in North America contained a certain amount of variability. Moreover three American NAN second vc group isolates and two European NAN isolates of a different vc group exhibit some typical EAN bands not found in any other NAN isolates tested; and two culturally unusual EAN isolates, H328 and H987 from the Ukraine and Georgia (Caucasus) respectively, exhibit a polymorphism found in every NAN isolate screened but absent from all other EAN isolates tested. The existence of such polymorphisms is best explained in evolutionary terms. We consider that such isolates may be very similar to isolates involved in the initial outbreak of *O. novo-ulmi* in North America, and may bear similar combinations of polymorphisms. Indeed, those or similar genotypes may have made up the postulated EAN *O. novo-ulmi* founder populations in the Illinois-Indiana area from within which the European NAN vc supergroup and the American NAN vc supergroup evolved. These preliminary investigations need to be confirmed with a much wider range of isolates and DNA probes.

Mitochondrial DNA polymorphisms

We have also studied mitochondrial DNA polymorphisms in *O. ulmi* and *O. novo-ulmi* to investigate these species' evolutionary relationship. In preliminary tests three restriction enzymes PvuII, EcoRI and BclI were found to give informative polymorphic fragments. The mtDNAs of ten *O. ulmi,* and twelve EAN and twelve NAN *O. novo-ulmi* isolates were therefore digested with these enzymes and the fragment sizes estimated by electrophoresis. Hierarchical analyses of each of the three resulting data sets were then carried out.

With the exception of NAN isolates H161 and H973 (considered below) the three dendrograms exhibited good clustering of *O. ulmi,* EAN and NAN. As might be expected, there was a separate grouping of *O. ulmi,* and a closer relationship of NAN and EAN within *O. novo-ulmi* confirming the relationships of these taxa proposed previously from biological and genetical studies. The most polymorphisms were shown by *O. ulmi* and by the NAN. Some clustering was also evident in relation to vc groups. Thus a North American NAN vc supergroup isolate from Maine, USA grouped closely with an isolate of the same vc group from Holland, and with two isolates of genetically related vc groups from the UK. This again suggests a high level of genetic conservation within a geographically widely distributed vc group. The EAN showed the least variability, all of the EAN isolates with the exception of R136 (Romania), M3 (Moscow) and H327 (Czechoslovakia) exhibiting identical patterns.

This is of particular interest since R136 and M3 are EAN isolates that also exhibit several typical *O. ulmi* nDNA polymorphisms, and may therefore be carrying residual *O. ulmi* DNA from an earlier evolutionary event (see above).

mtDNA sizes

Mitochondrial DNA sizes were estimated from PvuII, BclI and EcoRI digests. The estimates based on the three enzymes gave excellent agreement, and showed (with the exception of NAN isolates H161 and H973) that the mtDNA sizes of *O. ulmi,* the EAN and the NAN are distinct. Those of *O. ulmi* range from 74-78kb, of the EAN from 65-69kb, and of the NAN from 50-59kb. These estimates have been corroborated in the EAN and NAN by restriction maps (see below).

In their similar study of mtDNA polymorphism in six *O. ulmi* and twelve *O. novo-ulmi* isolates, Jeng et al. (ref. Jeng et al. 1990; and see Jeng, Chapter 18 this volume) also demonstrated characteristic differences between the two species, though they had an insufficient number of defined isolates to compare the EAN and NAN within *O. novo-ulmi.* They similarly recorded a ~40% larger mitochondrial genome in *O. ulmi* (their average 64.7kb) than in *O. novo-ulmi*

(38.4kb). However, their estimates for mtDNA sizes for individual isolates were rather variable over the three different restriction enzymes used, and are consistently some 6-20kb lower than our own estimates where the same isolate (H6, NAN) was used in both studies. This may be due to the fact that some large mtDNA fragments such as an additional 20kb EcoRI fragment observed by Rogers et al. (1987) and by us (Bates et al. 1990; 1993b) were undetected in their studies. In an extension of their study, Hintz et al. (1991) reported finding only three types of mtDNA restriction map when examining ten of the *O. novo-ulmi* isolates used by Jeng et al. (1990). The mtDNA sizes reported by Hintz et al. (1991) for EcoRI digests are however larger than those given by Jeng et al. (1990), and include a ~20kb fragment present in all isolates.

Restriction maps of EAN and NAN mtDNA in O. novo-ulmi.

Restriction maps of two NAN and two EAN isolates of *O. novo-ulmi* have been constructed and compared. The mtDNA sizes of these isolates at 54.6, 55.6, 63.4 and 65.4kb respectively are consistent with estimates for the NAN and EAN cited above. Large regions of similarity were found between the four isolates, and the mtDNA size differences between EAN and NAN appear to be due mainly to insertions (introns?) or deletions. These probably also account for many of the observed RFLPs. Thus one EAN isolate (H2023, 63.4kb) carries four insertions compared to the larger of the two NAN isolates (PG402, 55.6kb), and the latter also carries a single lkb insertion compared to NAN isolate W2TOL1 (54.6kb). MtDNA clones of *Podospora anserina* carrying known mtDNA genes hybridised to the same regions of mtDNA in each EAN or NAN isolate, allowing for introns. MtDNA gene order in *O. novo-ulmi* therefore appears to be the same in both EAN and NAN.

In their study Hintz et al. (1991) also observed mtDNA size differences in *O. novo-ulmi* due to introns or deletions. If the larger molecular size of the *O. ulmi* mtDNA genome (not yet mapped by us) proves also to be due to introns, then during the postulated evolutionary sequence *O. ulmi* → EAN → NAN there would have been a repeated loss of introns.

NAN isolates H161 and H973

An exception to some of the above mtDNA polymorphism observations were two North American NAN isolates H161 (Canada) and H973 (Maine, USA) both members of the American NAN vc supergroup. These isolates grouped closely to the *O. ulmi* isolates in the dendrograms, but were usually distinct from the main *O. ulmi* cluster. In some digests these two isolates also exhibited occasional *O. ulmi*-like bands. Their mtDNA sizes, at ~72kb, are intermediate between those of *O. ulmi* and *O. novo-ulmi.* However, their nuclear DNA profiles are quite typical for those of American NAN vc supergroup isolates. These isolates therefore appear to represent mtDNA recombinants between *O. ulmi* and *O. novo-ulmi,* indicating that some lateral mitochondrial gene transfer may occur between the two species in the absence of nuclear gene transfer. This would presumably result from transient hyphal fusions between the two species. Such events would be most likely to occur at recent epidemic fronts, when *O. novo-ulmi* is migrating into territory previously occupied only by *O. ulmi* (e.g. see Brasier 1987; 1990a).

Mitochondria As Targets For D-Factors

Disease- or d-factors are cytoplasmic virus-like genetic elements of *O. novo-ulmi* and *O. ulmi* that are transmitted mainly via hyphal fusions. They can severely debilitate these two fungal pathogens in terms of reducing their growth rate, colony stability, and spore viability. The general biology of d-factors has been reviewed elsewhere in some detail (Brasier 1983; 1986) and is discussed by Webber (Chapter 24, this volume). In nature, d-factors can prevent infection of elms via beetle feeding wounds, and therefore have the potential to break the cycle of Dutch elm disease (Webber 1987). D-infection can reach very high levels in *O. ulmi* and *O. novo-ulmi* populations, particularly at epidemic fronts where widespread and dominant vc clones occur. They are one of the few natural biological control agencies known to have influenced the outcome of Dutch elm disease epidemics, and may have brought about the decline in Europe during the 1940s of the first epidemic caused by *O. ulmi* (Brasier 1988a; 1990b). The molecular properties and modes of action of d-factors are being studied in our laboratories with a view to their genetic manipulation for disease control.

Most d-factors we have examined are associated with cytoplasmically transmissible, naked, multiple segment dsRNA profiles. Thus when 19 different d-infected isolates of *O. novo-ulmi* were examined, all were found to have dsRNA profiles of 4-14 segments ranging from 0.23-2.5kb (Rogers et al. 1988). Many healthy isolates often also contain some dsRNA, but commonly only between one and four bands (Rogers et al. 1988), and it is likely that only certain dsRNA segments in a profile are involved in d-infection, while others may be relatively benign. This is illustrated by the d^2-factor, one of a numbered series of d-factors (d^1-d^{13}) which are being investigated in more detail. The d^2-factor was first detected in a previously healthy isolate of the *O. novo-ulmi* European NAN vc supergroup, *W2tol*1, which became d-infected during a field experiment (Webber 1987). The healthy isolate already contained two dsRNA bands (of 2.40 and 0.61kb, ref. Bates et al. 1992a). The newly d^2-infected form of *W2tol*1 was found to contain ten dsRNA bands of from 2.40-0.23kb, which were numbered 1-10. These bands were shown to be faithfully cotransmitted along with the d-factor phenotype. However experiments on single conidial isolates during which some dsRNA bands were lost showed that only the segments 4, 7 and 10 (1.67, 0.65 and 0.23kb respectively) were specifically required for expression of the d^2-factor (Rogers et al. 1986).

Further studies on the d^2-factor showed that most of the dsRNA co-purified with the mitochondria. In mitochondrial preparations this dsRNA was protected against nuclease degradation and was therefore probably located within the mitochondria. Analyses of cytochrome profiles of newly d-infected recipient isolates showed that these had a reduction in cytochrome aa_3 (a mitochondrial gene product) compared to healthy isolates, and there appeared to be a correlation between low cytochrome aa_3 and d-infection levels (Rogers et al. 1987). A disturbed mitochondrial respiratory enzyme pathway may therefore be a main cause of the disease phenotype in d^2-infected isolates.

MtDNA plasmid involvement in the the d-factor disease phenotype

Further experiments were conducted to examine whether mitochondria could be transferred along with dsRNAs during cytoplasmic transmission of d-factors via hyphal fusions. In general mitochondria were found not to be transmitted with the dsRNA (Charter 1992; Charter et al. 1993). When experiments were conducted on transmission of the d^2-factor using recipient and donor isolates with mtDNAs distinguishable by their RFLPs, the profile of the newly d^2-infected recipient remained the same as that of the healthy original (i.e. unchanged). However, a DNA plasmid of ~2.2kb was also detected in the d^2-infected recipient, which was not present in the healthy form of the isolate.

Similar plasmids were also detected in mtDNA digests of the d^2-infected donor, and in other newly d^2-infected recipient isolates. The plasmids ranged in size from 4.1 to 0.8kb, and were again not found in the equivalent healthy isolates. Our results to date indicate that the plasmids, like the mtDNA, are not transmitted when the d^2-factor dsRNA is transferred. Rather the transmission of the d^2-factor results in the formation of a new DNA plasmid in the recipient (Charter et al. 1993).

When these d^2-factor associated plasmids were cloned and hybridised to enzyme restricted mtDNAs of recipient isolates, they were found to be homologous to certain specific mtDNA fragments, ie they represented partial mtDNA sequences. The plasmid DNAs were therefore mapped to the mtDNA genome and also compared with regions of the mtDNA map showing homology to *Podospora anserina* mtDNA clones. The plasmid DNAs were generally found to originate from one of two regions of the mtDNA, either from regions showing homology to *P. anserina* clones carrying large ribosomal RNA genes or clones carrying genes for the cytochrome sub-units. This indicates that the plasmids may contain sequences of these genes (Charter 1992).

Comparable studies with the d^5 (EAN) and d^{12} (NAN) factors (associated with 8 and 7 dsRNA segments respectively) show that the d^5 and d^{12} factors may be functionally similar to the d^2 factor. They are associated with cytoplasmically transmissible mitochondrially located dsRNAs, result in production of mitochondrially located DNA plasmids, and have altered cytochrome profiles. However, newly d^{12}-infected isolates possess mtDNA RFLPs of the donor rather than the recipient isolate of the d-reaction, which indicates that the mitochondria are transmitted with this d-factor. With the d^5-factor, some recipient/donor recombinant mtDNA profiles have been detected during d-factor transmission. Hence some mitochondrial transmission may also occur in this system (Charter 1992; Charter et al. 1993).

These results suggest that the d-factors d^2, d^5 and d^{12} all operate via a mechanism associated with debilitated mitochondria. The plasmids involved in the d^2-factor are now being sequenced to confirm which genes are present on a particular plasmid, and to investigate the mechanisms of plasmid excision. The possibility that the plasmids are generated as a result of the dsRNA encoded proteins interacting with the mtDNA, and the mode of plasmid replication, are also under investigation. It is postulated that such plasmids may affect mitochondrial function either by competing for replication factors or by producing aberrant transcripts or gene products (Charter 1992).

Concluding Remarks

The combination of traditional and molecular analytical techniques that led to the demonstration that there were two species of Dutch Elm disease pathogen is now providing fascinating insights into their behavioral differences and into the intriguing question of their evolutionary history. Parallel studies are rapidly enhancing our knowledge of the structure and cellular mechanisms of the virus-like d-factors of the pathogens themselves. They hold out the possibility that d-factors may one day be genetically manipulated for the control of Dutch elm disease.

References

Bates, M. R., Buck, K. W., and Brasier, C. M. 1990. Molecular variation in the Dutch elm disease fungus. Pages 171-178 in: Molecular Evolution. UCLA Symposium on Molecular and Cellular Biology New Series 122. M. T. Clegg and S. J. O'Brian, eds.

Bates, M. R., Buck, K. W., and Brasier, C. M. 1993a. Molecular relationships between *Ophiostoma ulmi* and the NAN and EAN races of *O. novo-ulmi* determined by restriction fragment length polymorphisms of nuclear DNA. Mycol. Res., in press.

Bates, M. R., Buck, K. W., and Brasier, C. M. 1993b. Molecular relationships of the mitochondrial DNA of *Ophiostoma ulmi* and the NAN and EAN races of *O. novo-ulmi* determined by restriction fragment length polymorphisms. Mycol. Res., submitted.

Brasier, C. M. 1977. Inheritance of pathogenicity and cultural characters in *C. ulmi*. Hybridisation of protoperithecial and non-aggressive strains. Trans. Br. Mycol. Soc. 68: 45-52.

Brasier, C. M. 1979. Dual origin of recent Dutch elm disease outbreaks in Europe. Nature 281:78-79.

Brasier, C. M. 1982. Occurrence of three sub-groups within *Ceratocystis ulmi*. Pages 298-321 in: Proceedings of the Dutch elm disease Symposium and Workshop Winnipeg, Manitoba, October 5-9, 1981. E. S. Kondo., Y. Hiratsuka and W. B. C. Denyer, eds. Manitoba Department of Natural Resources, Manitoba, Canada.

Brasier, C. M. 1983. A cytoplasmically transmitted disease of *Ceratocystis ulmi*. Nature 305:220-223.

Brasier, C. M. 1984. Inter-mycelial recognition systems in *Ceratocystis ulmi* : their physiological properties and ecological importance. Pages 451-497 in: The Ecology and Physiology of the Fungal Mycelium. D. Jennings and A. D. M. Rayner, eds. Cambridge University Press.

Brasier, C. M. 1986. The d-factor in *Ceratocystis ulmi:* its biological characteristics and implications for Dutch elm disease. Pages 177-208 in: Fungal Virology. K. W. Buck, ed. C.R.C. Press, Florida.

Brasier, C. M. 1987. Recent genetic changes in the *Ophiostoma ulmi* population: the threat to the future of the elm. Pages 213-226 in: Populations of Plant Pathogens. M. S. Wolfe and C. E. Caten, eds. Blackwells, Oxford.

Brasier, C. M. 1988a. Rapid changes in genetic structure of epidemic populations of *Ophiostoma ulmi.* Nature 332: 538-541.

Brasier, C. M. 1988b. *Ophiostoma ulmi,* cause of Dutch elm disease. Pages 207-223 in Genetics of Plant Pathogens. G. S. Sidhu, ed. Adv. Plant Pathol. Vol. 6. Academic Press, London and New York.

Brasier, C. M. 1990a. China and the origins of Dutch elm disease: an appraisal. Plant Pathol. 39:5-16.

Brasier, C. M. 1990b. The unexpected element: mycovirus involvement in the outcome of two recent pandemic events, Dutch elm disease and chestnut blight. Pages 289-308 in Pests, Pathogens and Plant Communities. J. J. Burdon and S. R. Leather, eds. Blackwell Scientific Publications, Oxford.

Brasier, C. M. 1991. *Ophiostoma novo-ulmi* sp. nov., causative agent of the current Dutch elm disease pandemics. Mycopathologia 115:151-161.

Brasier, C. M. 1993. The genetic system as a fungal taxonomic tool: gene flow, molecular variation and sibling species in the *'Ophiostoma piceae - Ophiostoma ulmi'* complex and its taxonomic and ecological significance. Pages 69-86 *In*: *Ophiostoma* and *Ceratocystis:* taxonomy, biology and pathology. M. J. Wingfield, K. Seifert and J. F. Webber, eds. American Phytopathological Society, St Paul, Minnesota.

Brasier, C. M., and Gibbs, J. N. 1973. Origin of the Dutch elm disease epidemic in Britain. Nature 242:607-609.

Brasier, C. M., and Gibbs, J. N. 1976. Inheritance of pathogenicity and cultural characters in *C. ulmi.* I. Hybridisation of aggressive and non-aggressive strains. Ann. Appl. Biol. 83:31-37.

Brasier, C. M., and Kirk, S. A. 1991. Dutch elm disease. The aggressive subgroup of *O. ulmi* is near clonal in North America. Report on Forest Research 1990, pp 52-54. HMSO, London.

Charter, N. W. 1992. Double-stranded RNA and mitochondrial DNA plasmids in diseased isolates of the Dutch elm disease pathogen *Ophiostoma novo-ulmi* . Ph.D. thesis. University of London. 289 pp.

Charter, N. W., Buck, K. W., and Brasier, C. M. 1993. Double stranded RNA and mitochondrial DNA plasmids in diseased isolates of the Dutch elm disease pathogen, *Ophiostoma novo-ulmi.* In: Viruses of Fungi and Simple Eukaryotes. Y. Koltin, M. J. Leibowitz, and V. Rubio, eds. In press. University of Delaware Press.

Gibbs, J. N., and Brasier, C. M. 1973. Correlation between cultural characters and pathogenicity in *Ceratocystis ulmi* from Europe and North America. Nature 241:381-383.

Hintz, W. E., Jeng, R. S., Hubbes, M., and Horgen, P. A. 1991. Identification of three populations of *Ophiostoma ulmi* (Aggressive subgroup) by mitochondrial DNA restriction-site mapping and nuclear DNA fingerprinting. Exp. Mycol. 15:316-325.

Jeng, R. S., Duchesne, L. C., Sabourin, M., and Hubbes, M. 1990. Mitochondrial DNA restriction fragment length polymorphisms of aggressive and non-aggressive isolates of *Ophiostoma ulmi.* Mycol. Res. 95:537-542.

Kile, G. A., and Brasier, C. M. 1990. Inheritance and inter-relationship of fitness characters in progeny of an aggressive x non-aggressive cross of *Ophiostoma ulmi.* Mycol. Res. 94:514-522.

Rogers, H. J., Buck, K. W., and Brasier, C. M. 1986. Transmission of double-stranded RNA and a disease factor in *Ophiostoma ulmi.* Plant Pathol. 35:277-287.

Rogers, H. J., Buck, K. W., and Brasier, C. M. 1987. A mitochondrial target for double-stranded RNA in diseased isolates of the fungus that causes Dutch elm disease. Nature 329: 558-560.

Rogers, H. J., Buck, K. W., and Brasier, C. M. 1988. Double stranded RNA in disease isolates of the aggressive subgroup of the Dutch elm disease fungus, *Ophiostoma ulmi.* Pages 327-351 in: Viruses of Fungi and Lower Eukaryotes. Y. Koltin and M. J. Leibowitz, eds. Marcel Dekker Inc. New York.

Webber, J. F. 1987. Influence of the d^2 factor on survival and infection by the Dutch elm disease pathogen, *Ophiostoma ulmi* Plant Pathol. 36:531-538.

24
D Factors and Their Potential for Controlling Dutch Elm Disease

J. F. Webber

Abstract

D factors are cytoplasmically transmitted agents that can have a deleterious effect on the growth and sporulation of the Dutch elm disease pathogens, *Ophiostoma ulmi* and *O. novo-ulmi.* Infected isolates are often characterized by slow and abnormal growth and impaired reproductive fitness. D factor transmission occurs readily in culture when d-infected and healthy isolates of the same vegetative compatibility group are paired together and hyphal anastomoses occur freely. However, the ease of d factor transfer declines sharply if the paired isolates share few common vegetative compatibility genes.

Introduction

Evidence suggests d factors are composed of dsRNA, with isolates of *Ophiostoma* infected by any particular d factor usually showing a characteristic profile of dsRNA segments. Profiles of many d factors have been characterized using polyacrylamide gel electrophoresis. However, much of the ecological work to assess their potential to act as biocontrol agents of Dutch elm disease has focused on a relatively small number of d factors, including the d^2, d^6 and d^{13} factors. These agents vary in the extent to which they impair the fitness of *O. novo-ulmi,* but can be extremely effective at controlling Dutch elm disease when targeted at the pathogen during the infection stage in beetle feeding grooves. Under these conditions, the number of spores necessary for successful infection may need to be increased as much as fifty-fold for d-infected individuals compared with their d-free counterparts.

Unfortunately, there are obstacles to be overcome in utilizing these potential biocontrol agents. The vegetative compatibility system of *O. ulmi* and *O. novo-ulmi* is effective at limiting the spread of d factors within pathogen populations. Sexual reproduction and transformation into the yeast phase during pathogenesis also allows *Ophiostoma* individuals to lose deleterious d factors.

Since the first fungal virus was positively identified by Hollings (1962) in the edible mushroom *Agaricus bisporus,* mycoviruses have been reported in a wide range of fungi, including representatives from all the major subdivisions (Buck 1986). Unlike viruses which attack other organisms, mycoviruses have no extracellular phase in their replication cycle (Lecoq et al. 1979) and are transmitted during cytoplasmic movement and exchange, via hyphal anastomoses and spore formation. Morphologically, the great majority are isometric particles consisting of a protein coat encapsidating double stranded (ds) RNA (Buck 1986). However, there are also several notable exceptions in which only dsRNA has been isolated from infected fungi, with no associated virus like particles (VLP's) being detected.

Hollings (1982) and Buck (1986) have speculated that as many as 5,000-15,000 species of fungi, out of an estimated total of 50,000, may eventually be found to harbor some type of viral infection. Currently, mycoviruses or other 'virus like' agents have been detected in only some 200 species, so clearly the majority of fungal viruses still await discovery! This comparatively small number of positive records may simply reflect the fact that most mycoviruses have a latent or cryptic lifestyle in their host, with no external symptoms to betray their presence. Nonetheless, some can and do induce a number of changes in their hosts such as reduced or aberrant growth, decreased spore viability, cold sensitivity and, in plant pathogens, a diminution in virulence. Recognition of these changes can be the trigger to examine affected individuals for the presence of dsRNA or virus like particles, although in the past probably many such fungal cultures were discarded because of their poor growth or atypical colony morphology.

Increasingly, however, plant pathogens are being screened for the presence of deleterious agents such as mycoviruses, in a bid to find effective biocontrol agents which act by reducing the pathogenic potential of these fungi (Dodds et al. 1988). This strategy may be particularly relevant to some of the species that have proved largely intractable to more traditional methods of control. At present, the action and control potential of deleterious cytoplasmic agents is being explored in depth for Dutch elm disease. Many aspects of this work are probably also applicable to other species of *Ophiostoma* and *Ceratocystis.*

Discovery of deleterious cytoplasmic factors

The Dutch elm disease pathogen, previously regarded as a single species with aggressive and nonaggressive subgroups, has recently been recognized as two clearly delineated sibling species, *Ophiostoma ulmi* and *Ophiostoma novo-ulmi* (Brasier 1991; Brasier this volume). *O. ulmi* corresponds to the non-aggressive subgroup responsible for the epidemics in Europe and North America during the first half of this century, while *O. novo-ulmi* describes the more recently evolved aggressive subgroup causing the current pandemics of Dutch elm disease.

The idea that mycoviruses or other cytoplasmic factors might influence the pathogenic behavior of *O. ulmi* was first predicted in 1973 (Brasier and Gibbs 1875). In particular, the way in which during the first epidemic of Dutch elm disease the relatively weak pathogen, *O. ulmi,* caused serious losses in elm populations throughout western Europe in the 1920's and 30's, followed by a decline in the 40's (Peace 1960), was suggestive that mycoviruses might have played a role in attenuating pathogenicity. However, examination by electron microscopy of both *O. ulmi and O. novo-ulmi* (then termed the non-aggressive and aggressive strains of *O. ulmi)* failed to reveal any VLP's to support this hypothesis (Hollings et al. 1974).

Following on from this, Pusey and Wilson (1981; 1982) assayed isolates of *O. ulmi* (non-aggressive) and *O. novo-ulmi* (aggressive) for dsRNA using serology and polyacrylamide gel electrophoresis, and were able to detect segments of dsRNA in isolates of both species. However, their initial hypothesis that the non-aggressive and aggressive groupings differed in pathogenicity purely because of the presence of dsRNA mycoviruses was not sustained, and they concluded that the presence of dsRNA alone was not consistently correlated with a reduction in pathogenicity.

A further clue to the existence of mycoviruses, or more strictly cytoplasmic factors, came with work on vegetative compatibility (v-c) in *O. novo-ulmi* (Brasier 1983). This involved pairing cultures on agar medium and assessing the nature of reaction that occurred along the line of contact. In some pairings, certain isolates consistently caused the development of sectors of pigmented and sunken mycelium in the adjacent 'recipient' colony; the region of abnormal growth often corresponding to a "butterfly" shape between the paired colonies (Fig. 1). Subcultures taken from these newly formed abnormal sectors were themselves able to act as transformants, indicating they had become infected with a transmissible factor. Moreover, the cultures that induced this change often tended to have an unstable 'amoeboid' type morphology (Fig. 2), and transformed previously normal colonies to a similar amoeboid form. Transformation only occurred, however, when paired isolates were of the same vegetative compatibility type allowing hyphal anastomoses to be formed between the previously separate individuals. If the isolates were vegetatively incompatible, a strong mycelial barrage reaction formed between the colonies, successful hyphal fusion failed and no transformation of the 'recipient' occurred (Brasier 1984).

Tests between paired cultures with genetic markets indicated that the transforming agent was not nuclear in origin but a cytoplasmic determinant, which could be transmitted between mycelia (Brasier 1983; 1986b). And because the transmission was usually characterized by the development of altered or abnormal growth in recipient isolates, the agent was termed the disease or 'd'-factor and the process of transformation a 'd reaction'. Not surprisingly, the number of the phenotypic similarities between d reactions and transmissible hypovirulence in *Cryphonectria parasitica,* which had produced the remission of chestnut blight in Europe (Grent and Sauret 1969), raised the possibility that d factors might contribute to the control of Dutch elm disease.

Cultural properties of d factors

D factors have now been identified among both major subpopulations of *O. novo-ulmi* (Brasier 1983; 1986b): the highly pathogenic North American (NAN) and Eurasian (EAN) subgroups; also in the 'old' non-aggressive species, *O. ulmi* (Brasier 1986). To discriminate between d factors originating from different *Ophiostoma* isolates, they were initially designated with a number, d^1, d^2 $d^{n,}$ in sequence of their discovery, although more recently, the numbering of d factors has been abandoned.

It is clear that some d factors have a detrimental effect on the growth and reproductive vigor of infected isolates. Infected cultures usually show a much reduced linear growth rate (Table 1) and an amoeboid colony morphology (Fig. 2), although the severity of a particular d factor can vary depending on the genetic background of the host isolate. This was illustrated most graphically when a previously healthy *O. novo-ulmi* isolate of the EAN subgroup was transformed by a d infected NAN isolate (Brasier 1983). The resulting d infection in the EAN isolate was extremely severe, leading Brasier (Brasier 1986a; 1986b) to suggest that d factors of each subgroup are likely to be co-evolved with their respective host population, thereby presumably minimizing the most damaging effects. However, not all d factors have a universally negative effect. Some appear to be latent, 'carried' by infected individuals of *O. ulmi* without causing any deleterious symptoms, while occasionally some may even increase the vigor of their host isolate although this effect can reverse spontaneously (Brasier 1986a; Rogers et al. 1986b).

The behavior of d factors during asexual and sexual reproduction also shows significant divergence in both *O. ulmi* and *O. novo-ulmi.* Just as d factors appear to move freely from infected to healthy individuals provided that viable hyphal anastomoses have been formed, transfer into asexual spores occurs frequently during conidiogenesis. In some instances, the majority of conidia formed by individuals carrying d factors are incapable of germinating (Brasier 1983, 1986a; Webber 1987) (Table 1), while others give rise to the typical amoeboid phenotype of d infected cultures. But a proportion of conidia derived from d infected isolates do appear to be free from d infection and no longer produce a d reaction when paired with other healthy isolates.

In contrast, all ascospore derived progeny appear to be substantially free of d factor infection. Despite tests on many crosses between pairs of isolates, where either one or both of the parents carried d factors none of the progeny ever showed any symptoms of d infection (Brasier 1983, 1986a; Rogers 1986). Thus transmission of d factors to sexually produced progeny appears in general not to occur, presumably in much the same way that transmissible hypovirulence in *Cryphonectria parasitica* also fails to incorporate into ascospores (Brasier 1986a; Rogers e al. 1986a, 1986b).

Structure of d factors

The altered morphology which signals d factor transfer between diseased and healthy isolates of *O. novo-ulmi is* invariably correlated with the transfer of multiple segments of dsRNA. Unlike mycoviruses, these do not appear to be encapsidated in a protein coat and presumably are transferred as naked nucleic acid. Polyacrylamide gel electrophoresis has shown that each d factor is associated with a unique profile of dsRNA, characterized by the number and size of dsRNA segments.

In a survey of more than forty isolates of *O. novo-ulmi,* including both the EAN and NAN subgroups, Rogers and co-workers (Rogers et al. 1988) found d infected individuals all had dsRNA profiles. These consisted of 4-14 segments ranging in molecular weight from 0.23-2.5 kb. Moreover, many apparently healthy isolates also contained some dsRNA (although rarely more than four segments) indicating that a d infected phenotype is not caused simply by the presence of dsRNA. It is probable that the deleterious effects associated with certain d factors are due to specific segments within the total dsRNA profile (Rogers et al. 1986a). Others may be neutral or inactive except when associated with particular 'enabling' segments; some may even confer advantages to infected *Ophiostoma* individuals. Where d infection is apparent, particularly in *O. novo-ulmi,* there is now strong evidence to indicate that the debilitation is mediated via mitochondrial disturbance; this aspect of d factor infection is discussed in more detail by Brasier (this volume).

In contrast to *O. novo-ulmi,* the biochemical nature and activity of d factors in *O. ulmi is* less clear-cut. It has not always proved possible to detect any associated dsRNA in transformed isolates following a typical d reaction (H J Rogers and A G Mitchell, personal communication). This may reflect very low quantities of dsRNA or it may suggest that d infection in *O. ulmi* has its origin in more than one type of transmissible cytoplasmic agent.

Occurrence of d factors

Routine isolation of *O. ulmi* and *O. novo-ulmi* from Dutch elm disease outbreaks throughout the northern hemisphere has revealed many isolates with the characteristic d infected amoeboid morphology. This implies that d factors are relatively common in nature although the incidence of infection is usually much greater in *Ophiostoma* isolates originating from scolytid breeding galleries in elm bark, compared with those derived from the pathogenic phase in elm xylem vessels.

That d factor transmission occurs quite frequently in bark, was first demonstrated during an experiment using a genetically marked isolate of *O. novo-ulmi* (Webber and Brasier 1984). The isolate was inoculated into a healthy elm, caused advanced crown symptoms and was later found to have established in the bark of the tree associated with other individuals of *O. novo-ulmi* introduced by breeding scolytid beetles. During this process it also became d infected, which could only have resulted from the marked individual acquiring the d factor from an adjacent d infected *O. novo-ulmi* individual. This particular d factor was designated as the d^2 factor and its biological, biochemical and ecological properties have been intensively researched.

Since then it has become clear that any d infected individuals of *O. ulmi* and *O. novo-ulmi* which infect elms and spread extensively within the xylem almost invariably lose the typical d infected phenotype and, in parallel, lose segments of dsRNA and eventually no longer transmit d factors (Brasier 1986a; Rogers et al. 1986b 19). Brasier (1986) has suggested that this is because replication by single-cell budding during pathogenesis results in a proportion of d free cells which are favored as the faster growing component in the elm xylem. The end result is that d factors tend to be lost during pathogenesis and also as a result of sexual reproduction, but apparently acquired during the saprotrophic phase of growth in the bark of dead and dying elms.

D factors are therefore likely to exert a degree of natural control during elm bark colonization associated with vector scolytid breeding galleries, particularly as this is a period of intense growth and sporulation when the fitness of individual genotypes is crucial to their survival and success (Webber et al. 1988). But the most powerful effect of d factors could occur during the dissemination of the pathogen by bark beetle vectors, a part of the Dutch elm disease cycle that can be a severe test for even the most vigorous individuals of *Ophiostoma*. Spores carried on the exoskeleton of vector beetles are subjected to extremes of desiccation and radiation, which can rapidly reduce the number of viable spore numbers available for infection (Webber and Brasier 1984). Not surprisingly, this process has the end result of limiting the ability of the pathogen to initiate xylem infection.

D factors and infection via feeding grooves

Evaluating the influence of d factors during pathogen invasion of elms via beetle feeding grooves has highlighted the process of elm infection. Initial work focused on the d^2 factor which infects *O. novo ulmi* (see page 7), comparing its ability to reduce spore germination in culture and the effectiveness of different spore loads of healthy and d^2 carrying conidia to infect moderately susceptible elms (Webber 1987). Spore loads of between 50 and 50,000 conidia were introduced into artificially made feeding grooves to mimic the conditions normally encountered during infection of elm. This took account of the fact that individual vector beetles disseminate spore loads which can vary markedly in the number of propagules they contain. Apparently only half of emerging *Scolytus scolytus* individuals carry more than 1000 spores of *O. novo-ulmi* or *O. ulmi,* whilst individuals of smaller sized vector species such as *S. multistriatus* only rarely carry as many as 1000 (Webber 1990; Webber and Brasier 1984; Webber and Gibbs 1989).

Cultural studies readily confirmed that d^2 infection could drastically reduce conidial germination from >90% to between 10 and 40%, depending on the genetic background of infected isolates and the prevailing conditions (Webber 1987, and Table 1). However, the influence of the d^2 factor was most striking during *Ophiostoma* infection via feeding grooves. Although the minimum spore threshold for successful xylem invasion was 1000 healthy conidia, at least 50,000 d^2 carrying conidia were required to achieve the same level of xylem penetration and cause systemic elm infection (Fig. 3). Apart from raising spore thresholds far above the number of spores that would typically be available from vector beetles, d2 infection also significantly reduced the persistence of *O. novo-ulmi* in feeding grooves (Fig. 2). It is clear that the reduced effectiveness of d^2 carrying inoculum is largely related to the number of spores unable to germinate following d infection, although low viability is not the complete explanation for reduced pathogenic potential. Probably decreased growth rate, inability to establish a vigorous mycelial phase and competition from other colonizing organizations all contribute to the loss of pathogenic potential in d infected *O. novo-ulmi* introduced into feeding grooves (Webber 1987).

Evaluation of other d factors has shown similar effects although none of those tested in feeding grooves have been quite as debilitating as the d^2 factor. For example, the d^{13} factor, originally found to be naturally widespread at elm disease epidemic fronts of the NAN subgroup in Spain (Brasier 1988), raised the spore threshold necessary for elm infection from 1000 conidia to 5,000 but only slightly decreased the ability of *O. novo-ulmi* to persist in feeding grooves. Nonetheless, this agent would be likely to play a very significant role in nature if we consider that probably less than 25% of the most effective vector species, *S. scolytus,* carry 5,000 or more spores (Webber 1990; Webber and Brasier 1984). Likewise, the d^6 factor which has been recorded in the EAN subgroup of *O. novo-ulmi* exerts a similar level of debilitation, necessitating a five-fold increase in spore numbers before successful infection can occur compared with the same spores free from d factor.

If indeed d factors can eventually be used to control Dutch elm disease on a wide scale, it seems likely that the most effective strategy will be to utilize mainly moderately deleterious d factors such as the d^6 and d^{13} factors. This should allow d infected individuals of *O. ulmi* and *O. novo-ulmi* to colonize elm bark and be maintained in the Dutch elm disease cycle during the saprotrophic phase. The d infected spores from these individuals would then contribute to the spore loads disseminated by vector beetles but nevertheless play a significant role in preventing new infections of healthy elms.

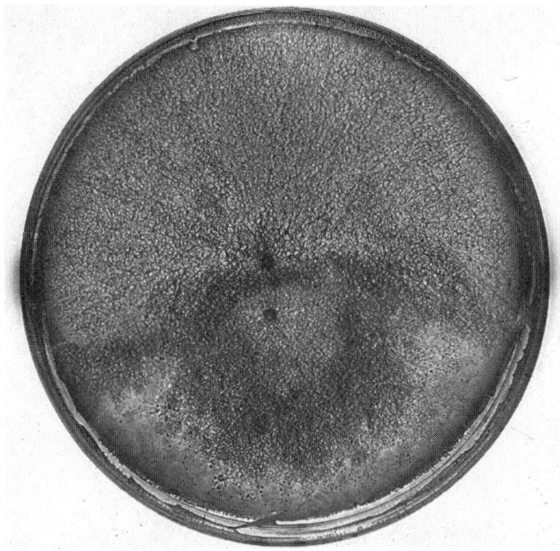

Figure 1. D reaction showing transformation of previously healthy culture of *Ophiostoma novo-ulmi* (top) with d^2-infected donor culture (below). Both isolates are of the same vegetative compatibility group.

Table 1. Effect of d factors on conidial germination and linear growth rate of *Ophiostoma novo-ulmi* and *O. ulmi*

Species and subgroup	Healthy		d infected	
	Percent germination[a]	Mean radial growth rate (mm)[b]	Percent germination[a]	Mean radial growth rate (mm)[b]
O. novo-ulmi, NAN				
W2 &w2 d^2	95.9	5.0±0.4[c]	14.2	2.9±2.5
H351 & H351 d^2	90.0	5.4±0.4	14.8	1.7±1.1
RDT/38 & RDT/38 d^2	93.0	5.2±0.3	34.0	1.8±0.5
BP6/1 & BP6/1 d^{13}	90.6	4.5±0.2	39.2	2.1±1.4
O. novo-ulmi, EAN				
P144 &P144 d^6	100.0	3.1±0.2	50.9	1.9±0.8
P144 &P144 d^9	100.0		28.8	1.8±0.3
O. ulmi				
C6/10 & C6/10 d	98.1	2.8±0.1	60.9	2.4±0.4

[a] Measured after 21h
[b] Expressed as mm $24h^{-1}$
[c] Standard deviation

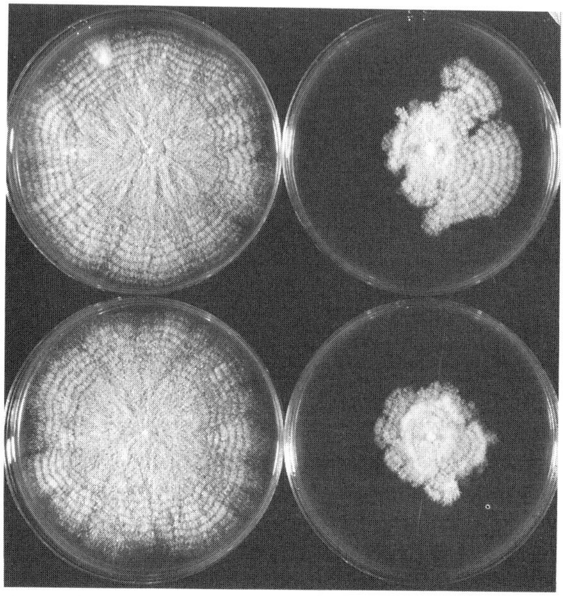

Figure 2. Healthy and d^{13} infected cultures of *Ophiostoma novo-ulmi*, NAN subgroup.

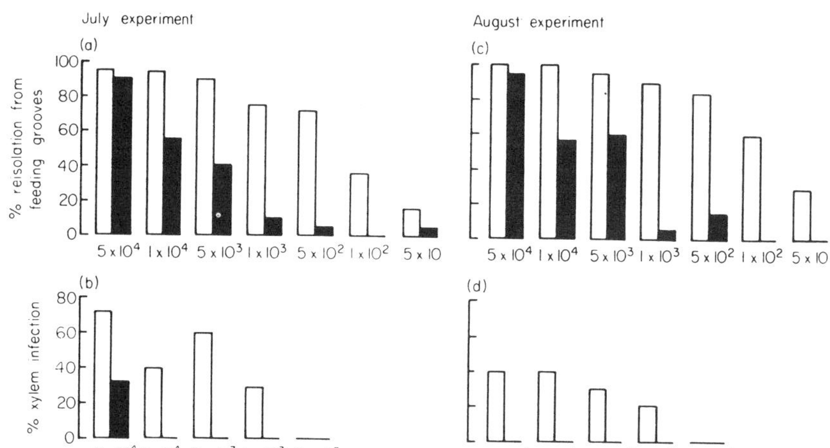

Figure 3. Comparison of the effectiveness of different quantities of healthy and d2-infected *Ophiostoma ulmi* conidia in artificial feeding grooves: (a) percentage of feeding grooves yielding *O. ulmi* 30 days after inoculation in July; (b) percentage of successful xylem infections following July inoculation of feeding grooves yielding *O. ulmi 32* days after inoculation in August; (d) percentage of successful xylem infections following August inoculation of feeding grooves. (■) d^2 infected conidia; (□) healthy conidia. Reproduced from Plant Pathology, after Webber (1987).

References

Brasier, C.M. 1983. A cytoplasmically transmitted disease of *Ceratocystis ulmi*. Nature 305:220-223.

Brasier, C.M. 1984. Inter-mycelial recognition systems in *Ceratocystis ulmi*: their physiological properties and ecological importance. Pages 451-497 in: The Ecology and Physiology of the Fungal Mycelium. D. Jennings and A.D.M. Rayner eds. Cambridge University Press. UK. pp.

Brasier, C.M. 1986a. The d factor in *Ceratocystis ulmi:* its biological characteristics and implications for Dutch elm disease. pp. 117-208 In: Fungal Virology. K.M. Buck, ed. CRS Press. Florida.

Brasier, C.M. 1986b. The population biology of Dutch elm disease: its principle features and some implications for other host-pathogen systems. Advances in Plant Pathology 5:55-118.

Brasier, C.M. 1988. Rapid changes in the genetic structure of epidemic populations of *Ophiostoma ulmi.* Nature 332:538-541.

Brasier, C.M. 1991. *Ophiostoma novo-ulmi* sp. nov., causative agent of current Dutch elm disease pandemics. Mycopathologia 115:151-161.

Brasier, C.M., and Gibbs, J.N. 1975. Variation in *Ceratocystis ulmi*: Significance of the aggressive and non-aggressive strains. pp. 53-56 in: Dutch Elm Disease. Proc. IUFRO Conf., Minn. St. Paul 1973. USDA Forest Service.

Buck, K.M. 1986. Fungal virology - an overview. pp. 1-84 In: Fungal Virology. K.M. Buck, ed. CRS Press. Florida.

Dodds, J.A., Valverde, R.A., and Mathews, D.M. 1988. Detection and interpretation of dsRNA. Pages 309-362 in: Viruses of Fungi and Simple Eucaryotes. Y Koltin and M.J. Leibowitz, eds. Marcel Dekker. New York and Basel.

Grent, J., and Sauret, S. 1969. L'hypovirulence exclusive phenomene original in pathologie vegetale. Comptes Rendus Hebdomadaires des Seances de l'Academie des Sciences 268:2347-2350.

Hollings, M. 1962. Viruses associated with a die-back disease of cultivated mushroom. Nature 196:962-96S.

Hollings, M. 1982. Mycoviruses and plant pathology. Plant Dis. 66: 1106.

Hollings, M., Stone, O.M., Barton, R.J., and Atkey, P.T. 1974. Dutch elm disease. Rep. Glasshouse Crops Res. Inst. 1973:124.

Lecoq, H., Boissonnet-Menes, M., and Delhotal, P. 1979. Infectivity and transmission of doublestranded RNA viruses. pp. 340-347 In: Fungal Viruses. H.P. Molitoris, M. Hollings and H.A. Woods. eds. Springer-Verlage, Heidelberg and New York.

Peace, T. 1960. The status and development of elm disease in Britain. For. Comm. Bull. 33:1-44.

Pusey, P.L., and Wilson, C.L. 1981. Possible role of fungal viruses in the distribution and spread of Dutch elm disease fungus. J. Arboriculture. 7:230-232.

Pusey, P.L., and Wilson, C.L. 1982. Detection of double stranded RNA in *Ceratocystis ulmi.* Pytopathology 72:423-428.

Rogers, H.J., Buck, K.W., and Brasier, C.M. 1986a. The molecular nature of the d factor in *Ceratocystis ulmi.* pp. 209-220 In: Fungal Virology K.W. Buck. ed. CRS Press. Florida.

Rogers, H.J., Buck, K.W., and Brasier, C.M. 1986b. Transmission of double stranded RNA and a disease factor in *Ophiostoma ulmi.* Plant. Pathol. 35:277-287.

Rogers, H.J., Buck, K.W., and Brasier, C.M. 1988. DsRNA and disease factors of the aggressive subgroup of *Ophiostoma ulmi.* pp. 327-352 In: Viruses of Fungi and Simple Eucaryotes. Y. Kottin and M.J. Leibowitz. eds. Marcel Dekker, New York and Basel.

Webber, J.F. 1987. The effect of the d^2 factor on survival and infection by the Dutch elm disease pathogen *Ophiostoma ulmi.* Plant Path. 36: 531-538.

Webber, J.F. 1990. Relative effectiveness of *Scolytus scolytus, S. multistriatus* and *S. kirschi* as vectors of Dutch elm disease. Eur. J. Fort. Path.

Webber, J.F., and Brasier, C.M. 1984. The transmission of Dutch elm disease: a study of the processes involved. pp. 271-306 in: Invertebrate-microbial Interactions. J.M. Anderson, A.D.M. Rayner and D.Walton. eds. Cambridge University Press, Cambridge.

Webber, J.F., Brasier, C.M., and Mitchell, A.G. 1988. The role of the saprophytic phase in Dutch elm disease. Pages 298-313 in: Plant Infecting Fungi. G.F. Pegg and P.G. Ayers. eds. Cambridge University Press, Cambridge.

Webber, J.F., and Gibbs, J.M. 1989. Insect dissemination of Fungal Pathogens of Trees. pp. 161-193 In: Insect-fungus Interactions. N.Wilding, N.M. Collins, P.M. Hammond and J.F. Webber. Academic Press, London.

Index